AKADEMIE DER WISSENSCHAFTEN UND DER LITERATUR

MATHEMATISCH-NATURWISSENSCHAFTLICHE KLASSE

MIKROFAUNA DES MEERESBODENS
61 (1977)

Redaktion: Peter Ax, Göttingen

The Meiofauna Species in Time and Space

Proceedings of a Workshop Symposium
Bermuda Biological Station · 1975

Editors: Wolfgang Sterrer and Peter Ax

AKADEMIE DER WISSENSCHAFTEN UND DER LITERATUR · MAINZ
IN KOMMISSION BEI FRANZ STEINER VERLAG GMBH · WIESBADEN

Redaktion: Prof. Dr. Peter Ax, D-3400 Göttingen
Berliner Straße 28, BRD

CIP-Kurztitelaufnahme der Deutschen Bibliothek

The meiofauna species in time and space:
proceedings of a workshop symposium, Bermuda Biolog. Station, 1975 / ed.: Wolfgang Sterrer and Peter Ax. – 1. Aufl. – Mainz: Akademie der Wiss. u. d. Literatur; Wiesbaden: Steiner (in Komm.), 1977.

(Mikrofauna des Meeresbodens; 61)
ISBN 3-515-02617-7

NE: Sterrer, Wolfgang [Hrsg.]; Bermuda Biological Station <Saint George, Hamilton>; Akademie der Wissenschaften und der Literatur <Mainz> / Mathematisch-Naturwissenschaftliche Klasse

Ausgegeben am 29. Juli 1977

Special Publication No. 15 from the Bermuda Biological Station.

Mikrofauna Meeresboden	61	1–316	1977

DRUCK: DR. ALEXANDER KREBS, 6944 HEMSBACH (BERGSTR.)
Printed in Germany

Foreword

In spite of the explosive growth in information which the study of benthic marine meiofauna has seen in its short history, the species, as the basic unit of communication among biologists, has largely remained elusive. While the earlier phases of research were dominated by taxonomic discoveries of the rank of orders, classes and phyla, an impressive amount of data regarding distributional ecology, biology and ecophysiology has been accumulated more recently; however, no major effort has been devoted to the species problem in meiofauna so far, nor do we understand the mechanisms of speciation and dispersal that have led to the global distribution and diversity as we know it today.

For the following reasons it can be expected that meiofauna research will continue to grow:

1) The high incidence of phylogenetic "key" forms makes meiofauna a unique data reservoir for our understanding of major phylogenetic problems. Ultrastructure techniques are only just beginning to be used, and will significantly contribute to the clarification of controversial issues such as the origin and early evolution of the Bilateria.

2) The role of meiofauna in the marine ecosystem, although suspected to be important, is still poorly understood. Its contribution to energy flow, position in the food chain, etc. remain essentially unexplored.

3) Our present knowledge of marine meiofauna is, with few exceptions, restricted to the North Atlantic and to shallow bottoms. Recent technical advances in deep-sea research, as well as the increasing accessibility of even the remotest shores will encourage many students to widen our knowledge in these directions; in addition, the interdisciplinary stimulus which emanated from the concept of plate tectonics has already begun to rejuvenate zoogeography into an exciting, dynamic discipline.

With this in mind, a workshop symposium was organized at the Bermuda Biological Station in September 24–30, 1975. "The Meiofauna Species in Time and Space" was attended by 33 scientists from 12 countries. A wide range of topics was covered, emphasizing the species-oriented approach in current meiofauna research, as is evident in the 19 papers and 7 abstracts which make up the present volume.

Our thanks are due to all of our colleagues who have participated in the Bermuda Workshop; to the staff of the Bermuda Biological Station; and finally to the Vollmer Foundation, the U.S. National Science Foundation, and the Akademie der Wissenschaften und der Literatur in Mainz who have made it possible to print these proceedings.

Wolfgang Sterrer,　　　　　　　　　　　　Peter Ax,
Bermuda Biological Station　　　　　　　　II. Zoologisches Institut
St. George's West, Bermuda.　　　　　　　der Universität
　　　　　　　　　　　　　　　　　　　　Göttingen, Fed. Rep. Germany.

Contents

Abstracts

Workshop Symposium "The Meiofauna Species in Time and Space"

September 24–30, 1975
Bermuda Biological Station

Organization
Wolfgang Sterrer

Participants

Bertil Åkesson,
University of Gothenburg,
Department of Zoology,
Fack., S-400 3300 Gothenburg,
Sweden.

Peter Ax and Renate Ax,
II. Zoologisches Institut und Museum
der Universität,
Berlinerstr. 28,
D-3400 Göttingen,
Fed. Rep. Germany.

Ian Ball,
Department of Entomology,
Royal Ontario Museum,
100 Queen's Park,
Toronto, Ontario,
Canada M5S 2C6

Patrick J. S. Boaden,
Department of Zoology,
The Queen's University of Belfast,
The Strand,
Portaferry, BT22 1PF,
Northern Ireland.

Claus Clausen,
Nordisk Kollegium för
Marinbiologi,
Biologisk Stasjon,
Espegrend, N-5065 Blomsterdalen,
Norway.

John O. Corliss,
Department of Zoology,
University of Maryland,
College Park, Maryland 20742,
U.S.A.

Bruce C. Coull,
Belle W. Baruch Institute for
Marine Biology and Coastal Research,
University of South Carolina,
Columbia, S. C., 29208, U.S.A.

Sebastian A. Gerlach,
Institut für Meeresforschung,
Am Handelshafen 12,
D-2850 Bremerhaven – G.,
Fed.Rep. Germany.

Olav Giere,
Zoologisches Institut und Museum,
Universität Hamburg,
Martin-Luther-King-Platz 3,
D-2000 Hamburg 13,
Fed. Rep. Germany.

Eike Hartwig,
Zoologisches Institut und Museum,
Universität Hamburg,
Martin-Luther-King-Platz 3,
D-2000 Hamburg 13,
Fed. Rep. Germany.

Carlo Heip,
Laboratorium voor Morfologie en
Systematik,
Ledeganckstraat 35,
B-9000 Gent, Belgium.

Robert P. Higgins,
Smithsonian Oceanographic Sorting
Center,
Smithsonian Institution,
Washington, D. C. 20560, U.S.A.

W. Duane Hope,
Department of Invertebrate Zoology,
Museum of Natural History,
Smithsonian Institution,
Washington, D. C. 20560, U.S.A.

William D. Hummon,
Department of Zoology,
Ohio University,
Athens, Ohio 45701, U.S.A.

M. Susan Ivester,
Belle W. Baruch Institute for
Marine Biology and Coastal Research,
University of South Carolina,
Columbia, S. C. 29208, U.S.A.

Tor G. Karling,
State Museum for Natural History,
Section for Invertebrate Zoology,
Stockholm 50, Sweden.

Ernst Kirsteuer,
American Museum of Natural History,
Central Park West,
New York, N. Y. 10024, U.S.A.

Brian M. Marcotte,
Department of Biology,
Dalhousie University,
Halifax, Nova Scotia, Canada.

Reinhard M. Rieger,
Department of Zoology,
University of North Carolina,
Chapel Hill, N. C. 27514, U.S.A.

Franz Riemann,
Institut für Meeresforschung,
Am Handeshafen 12,
D-2850 Bremerhaven-G,
Fed. Rep. Germany.

Edward E. Ruppert,
Department of Zoology,
University of North Carolina,
Chapel Hill, N. C. 27514, U.S.A.

Ernest R. Schockaert,
Universitaire Campus,
B-3610 Diepenbeek, Belgium.

Ernest Schoffeniels,
Laboratoire de Biochimie,
Université de Liège,
17 Pl. Delcour,
Liège, Belgium.

Thomas J.M. Schopf,
Marine Biological Laboratory
Woods Hole, Mass. 02543, U.S.A.

A. Sigurdsson,
Marine Research Institute,
P-O. Box 390,
Skulagata 4,
Reykjavik, Iceland.

Wolfgang Sterrer and Christiane Sterrer,
Bermuda Biological Station,
St. George's West, 1–15,
Bermuda.

John H. Tietjen,
Department of Biology,
City College of the City University
of New York,
New York, N. Y. 10031, U.S.A.

University of Maine at Orono
Dept. of Zoology
Murray Hall
Orono, Maine 04473, U.S.A.

Brigitte Volkmann,
Istituto di Biologia del Mare,
C. N. R. Riva 7 Martiti 1364/A,
30122 Venezia, Italy.

Wilfried Westheide,
II. Zoologisches Institut und
Museum der Universität,
Berlinerstr. 28,
D-3400 Göttingen, Fed. Rep. Germany.

Wolfgang Wieser,
Institut für Zoophysiologie,
Universität Innsbruck,
A-6020 Innsbruck, Austria.

Mikrofauna Meeresboden	61	11–18	1977

W. Sterrer & P. Ax (Eds.). The Meiofauna Species in Time and Space. Workshop Symposium, Bermuda Biological Station, 1975

Crossbreeding and Geographic Races: Experiments with the Polychaete Genus Ophryotrocha

by

Bertil Åkesson

Department of Zoology, University of Gothenburg, Fack, S–400 33 Gothenburg 33, Sweden

Abstract

Crossbreeding experiments have been performed with *Ophryotrocha labronica* and *O. puerilis*. *O. puerilis* is known as two subspecies, *O. puerilis puerilis* and *O. puerilis siberti*, which are partially interfertile.

In *O. labronica* a genetic incompatibility between geographically distant strains is revealed as an increased male ratio in the progeny and by the appearance of subvital larvae. All strains, however, are completely interfertile with 95–100 % development in the egg masses.

Ophryotrocha labronica is resistant to protracted inbreeding. Crosses at pilot scale between two substrains of the same origin, which had been inbred for 29 generations, as well as back crosses to the original stock culture indicated that genetic differences had arisen.

In *O. puerilis* no genetic incompatibility has been observed in intrasubspecies crosses between geographically distant populations. In intersubspecies crosses the reproductive potential, expressed as number of viable larvae per female per day, is drastically decreased compared to intrasubspecies crosses. Most embryos die at a very early stage of development, but surviving ones express hybrid vigor.

In both *Ophryotrocha* species reciprocal crosses yield different results. This may be due to maternal hereditary factors. Extranuclear DNA has been found during the vitellogenesis in *O. labronica*. The two species have also developed ethological isolation mechanisms which are reflected in a decreased mating propensity and in discrimination of alien males.

A. Introduction

Crossbreeding experiments have been performed with very few invertebrate species or species groups or genera. In order to be considered suitable material the organisms should have a rather short life cycle and also be easy to raise in the laboratory. Very often when extensive cultivation and crossbreeding have been performed in which strains of distant geographic origin have been employed, new species or various levels of incipient speciation have been discovered. The intensely studied fruit flies of the genus *Drosophila* provide the best example. Among marine animals, harpacticoid copepods and polychaetes, especially eunicid polychaetes of the genus *Ophryotrocha* (family Dorvilleidae), all small and morphologically uniform, have proved to provide an excellent material for crossbreeding experiments and speciation studies.

B. Experiments with Ophryotrocha labronica

My crossbreeding experiments started due to a disagreement about the sexual conditions in *O. labronica*. It had been described from Leghorn (Italy) as a protandrous hermaphrodite by La Greca & Bacci (1962). The hermaphroditic condition was confirmed by Parenti (1960) and Zunarelli-Vandini (1962, 1967), who both stated that isolated females can perform a secondary spermatogenesis and self-fertilize.

At that time I had two strains of *O. labronica*, both from Naples. They were both gonochoric with a male ratio of 30–40%.

I collected a strain of *O. labronica* from Leghorn which proved to be gonochoric and as incapable of self-fertilization as the other ones. The Naples and Leghorn strains were 100% interfertile. The details of these early crossbreeding experiments have already been published (Åkesson 1972 a, b).

The two Naples strains proved to be genetically different between themselves and both were different from the Leghorn strain. The differences were most conspicuous in the combination ♀ Leghorn x ♂ Naples II. The relative intraspecific incompatibility was revealed in several ways, as changes in sex ratios, by the appearance of larvae of a reduced viability in the progeny, by a decreased mating propensity, and, in discrimination experiments, as an almost complete ethological isolation.

In all these crossbreeding series reciprocal differences appeared, i. e., crosses with females from strain A and males from strain B yielded results different from those with females from strain B and males from strain A. This maternal hereditary factor has also been revealed in selection experiments towards increased and decreased male ratio (Åkesson 1972 b). The nature of the maternal factor is still unknown, but Emanuelsson (1969, 1971) found extranuclear DNA when he studied the vitellogenesis of *O. labronica*, and this DNA may form part of the genetic background of a cytoplasmic inheritance.

The expressions of reproductive isolation between the Naples and Leghorn strains of *O. labronica* are very similar to those reported by Battaglia and Volkmann-Rocco (1969, 1973) from crossbreeding experiments with the harpacticoid copepod *Tisbe clodiensis* (see also Volkmann 1977).

In subsequent crosses with other geographic strains of *O. labronica*, the male ratio approached 100% in the combination ♀ Los Angeles x ♂ Naples II. These strains and four others have been employed in interpopulation crosses where the egg mass production per unit of time (30 days) has been determined (Table 1). The number of egg masses is lower in interpopulation crosses than in intrapopulation matings. In some combinations this might be due to different mortality rates, female mortality being most important, but in others it reflects a decrease in mating propensity. No correlation seems to exist between mating propensity and sex ratio in the progeny.

Table 1. Variation in sex ratio and egg mass production in intra- and interpopulation crosses of *Ophryotrocha labronica*. Low egg mass production in interpopulation crosses reflects a decrease in mating propensity. The decrease is significant at the 1 % level ** or 0.1 % level *** in three combinations. Four bowls with 10 females and 10 males in each combination. Duration of the experiment is 30 days.

Crossing combination (♀ x ♂)	Male ratio, % ± S.E.	No. of egg masses	Survival after 30 days ♀	♂	Egg masses per ♀, X̄ ± S.E.	
Naples II x Naples II	41.2 ± 1.3	107	38	39	2.7 ± 0.2	
Naples II x Rovinj	49.9 ± 1.2	94	26	19	2.8 ± 0.1	
Rovinj x Naples II	39.0 ± 3.4	57	19	28	1.9 ± 0.2	
Rovinj x Rovinj	49.1 ± 0.7	75	24	26	2.3 ± 0.2	
Rhodos x Rhodos	42.8 ± 3.5	115	39	32	2.9 ± 0.2	
Rhodos x Faro	65.5 ± 6.2	40	38	37	1.0 ± 0.2	***
Faro x Rhodos	60.7 ± 10.1	64	34	30	1.7 ± 0.1	
Faro x Faro	57.5 ± 2.9	81	34	31	2.2 ± 0.1	**
Naples II x Naples II	41.2 ± 1.3	112	40	39	2.8 ± 0.1	
Naples II x Los Angeles	49.7 ± 1.3	83	37	33	2.2 ± 0.1	**
Los Angeles x Naples II	90.4 ± 6.2	88	34	24	2.4 ± 0.1	
Los Angeles x Los Angeles	33.3 ± 1.7	100	34	30	2.7 ± 0.1	

In a previously reported series of crosses between three populations from Cyprus (Åkesson 1975) three of the six combinations yielded very few or no egg masses in a 3-week period and no development at all in these egg masses. Subsequent crosses revealed that no sterility barriers existed. The result was due to a very high level of ethological isolation. In the three combinations (♀ x ♂), Limassol x Famagusta, Famagusta x Limassol, and Famagusta x Kyrenia, alien mating partners are not accepted in discrimination experiments and only occasionally in interpopulation crosses.

Table 2. Change of sex ratio during inbreeding of the Naples I strain of *Ophryotrocha labronica* through 46 generations. Eight substrains are employed.

Generation	(n)	Male ratio, % ± S.E.
P	8	33.7 ± 1.5
F 2	8	32.3 ± 1.4
F 3	8	35.2 ± 1.7
F 6	8	35.4 ± 2.2
F 12	8	41.2 ± 2.1
F 28	8	49.6 ± 1.8
F 29	64	50.3 ± 0.7
F 41	8	49.0 ± 1.3
F 46	8	50.5 ± 3.5

Ophryotrocha labronica is tolerant to protracted inbreeding. Eight substrains of the Naples I population have been inbred in brother-sister matings for 50 generations. After 46 generations one of the substrains was lost due to the appearance of 100% males in the progeny. The female-male ratio of the original stock culture was approximately 2:1. After 28 generations of inbreeding, maybe earlier, an 1:1 ratio was approached (Table 2). When crosses were performed between inbred sub-

Table 3. Crosses between two inbred substrains, A and B, and crosses between substrain C and the Naples I stock culture of *Ophryotrocha labronica*. The inbred substrains originate from Naples I.

Parent generation / Male ratio, % ± S.E.	F 1 generation / Male ratio, % ± S.E. (♀ x ♂)
Substrain A (F 29) / 53.4 ± 1.1	Substrain A x Substrain B (n = 2) / 65.8; 99.1
Substrain B (F 29) / 51.0 ± 2.4	Substrain B x Substrain A (n = 2) / 34.3; 62.8
Substrain C (F 29) / 51.1 ± 1.1	Substrain C x Naples I (n = 4) / 34,5 ±4.3
Naples I stock culture / 32.9 ± 1.6	Naples I x Substrain C (n = 4) / 61.2 ±4.5

strains of the F 29 generation, as well as back crosses to the stock culture Naples I, changes in the sex ratio of the progeny indicated genetic differences between the substrains and between them and the stock culture (Table 3).

All but two of my 13 cultivated strains of *O. labronica* were collected in the Mediterranean region. The exceptions are a strain from Faro in southern Portugal and another one from Los Angeles. There are reasons to doubt that the Los Angeles population is a "natural" one. The possibility of a transportation with ballast

Table 4. Discrimination experiments over three weeks with a new *Ophryotrocha* species from North Carolina and various strains of *O. labronica*. In each of 12 bowls one female and one male of the new species and one male of an *O. labronica* strain are kept together. The two species are intersterile when the female belongs to the North Carolina species.

Strains of *O. labronica*	No. of egg masses successful (+) / unsuccessful (−)	% homogamic matings (+)	Isolation index
Cyprus (Famagusta)	25 + / 0 −	100	1.00
Cyprus (Kyrenia)	24 + / 2 −	92.3	0.85
Crete (Kahnia)	24 + / 2 −	92.3	0.85
Italy (Naples II)	20 + / 9 −	69.0	0.38
Spain (Malaga)	25 + / 13 −	65.8	0.32
USA (Los Angeles)	19 + / 12 −	61.3	0.23
Cyprus (Limassol)	20 + / 17 −	54.1	0.08
Grece (Rhodos)	9 + / 11 −	45.0	− 0.10
Portugal (Faro)	15 + / 20 −	42.9	− 0.14
Italy (Naples I)	10 + / 14 −	41.7	− 0.17
Mallorca (Palma)	11 + / 17 −	39.3	− 0.21
Italy (Leghorn)	10 + / 20 −	33.3	− 0.33

water from the Mediterranean Sea (Gerlach 1977) should be considered. As is evident from the inbreeding experiments, a genetic difference may arise in rather few generations.

Populations of another *Ophryotrocha* species which is genetically different, but obviously closely related to *O. labronica* have been found in the Beaufort/Morehead region of North Carolina. The strains are completely intersterile in crosses with *O. labronica* if the female belongs to the North Carolina strains. The progeny of reciprocal crosses is subvital and the chances of a gene exchange in nature between populations of the new *Ophryotrocha* and *O. labronica* are extremely small. Results from more crosses are needed, however, before it is possible to decide whether to consider the North Carolina strains as representatives of a new sibling species or as an American subspecies of *O. labronica*. In Table 4 the variations in ethological isolation between the North Carolina species and various strains of *O. labronica* are summarized.

C. Experiments with Ophryotrocha puerilis

Another *Ophryotrocha* species, the protandrous hermaphrodite *O. puerilis*, is known as two subspecies, the Mediterranean *O. puerilis puerilis,* and the Atlantic *O. puerilis siberti*. They are almost reproductively isolated. The level of interfertility differs in reciprocal crosses between the two subspecies. When the female belongs to the Mediterranean subspecies the successful egg masses are 5.2% of the number produced. In the reciprocal crosses the percentage is 42.2 (Åkesson 1975). More than 90% of the egg masses are successful in intrapopulation matings. The same is true of crosses between strains of the same subspecies independent of geographic distance between them. An egg mass is considered successful if one or more viable larvae are released from it. Similar reciprocal differences exist in mean number of viable larvae per successful egg mass.

The reproductive potential is often expressed as reproductive rate, i. e., number of eggs per female per day. As only a fraction of the eggs develop in egg masses resulting from interpopulation matings between the two subspecies, a calculation of number of viable larvae per female per day should be preferred. When calculated on 150—200 egg masses in each combination the figures are as follows:

($♀$ x $♂$)	No. of viable larvae per female per day
Atlantic x Atlantic	36.6
Atlantic x Mediterranean	3.3
Mediterranean x Atlantic	0.1
Mediterranean x Mediterranean	27.8

In Table 5 A those crossing combinations between the two subspecies have been selected in which few if any egg masses are successful and the percentage of viable larvae is low. As a contrast some interpopulation crosses between geographically distant strains of the same subspecies are included in Table 5 B. In the selected intrasubspecies crosses only five egg masses are unsuccessful. When calculated on the whole available material, 3% of the egg masses from the Atlantic intrasubspecies matings and 9% of the Mediterranean ones are unsuccessful.

Table 5. Crosses between the two subspecies of *Ophryotrocha puerilis,* (A), and between strains of the same subspecies, (B). The experiments run for 30 days with one female and one male in each bowl. An egg mass is considered successful if one or more viable larvae are released.

Crossing combination (♀ x ♂)	No. of couples	No. of egg masses successful (+)/ unsuccessful (−)	% successful egg masses	% eggs developing into viable larvae in (+)
(A)				
Roscoff x Leghorn	6	7+ / 14−	33	< 3
Leghorn x Roscoff	6	1+ / 24−	4	< 10
Lagos x Leghorn	6	4+ / 13−	24	< 5
Leghorn x Lagos	6	0+ / 17−	0	−
Naples x Plymouth	5	0+ / 13−	0	−
Leghorn x Plymouth	6	1+ / 20−	5	< 10
Malaga x Roscoff	6	1+ / 23−	4	< 1
Malaga x Lagos	6	0+ / 25−	0	−
(B)				
Plymouth x Lagos	6	26+ / 0−	100	100
Lagos x Plymouth	6	23+ / 0−	100	100
Roscoff x Lagos	6	22+ / 1−	96	100
Lagos x Roscoff	6	30+ / 0−	100	> 90
Leghorn x Malaga	6	22+ / 2−	92	100
Malaga x Leghorn	6	28+ / 2−	93	> 90

Table 6. Discrimination experiments over 30 days with the Atlantic and Mediterranean strains of *Ophryotrocha puerilis.* In each bowl one female and one male from the same strain and one male from a strain of the other subspecies. In this table an egg mass is considered successful if more than 10 % of the eggs develop into viable larvae. Further explanations in the text.

Crossing combination (♀ x ♂) x ♂	No. of replicates	No. of egg masses successful (+) / unsuccessful (−)	% successful egg masses	% homogamic matings	Isolation index
Roscoff x Leghorn	32	26+ / 41−	38.8	40.3	− 0.19
Leghorn x Roscoff	40	95+ / 45−	67.9	74.3	0.49
Lagos x Leghorn	24	48+ / 23−	67.6	69.0	0.38
Leghorn x Lagos	44	96+ / 46−	67.6	73.9	0.48
Naples x Plymouth	16	17+ / 28−	37.8	42.2	− 0.16
Leghorn x Plymouth	16	28+ / 20−	58.3	64.6	0.29
Malaga x Roscoff	52	114+ / 64−	64.0	69.7	0.39
Malaga x Lagos	64	156+ / 58−	72.9	79.4	0.59

The crossing combinations from Table 5 A have been employed in a discrimination experiment in which the females had free mating choice between one male of her own strain and one from a strain of the other subspecies (Table 6). As in *O. labronica* the females mate with only one male. The 3% and 9% of unsuccessful homogamic matings mentioned above have been considered when the percentage of homogamic matings and the isolation index were calculated. The isolation index is calculated as homogamic matings minus heterogamic matings, divided by the total number of matings.

The hybrid larvae of *O. puerilis* are different frdm *O. labronica* in that they express hybrid vigor (heterosis). When newly released vital larvae were cultivated for one month the mortality rates were not higher in the hybrids than in the intrapopulation crosses. The growth rate of the hybrids exceeded that of the parent strains (Table 7).

Table 7. Growth and mortality in intra- and interpopulation crosses of *Ophryotrocha puerilis puerilis* and *O. puerilis siberti.* Fifty newly released vital larvae were transferred to each bowl and cultivated for one month at 20 °C.

	Crossing combinations ($\female$ x $\male$)			
	Leghorn x Leghorn	Plymouth x Plymouth	Leghorn x Plymouth	Plymouth x Leghorn
No. of bowls	16	16	8	13
Mortality in 30 days (%)	10.8 ± 2.8	23.3 ± 4.3	18.0 ± 4.0	21.8 ± 3.3
Size classes (%)				
0 – 9 segments	8.9	14.2	9.4	12.1
10 – 15 segments	65.5	58.7	45.0	43.3
16 – 24 segments	25.6	27.1	45.6	44.6

D. Discussion

In the early experiments with interpopulation crosses of *O. labronica* there was a good correlation between high male ratio and low mating propensity as expressions of genetic incompatibility. In subsequent series this correlation has not been found. These isolation factors seem to be independent.

For *O. puerilis* the highest isolation index, 0.59, is found in crosses between $\female$ Malaga x $\male$ Lagos where a complete intersterility was also found (Tables 5, 6). This is in good agreement with the concept, as discussed by Mayr (1963) and others, that a reinforcement of the isolation along secondary boundary regions will occur between two subspecies.

The two *Ophryotrocha* species employed in the experiments have in common the discrimination of alien males and the lowered mating propensity together with an alien mating partner. They differ in other respects. In *O. labronica* the early embryonic development is quite normal in crosses between distant strains. The genetic incompatibility is revealed as an increase of the frequency of subvital larvae. In *O. puerilis* geographic distance does not mean anything in crosses between populations of the same subspecies except perhaps a certain heterosis effect. In intersubspecies crosses most embryos die during the early cleavage stages. The surviving ones are released as quite normal larvae which grow unusually fast, an indication of a heterosis effect.

Bibliography

Åkesson, B.: Incipient reproductive isolation between geographic populations of *Ophryotrocha labronica* (Polychaeta, Dorvilleidae). Zool. Scripta 1, 207–210 (1972a).

—: Sex determination in *Ophryotrocha labronica* (Polychaeta, Dorvilleidae). In Battaglia (Ed.), Fifth European Marine Biol. Symposium. Piccin Editore, Padova (1972b).

—: Morphology and life history of *Ophryotrocha maculata* sp. n. (Polychaeta, Dorvilleidae). Zool. Scripta 2, 141–144 (1973a).

—: Reproduction and larval morphology of five *Ophryotrocha* species (Polychaeta, Dorvilleidae). Zool. Scripta 2, 145–155 (1973b).

—: Reproduction in the genus *Ophryotrocha* (Polychaeta, Dorvilleidae). Pubbl. Staz. Zool. Napoli 39 (Suppl.), 377–398 (1975).

Battaglia, B. & Volkmann-Rocco, B.: Gradienti di isolamento riproduttivo in popolazioni geografiche del copepode *Tisbe clodiensis*. Atti Ist. veneto Sci. 127, 371–381 (1969).

—: Geographic and reproductive isolation in the marine harpacticoid copepod *Tisbe*. Mar. Biol. 19 (2), 156–160 (1973);

Emanuelsson, H.: Electronmicroscopical observations on yolk and yolk formation in *Ophryotrocha labronica* La Greca & Bacci. Z. Zellforsch. 95, 19–36 (1969).

—: Metabolism and distribution of yolk DNA in embryos of *Ophryotrocha labronica* La Greca & Bacci. Z. Zellforsch. 113, 450–460 (1971).

Gerlach, S. A.: Means of meiofauna dispersal. Mikrofauna Meeresboden 61, 89–103 (1977).

La Greca, M. & Bacci, G.: Una nuova specie di *Ophryotrocha* delle coste tirreniche. Boll. Zool. 29, 13–23 (1962).

Mayr, E.: Animal Species and Evolution. The Belknap Press of Harvard University Press, Cambridge, Mass. (1963).

Parenti, U.: Self-fertilization in *Ophryotrocha labronica*. Experientia 16, 413–414 (1960).

Volkmann, B.: Geographic and reproductive isolation in the genus *Tisbe* (Copepoda, Harpacticoida). Mikrofauna Meeresboden 61, 313–314 (1977).

Zunarelli, R.: Il differenziamento citosessuale di *Ophryotrocha labronica*. Atti Accad. naz. Lincei Rc., 8 Ser., 32, 703–706 (1962).

Zunarelli-Vandini, R.: Azioni reciproche sulle gonadi in coppie omeospecifiche ed eterospecifiche di *Ophryotrocha puerilis siberti* ed *Ophryotrocha labronica*. Arch. zool. ital. 52, 177–192 (1967).

Mikrofauna Meeresboden	61	19—28	1977

W. Sterrer & P. Ax (Eds). The Meiofauna Species in Time and Space. Workshop Symposium, Bermuda Biological Station, 1975

Parasite-Host Relationships and Phylogenetic Systematics. The Taxonomic Position of Dinophilids

by

Bertil Åkesson

Department of Zoology, University of Gothenburg, Fack, S–400 33 Gothenburg 33, Sweden

Abstract

Coelomic coccidian parasites from polychaetes usually have a life cycle without schizogony. They seem to be extremely host specific. With a few exceptions they are only found in one natural host. Infection experiments have been performed with three members of the coccidian genus *Grellia*. Two are found in polychaetes of the genus *Ophryotrocha*, one in the archiannelid *Dinophilus gyrociliatus*. In infection experiments the coccidian from *Dinophilus* could also infect one of the *Ophryotrocha* species. One of the coccidians from *Ophryotrocha*, but not the other one, could infect *D. gyrociliatus*. When other polychaetes were exposed to spores of the coccidians none were infected by the coccidian from *Dinophilus*. Only members of the family Dorvilleidae, to which the genus *Ophryotrocha* also belongs, were infected by the coccidians from *Ophryotrocha*.

On the basis of the results from the infection experiments it is stated that the archiannelid family Dinophilidae is closely related to the polychaete family Dorvilleidae. This is in good agreement with previous statements based on similarities in morphology and embryology. Within dorvilleid polychaetes it is possible to establish series with increased structural reductions as a result of adaptations to a life in the mesopsammon. A recent discussion of the phylogeny of archiannelids and in particular of the taxonomic position of dinophilids is reviewed. The statement that dinophilids and dorvilleid polychaetes are related is discussed in the light of "Fahrenholz'rule" according to which the phylogeny of parasites parallels that of their hosts. As a result of the present study the concept of archiannelids as a monophyletic group should be rejected.

A. Introduction

Natural populations of *Ophryotrocha* species are often infected with sporozoans, the most obvious of which are coccidians. Coccidians are found in various polychaete families. Vivier & Henneré (1964) listed eight species. Since then a few more have been added and now the list includes 15 species distributed among 9 genera (Table 1). In the list of hosts 11 polychaete families and the dinophilids are represented. Some descriptions of the coccidian life cycles are old and incomplete and, as discussed by Vivier & Henneré (l. c.), there is also some risk of confusion between species.

Table 1. Coccidian parasites in polychaetes.

Host		Parasite
Family	Species	
Orbiniidae	*Scoloplos armiger*	*Eleutheroschizon duboscqui* Brasil, 1906
Opheliidae	*Ophelia limacina*	*Eleutheroschizon murmanicum* Averintsev, 1908
Spionidae	*Scolelepis fuliginosa**	*Dorisiella scolelepidis* Ray, 1930
”	*Polydora ciliata*	*Myriospora polydorae* Ganapati, 1953
Cirratulidae	*Audouinia tentaculata*	*Angeiocystis audouiniae* Brasil, 1904
”	*Cirratulus filiformis*	*Myriospora gopalai* Ganapati, 1945
Flabelligeridae	*Trophonia plumosa*	*Myriospora trophoniae* Lermantoff, 1913
Capitellidae	*Notomastus latericeus*	*Coelotropha vivieri* Henneré, 1963
Terebellidae	*Polymina nebulosa**	*Caryotropha mesnili* Siedlecki, 1902
Sabellidae	*Amphiglena mediterranea*	*Myriosporides amphiglenae* Henneré, 1966
Phyllodocidae	*Eulalia viridis*	*Defretinella eulaliae* Henneré, 1966
Nereidae	*Nereis diversicolor*	*Coelotropha durchoni* (Vivier, 1963)
Dorvilleidae	*Ophryotrocha puerilis* *Ophryotrocha labronica*	*Grellia ophryotrochae* (Grell, 1960)
”	*Ophryotrocha macrovifera* *Ophryotrocha labronica*	*Grellia* sp. (new species to be described)
Dinophilidae	*Dinophilus gyrociliatus*	*Grellia dinophili* (Grell, 1953)

* Schizogony and intracellular development reported. Both statements doubted by Vivier & Henneré (1964)

Table 2. Classification of the coccidian order Protococcidiorida Kheisin, 1956. From Levine (1973).

	Host Species
Family Eleutheroschizonidae Chatton & Villeneuve, 1936	
Genus *Eleutheroschizon* Brasil, 1906	
E. dubosqui Brasil, 1906	*Scoloplos armiger*
E. murmanicum Averintsev, 1908	*Ophelia limacina*
Family Grellidae Levine, 1973	
Genus *Grellia* Levine, 1973 (syn., *Eucoccidium* Grell, 1953)	
G. dinophili (Grell, 1953)	*Dinophilus gyrociliatus*
G. ophryotrochae (Grell, 1960)	*Ophryotrocha puerilis, O. labronica*
G. species	*Ophryotrocha macrovifera, O. labronica*
Genus *Coelotropha* Henneré 1963	
C. durchoni (Vivier & Henneré, 1964)	
(syn., *Eucoccidium durchoni* Vivier, 1963)	*Nereis diversicolor*
C. vivieri Henneré, 1966	*Notomastus latericeus*
Genus *Myriosporides* Henneré, 1966	
M. amphiglenae Henneré, 1966	*Amphiglena mediterranea*
Genus *Defretinella* Henneré, 1966	
D. eulaliae Henneré, 1966	*Eulalia viridis*

Schizogony is generally absent in coccidians from polychaete hosts. It has been reported from two genera only, *Dorisiella* and *Caryotropha*, and both observations have been called into question by Vivier & Henneré (l. c.). The same is true of statements that the life cycle is intracellular. In the other genera most of the life cycle is extracellular.

Recently Levine (1973) has discussed the taxonomy of coccidian parasites in polychaetes. He excludes the incompletely known and disputed genera and suggests a subdivision of the order Protococcidiorida into two families, Eleutheroschizonidae and Grellidae (Table 2). The two species of the genus *Eleutheroschizon* are intestinal parasites, whereas the members of the genera within the family Grellidae are found in the coelom.

All members of the family Grellidae have the same type of life cycle. The sporozoites penetrate the intestinal wall and develop into macro- and microgamonts in the coelom. The subdivision into genera is based on a variation in number of microgametes, number of sporocysts per oocyst, number of sporozoites in the sporocysts, as well as slight morphological variations in different life cycle stages. It is assumed that the family Grellidae is a monophyletic group.

B. The coccidian genus Grellia

For the present discussion of parasite-host relationship and its bearing upon taxonomy, the genus *Grellia* is of particular interest. The type species, *G. dinophili*, has been studied in detail by Grell (under the name of *Eucoccidium dinophili*) (1953 a, b, 1960). A second member of the genus, *G. ophryotrochae*, has also been named by Grell (1960), who could demonstrate that in the laboratory *Dinophilus gyrociliatus* can also be infected by *G. ophryotrochae*. This coccidian, however, is not as fatal for *Dinophilus* as the natural parasite. The reciprocal infection experiments with *G. dinophili* and *Ophryotrocha puerilis* were not successful.

The parasite-host relationship between *G. ophryotrochae* and *O. labronica* was described by Åkesson (1968). A third coccidian of the same genus, not yet described, was found by Åkesson in 1971 in *O. labronica* and *O. macrovifera* from Cyprus. Simon & Friedl (personal communication) have found the same parasite in a strain of *O. macrovifera* from Tampa, Florida. Of all coccidians reported from polychaetes only *Grellia ophryotrochae* and *Grellia sp.* have been found in more than one natural host (Table 2).

In an attempt to evaluate the taxonomic relations between the hosts, I have continued and extended Grell's infection experiments. The results are summarized in Table 3. In general the *Ophryotrocha* species are much more affected by *G. ophryotrochae* than by *Grellia sp.* The latter is only fatal for the two natural hosts. *Dinophilus gyrociliatus* can be infected by *G. dinophili* and *G. ophryotrochae*, but not by *Grellia sp.* With the exception of *O. notoglandulata*, which can be slightly infected, the *Ophryotrocha* species are resistant to infection by *G. dinophili.*

Table 3. Results from infection experiments in the laboratory. The plus signs indicate various degrees of infection, the minus sign means no infection.

Host	Parasite		
	Grellia ophryotrochae	*Grellia species*	*Grellia dinophili*
Ophryotrocha puerilis	++	+	−
Ophryotrocha labronica	+++	+++	−
Ophryotrocha macrovifera	+++	+++	−
Ophryotrocha notoglandulata	++	+	+
Ophryotrocha robusta	+++	+	−
Ophryotrocha hartmanni	++	+	−
Ophryotrocha adherens	++	+	−
Ophryotrocha diadema	+	+	−
Dinophilus gyrociliatus	++	−	+++

When a culture of *D. gyrociliatus*, which was infected with *G. dinophili*, was given as food to *O. puerilis* for more than six weeks, the polychaete was not infected. At the end of the period, when the same individuals of *O. puerilis* were given food contaminated with spores of *G. ophryotrochae*, they became infected to the same degree as a control group. Obviously there is no immunity reaction caused by the spores of the "wrong" coccidian.

Before the results of the infection experiments could be evaluated it proved necessary to extend them to other possible hosts. Spores of *G. dinophili* and *G. ophryotrochae* were given to a *Nerilla sp., Nereis diversicolor, Polydora ciliata, Pholoe minuta, Schistomeringos coeca,* and a *Dorvillea sp.* in parallel series. In addition, the chaetognath *Spadella cephaloptera* was given *Ophryotrocha* infected by *G. ophryotrochae.* Not a single one of these potential hosts was infected by *G. dinophili. Grellia ophryotrochae* could cause a slight infection in *Schisto- meringos coeca* and the *Dorvillea sp.* Like *Ophryotrocha*, both genera belong to the family Dorvilleidae, and they are, according to Day (1967), very closely related.

It is of particular interest that *Nereis diversicolor* was not infected. The natural parasite of *N. diversicolor, Coelotropha durchoni,* is so similar in morphology and life cycle to the *Grellia* species that it was originally described as *Eucoccidium durchoni.* The generic name of *Eucoccidium* was preoccupied and changed to *Grellia* by Levine (1973).

The infection experiments support the conclusion which can be drawn from the distribution of coccidians among polychaetes, that these parasites are host specific to a very high degree. The successful reciprocal infections add new evidence of a close relationship between dinophilids and eunicid polychaetes, a relationship which has been repeatedly suggested, above all by Russian zoologists.

C. Phylogeny of archiannelids

Hermans (1969) recently published an exhaustive review of the phylogeny and systematic position of archiannelids. Obviously, there has always been, and still exists, a wide disagreement concerning the phylogenetic position of the archiannelids. Do they constitute a natural group or are they "a heterogenous assemblage of genera" (Dales 1963)? Are the archiannelids primitively simple or secondarily adapted to a life in the mesopsammon? In his discussion Hermans came to the conclusion that the archiannelids represent a monophyletic group which is not closely related to any of the polychaete orders. According to Hermans (l. c., p. 98), "The question of whether the archiannelids are primitively or secondarily simple cannot be resolved on the basis of present knowledge".

Hermans' arguments have been discussed and criticized by Orrhage (1974) and Fauchald (1974). Both came to the conclusion that the archiannelids cannot be considered a phylogenetic-systematic taxon. "No features indicate a particularly close relationship between the group Archiannelida and any given polychaete family or between the five archiannelidan families . . . . Archiannelida is here considered a polyphyletic group, secondarily adapted to life in the mesopsammon". (Fauchald 1974, p. 497)

Table 4. Review of the concept of dinophilid affinities. Extracted from Hermans (1969).

Schmidt 1848	Rhabdocoel turbellaria
Van Beneden 1851	Nemertines
Schmarda 1861	Naid oligochaeta
Hallez 1879	Nemertines
Korschelt 1882	Turbellaria
Graff 1882	Rotifera or archiannelida
Lang 1884	Rotifera or archiannelida
Repiachoff 1886	Between rotifera and primitive annelida
Korschelt 1887	Related to annelida
Weldon 1887	Between turbellaria and archiannelida
Hatschek 1888	Side branch of archiannelida
Harmer 1889	A true archiannelid
Schimkevitsch 1895	Archiannelida or rotifera
Benham 1896	Ancestrally simple archiannelida
Salensky 1907	Secondarily primitive and degenerate polychaeta
Goodrich 1912	Degenerate archiannelida
Nachtsheim 1919	Neotenic annelida also related to rotifera
Heider 1922	Archiannelida
Iwanoff 1928	Oligomerous annelida
Remane 1932	Simplified polychaeta
Livanov 1940	Eunicemorph polychaeta
Ruebush 1940	Rotifera or turbellaria
Beklemischev 1952	Secondarily reduced eunicemorph polychaeta
Sveshnikov 1958	Eunicemorph polychaeta
Ruttner-Kolisko 1963	Primitive to annelida and rotifera, connecting both to turbellaria

If we confine ourselves to the dinophilids, this family has been connected with almost everything from turbellaria and rotifers to eunicemorph polychaetes. In Table 4 the various opinions about dinophilid relationship have been summarized. The Russian zoologists Livanov, Beklemischev, and Sveshnikov all relate dinophilids to eunicemorph polychaetes and emphasize the structural similarities in the ventral proboscis (muscle bulb) and in the larval morphology. Bubko (1973) expresses another opinion. He considers the dinophilids as a separate class of Annelida, equal to Chaetopoda. Polychaeta and (the other) Archiannelida constitute subclasses in the class of Chaetopoda.

The presence of closely related coccidians in the genera *Dinophilus* and *Ophryotrocha* as well as the successful experiments with reciprocal infection, speak in favour of a close relationship between them. The evolution of the parasites has been parallel to that of their hosts. If the two genera had a very remote common ancestor the divergent evolution of the parasites would have passed far beyond the point where reciprocal infection is possible.

D. Systematics of the family Dorvilleidae

Hartmann-Schröder (1971) includes the genus *Parapodrilus* in the family Dorvilleidae. She also erects a new genus *Parophryotrocha* with *P. isochaeta* as genotype. Jumars (1974) is of the opinion that the placement of *Parapodrilus* in the family Dorvilleidae is not justified. The available evidence is too scanty. According to Jumars the lack of prostomial appendages and jaws might indicate separate familiar status for *Parapodrilus*.

When Westheide described *Parapodrilus psammophilus* in 1965 he did not make any final decision about its systematic position. He emphasized the similarities

Table 5. Various levels of structural reduction in *Ophryotrocha, Parophryotrocha, Parapodrilus,* and *Dinophilus.*

	Parapodia	Setae	Pharyngeal muscle bulb	Jaws	Eyes	Palps	Antennae
Ophryotrocha puerilis	well developed, dors. & ventr. cirri	5–9 of 3 types, simple and composite	yes	P-type & K-type	yes	yes	yes
Ophryotrocha gracilis	simple, no cirri	5–9 of 3 types, simple and composite	yes	P-type	vestigial, no pigment	vesti-gial	yes
Parophryotrocha isochaeta	none	10–25 of 1 type, simple	yes	P-type & K-type	no	no	no
Parapodrilus psammophilus	simple, rud.of dorsal cirri	3 of 1 type, simple	no	none	no	no	no
Dinophilus gyrociliatus	none	none	yes	none	yes	no	no

with the dinophilids from which *P. psammophilus* differs by having simple setae and by its lack of a ventral pharyngeal muscle bulb. According to Hartmann-Schröder the connection with the dorvilleids goes through *Parophryotrocha isochaeta* which has simple setae and reduced parapodia.

Although Hartmann-Schröder provides few arguments in support of her opinion, the placement of *Parapodrilus* in the family Dorvilleidae seems reasonable to me. To those similarities emphasized by Westheide and Hartmann-Schröder it should be added that the organization of the pygidium in *Parapodrilus* resembles that of *Ophryotrocha* larvae. Both have one median and two lateral appendages (Westheide 1965; fig. 1; Åkesson 1973, fig. 2). It is possible to present a series with increased reduction as follows: (1) *Ophryotrocha puerilis* with three types of setae in well developed parapodia, a jaw apparatus where the P-type juvenile maxillae are replaced by an adult set of K-type maxillae, well developed eyes, palps, and antennae. (2) *O. gracilis,* which retains the juvenile P-type maxillae throughout life and has reduced eyes and palps. (3) *Parophryotrocha*, which has lost parapodia and composite setae as well as eyes, antennae, and palps. (4) *Parapodrilus* which retains simple setae in simple parapodia, but has lost eyes, antennae, and palps as well as jaw apparatus and the ventral muscle bulb. (5) *Dinophilus*, which has lost setae, parapodia, antennae, palps, and jaws, but still has eyes and the ventral pharyngeal muscle bulb (Table 5).

E. Other parasites in Ophryotrocha

In the genus *Ophryotrocha* there is a gonochoric group of four sibling species which are completely intersterile. They can be distinguished by variation in reproductive patterns and differences in maximum segment number and maximum length as well as in pigmentation and other inconspicuous morphological details.

In a recently published paper (Åkesson 1975) I reported some experiments with an intestinal sporozoan parasite from *Ophryotrocha labronica*. Originally I believed it to be a gregarine, but now I am more and more convinced that it is a coccidian of the enigmatic family Eleutheroschizonidae. In the experiments the four members of the sibling group were given food which was contaminated with parasite spores. The seemingly harmless parasite proved to reduce the reproductive rate by about 50 per cent in two of the *Ophryotrocha* species. The other two species were not affected; one had a slight infection, the second one was not infected at all. Obviously even closely related sibling species are physiologically different. In parallel experiments it could be shown that no *Ophryotrocha* species outside the sibling group could be infected.

Another parasite, which is morphologically identical with the intestinal parasite of *O. labronica,* has been found in the gut of *O. adherens* (a provisional name of a new species from Cyprus), which is one in a group of six simultaneous hermaphrodites with anterior male and posterior female segments. This sporozoan cannot be transferred to any of the other *Ophryotrocha* species which I keep in culture.

An *Ophryotrocha sp.* from Beaufort, North Carolina, which presumably is a fifth member of the sibling *labronica* group, is often infected by an ectocommensal ciliate. In laboratory experiments this ciliate only accepted the members of the sibling group as hosts, but never attached to any of the other species.

F. Discussion

The statement of a connection between dorvilleid polychaetes and dinophilids on the basis of parasite-host relationships is in good agreement with the "Fahrenholz'rule" (Fahrenholz 1913), which was discussed and given its name by Eichler (1941). According to this rule the phylogeny of the parasites parallels that of the hosts. Fahrenholz' rule and its implications has been reviewed and discussed by Stammer (1957) and Hennig (1966). Stammer gives numerous examples of the validity of the rule as well as examples of parasite-host relationships where no parallel evolution has occurred. Hennig provides a series of diagrams in order to demonstrate the pitfalls which might occur in the phylogenetic evaluation of parasite-host relationships. He also gives examples where the phylogenetic systematics have been better understood in the light of parasite-host relationships.

Mayr (1963, p. 54) reviews cases where sibling species have been discovered due to the presence of different parasites. Similarly the sibling *Ophryotrocha* species of the *labronica* group could have been recognized by their different susceptibility to the intestinal sporozoans. The complete intersterility and rank of sibling species had been revealed, however, in crossbreeding experiments ahead of the infection studies.

As far as endoparasites are concerned the variations in parasite-host relationships can be easily referred to physiological differences in the hosts. It is more difficult to understand why the ectocommensal ciliate, which was found on the *Ophryotrocha sp.* from Beaufort accepts only members of the *labronica* group as hosts. Some of the other *Ophryotrocha* species are morphologically so similar that they have often been confused (Pfannenstiel 1972, Åkesson 1973 a,b).

In his discussion of Fahrenholz' rule Hennig points out that the phylogeny of parasites is often retarded compared to that of their hosts. This is true as far as parasite morphology is concerned, but the example of extremely host-specific intestinal sporozoans in *O. labronica* and *O. adherens* indicates that morphological differences can be preceeded by physiological adaptations to the hosts, and as a result a subdivision into sibling parasites.

My concept of a structural reduction within the genera *Ophryotrocha, Parophryotrocha,* and *Parapodrilus* as adaptations to a life in the mesopsammon as well as similar and perhaps even more far-reaching reductions in the family Dinophilidae, are in good agreement with Jumars' tentative phylogeny of the dorvilleid genera

(1974, fig. 14). Originating from a hypothetical ancestral dorvilleid he presents two major branches, both with gradual reductions.

Even if the present study adds new evidence of a relationship between dorvilleid polychaetes and dinophilids, we still do not know how closely related they might be. If the family Dinophilidae is removed from the archiannelids and placed as a family among eunicemorph polychaetes, we also have to reject Hermans' (1969) concept of archiannelids as a monophyletic group. No affinities have been demonstrated between other archiannelid families and eunicemorph polychaetes.

Bibliography

Åkesson, B.: The parasite-host relation between *Eucoccidium ophryotrochae* Grell and *Ophryotrocha labronica* La Greca & Bacci. Oikos **19**, 158–163 (1968).

–: Morphology and life history of *Ophryotrocha maculata* sp. n. (Polychaeta, Dorvilleidae) Zool. Scripta **2**, 141–144 (1973a).

–: Reproduction and larval morphology of five *Ophryotrocha* species (Polychaeta, Dorvilleidae). Zool. Scripta **2**, 145–155 (1973b).

–: Reproduction in the genus *Ophryotrocha* (Polychaeta, Dorvilleidae). Pubbl. Staz. Zool. Napoli **39** (Suppl.), 377–398 (1975).

Bubko, O. V.: On systematic position of Oweniidae and Archiannelida (Annelida). Zool. Zhur. **52** (9), 1286–1296 (1973).

Dales, R. P.: Annelids. London: Hutchinson Univ. Library (1963).

Day, J.: A monograph on the Polychaeta of southern Africa. Part 1. Errantia. London: Brit. Mus. (1967).

Eichler, W.: Wirtsspezifität und stammesgeschichtliche Gleichläufigkeit (Fahrenholzsche Regel) bei Parasiten im allgemeinen und bei Mallophagen im besonderen. Zool. Anz. **132**, 254–262 (1941).

Fahrenholz, H.: Ektoparasiten und Abstammungslehre. Zool. Anz. **41**, 371–374 (1913).

Fauchald, K.: Polychaete phylogeny: A problem in protostome evolution. Syst. Zool. **23** (4), 493–506 (1974).

Grell, K. G.: Entwicklung und Geschlechtbestimmung von *Eucoccidium dinophili*. Arch. f. Protist. **99**, 156–186 (1953a).

–: *Eucoccidium dinophili*, n. g., n. sp., und das System der Coccidien. Naturwiss. **40** (7), 227–228 (1953b).

–: Reziproke Infektion mit Eucoccidien aus verschiedenen Wirten. Naturwiss. **47**, 47–48 (1960).

Hartmann-Schröder, G.: Annelida, Borstenwürmer, Polychaeta. Tierwelt Dtl. **58**, 1–594 (1971).

Hennig, W.: Phylogenetic Systematics. Urbana, Chicago, London: Univ. of Illinois Press (1966).

Hermans, C. O.: The systematic position of the Archiannelida. Syst. Zool. **18** (1), 85–102 (1969).

Jumars, P. A.: A generic revision of the Dorvilleidae (Polychaeta), with six new species from the deep North Pacific. Zool. J. Linn. Soc. **54**, 101–135 (1974).

Levine, N. D.: *Grellia* gen. n. for *Eucoccidium* of Grell (1953) preoccupied. J. Protozool. **20** (5), 548–549 (1973).

Mayr, E.: Animal Species and Evolution. The Belknap Press of Harvard University Press, Cambridge, Mass. (1963).

Orrhage, L.: Über die Anatomie, Histologie und Verwandtschaft der Apistobranchidae (Polychaeta Sedentaria) nebst Bemerkungen über die systematische Stellung der Archianneliden. Z. Morph. Tiere **79**, 1–45 (1974).

Pfannenstiel, H. D.: Eine neue *Ophryotrocha*-Art (Polychaeta, Eunicidae) aus Japan. Helgoländer wiss. Meeresunters. **23**, 117–124 (1972).

Stammer, H. J.: Gedanken zu den parasitophyletischen Regeln und zur Evolution der Parasiten Zool. Anz. **159**, 255–267 (1957).

Vivier, E., Henneré, E.: Cytologie, cycle et affinités de la coccidie *Coelotropha durchoni*, nomen novum (=*Eucoccidium durchoni* Vivier), parasite de *Nereis diversicolor* O. F. Müller (Annélide Polychète). Bull. Biol. France Belg. **98**, 153–206 (1964).

Westheide, W.: *Parapodrilus psammophilus* nov. gen. nov. spec., eine neue Polychaeten-Gattung aus dem Mesopsammal der Nordsee. Helgoländer wiss. Meeresunters. **12**, 207–213 (1965).

Mikrofauna Meeresboden	61	29—43	1977

W. Sterrer & P. Ax (Eds.). The Meiofauna Species in Time and Space. Workshop Symposium, Bermuda Biological Station, 1975

Problems of Speciation in the Interstitial Fauna of the Galapagos

by

Peter Ax

II. Zoologisches Institut und Museum der Universität, D-3400 Göttingen, Berliner Str. 28, Fed. Rep. Germany

Abstract

Our research on the interstitial fauna of the Galapagos Islands (1972—1973) identified about 400 species. The fact that the majority of the newly described species belonged to well known genera gave a decisive starting point for the analysis of speciation processes in the Galapagos.

In genera represented by a single Galapagos species clear sister species relationships to certain taxa in the continental interstitial fauna can often be established. In these cases, speciation in the Galapagos Archipelago is very probable.

In genera with two or more species in the Galapagos one must decide: did these species descend from a common stem species in the Galapagos, or did they reach the islands independently? In the turbellarians *Promesostoma* and *Ceratopera* two species from each genus settled independently in the Galapagos Islands. On the other hand the phylogenetic development of 4 closely related species of the genus *Duplominona* probably has taken place within the archipelago.

A. Introduction

The Galapagos Archipelago has ideal conditions for the study of speciation: (1) the islands are volcanic in origin; (2) their maximum age is probably three million years (Bailey 1976); (3) a land bridge between the Galapagos and the American continent never existed. In other words, the fauna and flora of the islands must have recently crossed the broad barrier of the Pacific.

In spite of the recent formation of the islands, geologically speaking, the terrestrial fauna has undergone extensive speciation. Just to mention a few classic examples, there are the Darwin's finches (Geospizinae) with 13 endemic species and numerous subspecies, the lava lizards of the genus *Tropidurus* with 7 species, and the giant tortoise *Geochelone elephantopus* with 15 subspecies. The Galapagos Islands are also a good site for testing the question of whether marine fauna can undergo a similar evolution. This may well be the case for the interstitial fauna of marine sand, as suggested by the following facts: (1) this fauna is bathymetrically limited to the littoral sand bottom; (2) it is strongly dependent on the interstitial spaces of sand, practically speaking it is a sessile fauna in a mobile substrate;

(3) with only a few, unimportant exceptions, pelagic larvae do not exist (Sterrer 1973, Ax & Schmidt 1973).

As a result of these adaptations to the mesopsammic biotope, the interstitial fauna cannot cross the barrier represented by deep and wide oceans, either actively on the sea bottom or passively in the pelagic zone as planktonic larvae. In order to settle geographically isolated areas the marine interstitial fauna, like terrestrial animals, must depend on accidental transport on drifting objects. Because of this, the chance of regular gene flow between populations of widely separated continents and islands is minimal.

These facts form the basis for the evolutionary aspects of our Galapagos Project in 1972 and 1973. In the course of 13 months, 7 members of our Göttingen research team (P. & A. Schmidt, U. Ehlers, S. Hoxhold, W. Westheide, P. & R. Ax) conducted studies at the Charles Darwin Station on Santa Cruz Island.

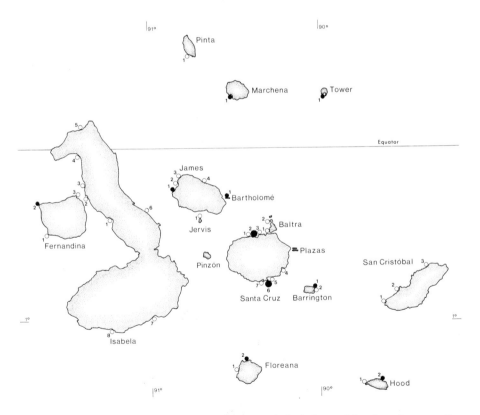

Fig. 1. Galapagos Archipelago showing our stations. Black circles mark beaches where quantitative investigations were made. Open circles indicate beaches where qualitative samples were taken (after Ax & Schmidt 1973).

B. Results

Almost all of the characteristic groups of the interstitial fauna were found at our stations on 14 islands of the archipelago (Fig. 1). All together we identified about 400 species. Most of these species are new to science, but the overwhelming majority belong to well known genera. This important discovery gave us a decisive starting point for our analyses.

I would like to illustrate the problem of understanding speciation in the Galapagos Archipelago with four examples from the turbellarians.

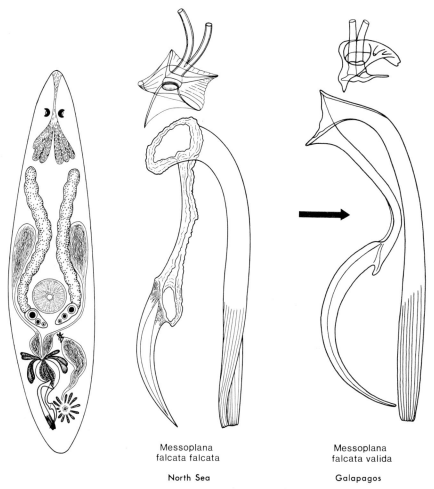

Messoplana
falcata falcata

North Sea

Messoplana
falcata valida

Galapagos

Fig. 2. *Messoplana falcata* (Ax, 1953), length 1–1.2 mm. To the right of the habitus picture are the mouth pieces of the receptaculum seminis and the male copulatory organ of the continental subspecies *Messoplana falcata falcata* (Ax, 1953) and the Galapagos subspecies *Messoplana falcata valida* Ehlers & Ax, 1974

I. Differentiation of subspecies

The Galapagos population of *Messoplana falcata*, order Typhloplanoida, is nearly identical with that of the North and Baltic Seas, but there are consistent differences in the habitus, the details of the cuticular organs, and the structure of the mouthpieces of the seminal receptacle (Fig. 2). Are these differences the result of the formation of a subspecies or do they represent the formation of a separate species?

As a rule, crossbreeding experiments with the hermaphroditic turbellarians and other groups of the fragile, or soft-bodied interstitial fauna are impossible; the problem of differentiating species from subspecies must be solved indirectly. In general we use the following criteria. If the differences between insular and continental populations are as great as between similar sympatric species from the continental fauna, then the two populations are determined to represent separate species. When the differences are not this great, then the insular and the continental populations are interpreted to be subspecies.

Messoplana falcata represents the second alternative. We regard *Messoplana falcata valida* from the Galapagos and *Messoplana falcata falcata* from the North Sea to be subspecies. These two populations still appear to have the potential to interbreed.

II. Sister group relations between individual species of the Galapagos and continental faunas

In many cases, in which genera are represented by a single species in the Galapagos Islands, we were able to establish a clear sister species relationship to a certain species in the continental interstitial fauna. Such a demonstration is especially meaningful when the closest relative inhabits the continental coast nearest the Galapagos.

These prerequisites are met by two species of the Otoplanid genus *Philosyrtis*. In this genus, consisting of 5 species, *Philosyrtis sanjuanensis* from the North American Pacific coast and *Philosyrtis santacruzensis* from the Galapagos are undoubtely sister species (Fig. 3). The basis for this conclusion is the great similarity in the construction of the stylet apparatus.

In these two species, the cuticular copulatory organ is differentiated into a dorsal and a ventral group of needles, clearly a synapomorphic character. There are very small but consistent differences in the form and number of these needles. Other minor variations include the number of testicular follicles and the form of the posterior end where the adhesive plate has a small neck in *Philosyrtis sanjuanensis* and is conical in *Philosyrtis santacruzensis*. In this example we could trace the descent of *Philosyrtis santacruzensis* back to the Eastern Pacific species, *Philosyrtis sanjuanensis*. The differences between the two sister species are obviously the result of speciation in the Galapagos Archipelago.

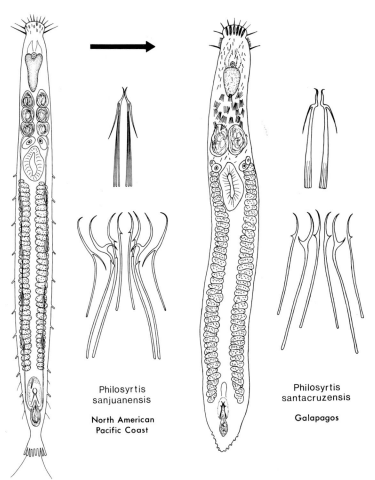

Philosyrtis
sanjuanensis

North American
Pacific Coast

Philosyrtis
santacruzensis

Galapagos

Fig. 3. *Philosyrtis sanjuanensis* Ax & Ax, 1967, from the North American Pacific Coast and *Philosyrtis santacruzensis* Ax & Ax, 1974, from the Galapagos, length 1.2–1.5 mm. Left, habitus pictures. Right, cuticular copulatory organ, where differences are especially noticeable in the lower group of needles (x 1000). The arrow shows the direction of phylogenetic derivation.

III. Sister group relations in genera represented by two Galapagos species

We frequently found two species of the same genus in the beaches of the Galapagos Islands. When this happened, our first question was, "Are these two species descended from a common stem species here in the Galapagos, or did they reach the islands independently? "

Our method of answering this question can be illustrated by two well analyzed examples from the turbellarian order Typhloplanoida. *Promesostoma sartagine* and *Promesostoma tenebrosum* are two Galapagos species from a genus which is especially well represented by numerous species in the North Atlantic and adjacent seas. Three species are also known from the Pacific Coast of North America. There are large differences in the form and size of the cuticular stylet in *P. sartagine* and *P. tenebrosum*; they are surely not sister species. By means of the same character we can rule out a close relationship between the Galapagos species and the 3 species on the North American Coast. But each of these two Galapagos species exhibits striking similarities to a different species of *Promesostoma* from Europe. We therefore concluded that *P. nynaesiensis* is the sister species of *P. sartagine* and *P. balticum* is the sister species of *P. tenebrosum* (Fig. 4 and 5).

The situation in the genus *Ceratopera*, a warm water taxon, is exactly the same. *Ceratopera paragracilis* and *Ceratopera bifida* from the Galapagos are not closely

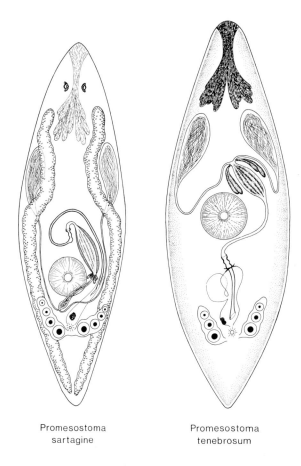

Promesostoma
sartagine

Promesostoma
tenebrosum

Fig. 4. *Promesostoma sartagine* Ax & Ehlers, 1973, and *Promesostoma tenebrosum* Ax & Ehlers, 1973 from the Galapagos, habitus and organization, length 0.5–0.7 mm.

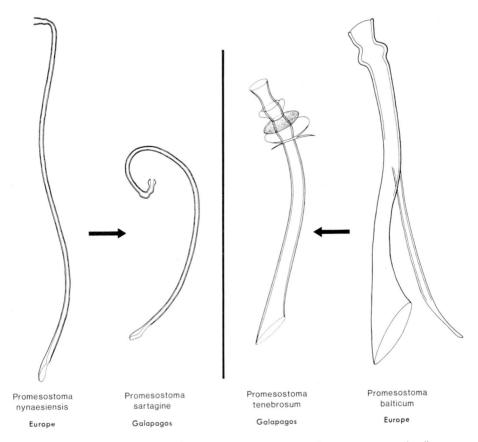

Promesostoma   Promesostoma   Promesostoma   Promesostoma
nynaesiensis   sartagine   tenebrosum   balticum

Europe   Galapagos   Galapagos   Europe

Fig. 5. Enlargement of the cuticular stylet of both Galapagos *Promesostoma* species (inner illustrations) and their European sister species *Promesostoma nynaesiensis* Karling, 1957 and *Promesostoma balticum* Luther, 1918, (outer illustrations). The arrows indicate the separate derivation of the Galapagos species.

related. The closest relatives of these two species are known at present only from the Mediterranean Sea. *C. paragracilis* and *C. gracilis* must be derived from one common stem species, *C. bifida* and *C. axi* from another (Fig. 6 and 7).

In both these examples we could state that two species from the same genus settled independently in the Galapagos Islands. The small differences between the sister species may have arisen through speciation in the Galapagos. The latter suggestion should be tested by studying related populations from the Pacific Coast of the Americas. Theoretically we must consider three possibilities: (1) the related continental populations of the Eastern Pacific could be identical with the corresponding European species; (2) the related continental populations of the Eastern Pacific could be identical with the species of Galapagos; (3) the Eastern Pacific coastal

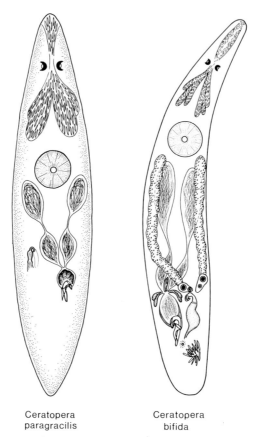

Ceratopera
paragracilis

Ceratopera
bifida

Fig. 6. *Ceratopera paragracilis* Ehlers & Ax, 1974 and *Ceratopera bifida* Ehlers & Ax, 1974,
from the Galapagos, habitus and organization, length 1.5 mm.

populations could form separate species which are morphologically intermediate
between the pairs which are presently considered to be sister species. Speciation in
the Galapagos Islands can only be demonstrated if case 1 or case 3 proves to be true.

IV. Species multiplication within the Galapagos Archipelago

The presence of genera represented by several species in the Galapagos Archipelago raised an especially difficult question: has a species multiplication, comparable to what occurred among Darwin's finches, also taken place in the interstitial
fauna?

Obviously this was rarely the case, but I can discuss an instructive example in
the genus *Duplominona* Karling from the family Monocelididae (Proseriata). We
discovered four new, very similar species of this genus in the Galapagos: *D. galapagoensis*, *D. karlingi*, *D. krameri* and *D. sieversi* (Fig. 8). The four species live next to

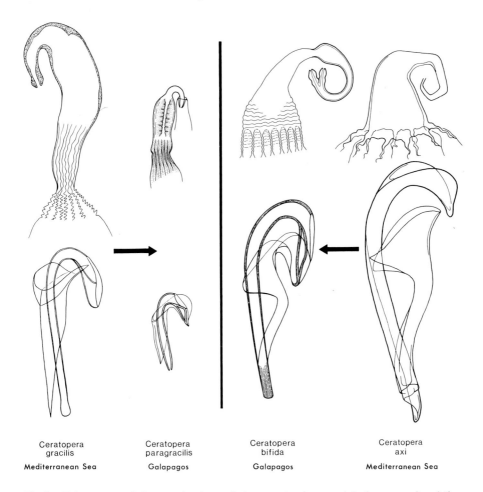

Ceratopera	Ceratopera	Ceratopera	Ceratopera
gracilis	paragracilis	bifida	axi
Mediterranean Sea	Galapagos	Galapagos	Mediterranean Sea

Fig. 7. Enlargement of the mouth pieces of the receptaculum seminis (upper row) and the copulatory organ of both Galapagos *Ceratopera* species and their sister species, *Ceratopera gracilis* (v. Graff, 1882) and *Ceratopera axi* (Riedl, 1954), from the Mediterranean Sea. The arrows indicate the separate derivation of the Galapagos species from their continental species.

each other in the littoral region of the Islands of Santa Cruz. All four have a cirrus with a central, enclosed stylet. Very small, but consistent, differences exist in the form and size of the stylet as well as in the formation of the cirral spines and in the existence or absence of granule glands. First, we had to find out wether these four *Duplominona* species are more closely related to each other than to any other species outside of the island group. This could only be determined by a carefully detailed kinship analysis of the whole genus using the methods of phylogenetic systematics (Hennig).

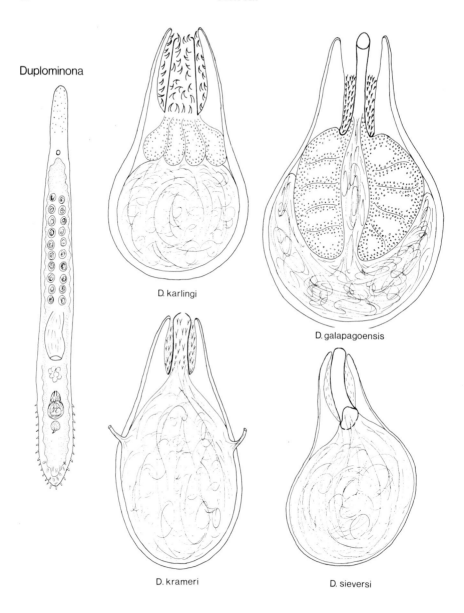

Duplominona

D. karlingi

D. galapagoensis

D. krameri

D. sieversi

Fig. 8. Genus *Duplominona* Karling from the Galapagos. Left, habitus and organization of *Duplominona karlingi*. Right, copulatory organs of the 4 species *D. karlingi* Ax & Ax 1977, *D. galapagoensis* Ax & Ax 1977, *D. krameri* Ax & Ax 1977, and *D. sieversi* Ax & Ax 1977. Differences among the species exist, for example, in the length and form of the stylets and the form of the cirral spines.

Let me briefly explain how this analysis was conducted (Fig. 9). In the family Monocelididae the genus *Monocelis* is a primitive taxon. It has a copulatory organ of the simplex type and three separate genital openings: vagina, male pore and female pore.

From this level of organization the so called *Minona* group, including the genera *Minona, Preminona, Duplominona, Peraclistus* and *Ectocotyla,* can be derived. A definite synapomorphy of this group of genera is the accessory prostatoid organ with glands and a stylet. The most primitive condition in the *Minona* group is

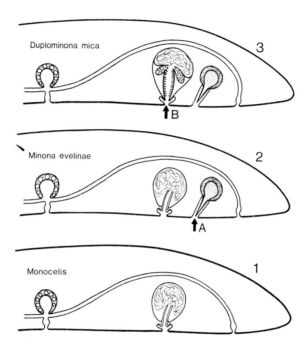

Fig. 9. Scheme of the organization of the sex organs of members of the genera *Monocelis, Minona,* and *Duplominona.* The gradual evolution of synapomorphic characters is indicated with arrows and capital letters.

(1) Ground plan of *Monocelis* Ehrbg., primitive genus of the family Monocelididae. Copulatory organ of the simplex type. Three separate genital openings: vagina, male pore and female pore.

(2) Ground plan of the genus *Minona* Marcus as illustrated by *Minona evelinae* Marcus. Apomorphy A (= Synapomorphy for the *Minona* group of the Monocelididae): development of an accessory prostatoid organ with muscular secretory bladder and cuticular stylet. Plesiomorphies: copulatory organ of the simplex type; four separate genital openings: vagina, male pore, female pore and pore of the accessory organ.

(3) Ground plan of the genus *Duplominona* Karling as illustrated by *Duplominona mica* (Marcus). Apomorphy B (= Synapomorphy of all *Duplominona* species): copulatory organ of the duplex type with cirral spines. Plesiomorphy: four separate genital openings.

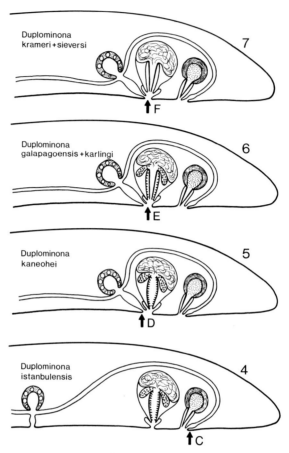

(4) *Duplominona istanbulensis* Ax. Apomorphy C (= Common synapomorphy with *D. kaneohei* and the *galapagoensis* group): union of the female pore with the pore of the accessory organ. Plesiomorphy: separation of the vagina and male pore.

(5) *Duplominona kaneohei* Karling, Mack-Fira & Dörjes. Apomorphy D (= Synapomorphy with the *galapagoensis* group): union of the vagina with the male pore. Plesiomorphy: cirrus of the copulatory organ without a stylet.

(6) *Duplominona galapagoensis* and *D. karlingi*. Apomorphy E (= Synapomorphy with *D. krameri* and *D. sieversi*): development of an enclosed cuticular stylet.

(7) *Duplominona krameri* and *D. sieversi*. Apomorphy F (= Synapomorphy of both species): reduction of the granule glands and the cirral spines.

realized in the genus *Minona.* In this genus we find a copulatory organ of the simplex type, an accessory organ positioned behind the copulatory organ, and four separate genital openings.

As a synapomorphic character, the members of the more evolved genus *Duplominona* have a copulatory organ of the duplex type containing a cirrus armed with spines. This genus contains 8 species to date. The most primitive taxon is *D. mica* from the Atlantic Coast of South America; it has four separate genital openings. Within the genus *Duplominona,* evolution proceeded to the condition found in *D. istanbulensis* where there is a common opening for the femal genital pore and the accessory prostatoid organ. The next apomorphic evolutionary step was the combination of the vagina and the male pore, the condition found in *D. kaneohei* from Hawaii. Finally, a stylet developed inside the cirrus; this character is clearly a synapomorphy for the species from the Galapagos Islands. For two of these species we have to postulate the additional reduction of the cirral spines and the granule glands.

The kinship diagram resulting from this analysis (Fig. 10) shows that the four Galapagos species are the most evolved members of the genus *Duplominona* and that they are more closely related to each other than to the other members of the genus. Furthermore, we can separate this Galapagoensis group into two pairs of sister species: *D. galapagoensis* and *D. karlingi* in one group, *D. krameri* and *D. sieversi* in the other. Finally, from this analysis we can conclude that the phylogenetic development of these four species of *Duplominona* has taken place within the Galapagos Archipelago. They evolved from a common stem species, which must have had a cuticular stylet inside of its cirrus. To find the most closely related taxon on the Pacific Coast of the Americas, however, is the task of additional investigations.

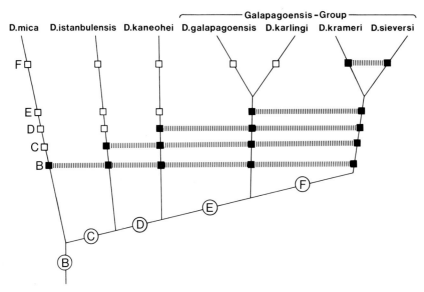

Fig. 10. Kinship diagram of the genus *Duplominona* Karling. From the analysis of apomorphic characters shown in figure 9 the *galapagoensis* group is concluded to be the most highly altered species group of the genus *Duplominona* (The problematic position of *D. westbladi* Karling is not considered).

C. Concluding remarks

I would like to stress that the foregoing interpretations suffer from the unfortunate lack of knowledge of the interstitial fauna of the Eastern Pacific. What we need most are similar investigations of the fauna of the Central American and South American coasts. I will not wait for these investigations, but will present the currently acceptable hypotheses, in the hope that they will be of heuristic value, that they will stimulate new goal oriented research.

The apparent speciation in the Galapagos interstitial fauna within the last 2—3 million years seems to contradict the hypothesis of Sterrer (1973) which states that the interstitial species of the continental coasts are between 50 and 200 million years old. This estimate was based on the modern concept of "continental drift." Sterrer interpreted the far reaching similarities in the species structure of certain interstitial taxa on both sides of the Atlantic as being the result of the genesis of that ocean. He hypothesized that the opening of the Atlantic, about 200 million years ago, separated the old "pan — Atlantic" fauna into two identical sets of species.

I believe that this contradiction can be eliminated by a careful examination of the conditions involved. There is a basic difference between the settlement of extensive coasts and immigration to isolated islands, and this fact must have a great influence on the speed of speciation. In the separation of complete communities, with several hundred, coexisting species, by "continental drift," the community structure remained intact. In a well established system of complicated interspecific competition, there surely exist only minimal possibilities for new speciation. In the settlement of a newly created group of islands, without a shallow connection to the continents, only a few, randomly selected, interstitial organisms would reach the insular region. As with the terrestrial fauna, the relatively small number of species of interstitial turbellarians from the different taxa in the Galapagos is in good agreement with this assumption. A comparison of the number of species of the interstitial Turbellaria, in all sandy tidal beaches in the Galapagos Islands, with the number of eulittoral species of only one middle-lotic sandy beach on the east side of the island of Sylt in the North Sea, shows the following remarkable differences: Acoela 14 : 23, Macrostomida 6 : 13, Proseriata 11 : 33, Typhloplanoida 9 : 60, Kalyptorhynchia 15 : 68.

The abiotic conditions of a new mesopsammic island biotope could be very similar to those of the closest continental coast. In any case, the accidentally transported fragments of such an interstitial community would at first experience only minor interspecific competition. This initial lack of competition could be the important prerequisite for rapid speciation.

Bibliography

Ax, P.: Zur Evolution der marinen Mikrofauna von Galapagos, Akad. d. Wiss. u. d. Lit. Mainz 1949–1974, 90–105 (1974).

– & R. Ax: Interstitielle Fauna von Galapagos. V. Otoplanidae (Turbellaria, Proseriata). Mikrofauna Meeresboden **27**, 1–28 (1974).

– & R. Ax: Interstitielle Fauna von Galapagos. XIX. Monocelididae (Turbellaria, Proseriata). Mikrofauna Meeresboden **64** (1977, in press).

– & U. Ehlers: Interstitielle Fauna von Galapagos. III. Promesostominae (Turbellaria, Typhloplanoida). Mikrofauna Meeresboden **23**, 1–16 (1973).

– & P. Schmidt: Interstitielle Fauna von Galapagos. I. Einführung. Mikrofauna Meeresboden **20**, 1–38 (1973).

Bailey, K.: Potassium-Argon ages from the Galapagos Islands. Science **192**, 465–467 (1976).

Ehlers, U. & P. Ax: Interstitielle Fauna von Galapagos. VIII. Trigonostominae (Turbellaria, Typhloplanoida). Mikrofauna Meeresboden **30**, 1–33 (1974).

Hennig, W.: Phylogenetic systematics. Univ. Illinois Press, Urbana, Chicago, London (1966).

Karling, T. G.: Marine Turbellaria from the Pacific Coast of North America. IV. Coelogynoporidae and Monocelididae. Ark. Zool. (2) **18**, 493–528 (1966).

Karling, T. G.; V. Mack-Fira & J. Dörjes: First report on marine Microturbellarians from Hawaii. Zool. Scripta **1**, 251–269 (1972).

Sterrer, W.: Plate tectonics as a mechanism for dispersal and speciation in interstitial sand fauna. Netherl. Journ. Sea Res. **7**, 200–222 (1973).

| Mikrofauna Meeresboden | 61 | 45–63 | 1977 |

W. Sterrer & P. Ax (Eds.). The Meiofauna Species in Time and Space. Workshop Symposium,
Bermuda Biological Station, 1975

Thiobiotic Facts and Fancies
(Aspects of the Distribution and Evolution of Anaerobic Meiofauna)

by

P. J. S. Boaden

The Queen's University Marine Biology Station, Portaferry, BT22 1PF
Northern Ireland

Abstract

The thiobios is defined as the entire biota of sulphide rich and redox potential discontinuity layer biotopes. Phylogenetic terms are extended to habitat evolution. Four main types of vertical distribution of beach sand meiofauna are apparent: individual species being either limited to aerobic, RPD layer or anaerobic sand or else occurring mainly near the surface but extending into the anaerobic layer. Vertical habitat partitioning is associated with sympatric speciation for example in the turbellarian genera *Neoschizorhynchus, Proschizorhynchus* and *Schizochilus* and the nematodes *Microlaimus* and *Theristus*. Short term changes in vertical distribution for example of the gastrotrich *Thiodasys sterreri* are correlated with diurnal, tidal and biological factors. Annual changes are apparently associated with the oxygen and temperature regime.

Pigmented species often occur in the RPD layer and may be useful as "bioprobes". The red colour of some thiobiotic turbellarians is due to haem compounds; the function may be respiratory or anti-oxidative; in the gnathostomulid genus *Haplognathia* red pigmentation is probably plesiomorphic and protective; in the turbellarians *Paratomella, Subulagera, Diascorhynchus* and *Karkinorhynchus* little phylogenetic evidence is available. In the latter two genera karyological investigations may prove useful since meiotic figures indicate polyploidy. Sympatric speciation and genetic isolation by this means would help explain the common occurrence of congeneric species in meiofaunal samples.

Elongation, sparse ciliation, poor adhesion and unusual reproductive features are *lebensform* characters of thiobiotic metazoans. Thiobiotic taxa are both widely distributed and inhabit a very stable environment. However the metazoan thiobios has a low species diversity in spite of its extreme age. This is due to restriction of food and respiratory resources leading to a lack of available energy in the system.

These restrictions did not exist in the Precambrian Era. It is probable that the original metazoan was an anaerobic, holobenthic, interstitial organism feeding by epidermal absorption although this model — the Thiozoon — conflicts with several current zoological theories. Helminth types of organization became established with the need to exploit new food sources even before aerobic metabolism had evolved. Poriferan, coelenterate, annelid and other types of organization arose subsequently as aerobic apomorphic habitats became available. These features are incorporated in a scheme linking the major innovative advances in lower Metazoa and other taxa to changes in habitat and niche availability in the Precambrian Era.

A. Introduction

I. Some Definitions

Baas Becking (1925) stated "As the sulphur bacteria play an important role in the cycle of sulphur in nature an attempt will be made to ascertain their true position in that cycle. The natural ecological community of these bacteria is a miniature cycle in itself, and will be called a Sulphuretum." He listed nematodes as one of the seven types of members of this community, the rest being prokaryotes. Fenchel & Riedl (1970) reported briefly on representatives of "12 to 18 phyla" including gnathostomulids, turbellarians, gastrotrichs and nematodes from "the living system of the sulphide biome". In the interim period there had been very little information published on sulphide system metazoans, with the major exception of the discovery by Ax (1956) and subsequent work on the new phylum Gnathostomulida — see Sterrer (1972), and the term sulphuretum had become the prerogative of bacteriologists and microflorists. Accordingly Boaden & Platt (1973) introduced the term "Thiobios" to reinstate meiofauna into sulphide system terminology. Thiobios (Greek, *theion*, sulphur; *bios*, life) may be defined as the entire biota of sulphide-rich and redox-potential discontinuity (RPD) layer biotopes.

II. Some Problems

The primary aim of this symposium was to discuss "the meiofauna species in space and time". Because of the need to conserve both these commodities and the lack of other factual information, my discussion will be limited to the meiofaunal thiobios of marine sands. A recent review of marine sulphur biogeochemistry is given by Goldberg & Kaplan (1974).

A thiobiotic meiozoan will normally (by definition of the thiobios) be limited in occurrence to areas which are deficient in free oxygen but it may make temporary excursions to, or temporarily encounter, oxygen-rich conditions. Some thiobionts presumably have aerobic metabolism adapted to very low levels of dissolved oxygen and as a result would be mainly limited to the RPD layer, but I fancy that many thiobiotic metazoans will prove to be true anaerobes. Whether such forms are obligatory or facultative anaerobes and whether these conditions are primary or secondary will be linked with the evolutionary history of the taxon concerned. In trying to trace this we are of course faced with the usual problem of deciding which taxonomic features are primary and which secondary, that is which conditions are original and which derived or, in terms of Hennig's (1966) phylogenetic systematics, which are plesiomorphic and which apomorphic.

I can see no reason why Hennig's concepts should not be extended to include the totality of expression of a taxon's being; they could then embrace not only morphology but biochemistry, physiology, behaviour and ecology. Indeed, Hennig himself quotes the dictum "It will also be characteristic of species in general to exhibit morphological, physiological, and ecological peculiarities which are the pheno-typic expression of the unique hereditary configurations present in the gene pool." (Meglitsch 1954). The important idea of Lebensort-Typen (Riedl 1963) may be re-expressed as taxa in which habitat is a plesiomorph. Use of phylogenetic ter-minology may help correct the subjective impressions gained from application of terms such as advanced and primitive or higher and lower which we often apply to present day metazoan species — regardless of the fact that they have all had the same length of time (from the origin of life until now) to evolve.

Since it is the general rule at the present day for anaerobic sand to be covered by an aerobic layer, synapomorphy of anaerobiosis can be taken to imply that the stem form of the monophyletic group concerned was an interstitial aerobe; equally symplesiomorphy of anaerobiosis must usually imply common ancestry within anaerobic sand although there may be some possibility of lateral invasion from anaerobic mud. Riedl (1963, 1966) presents diagrams showing habitat evolution in marine Turbellaria with a lateral or depth progression from mud, through sand, to rock; this however predates awareness of the thiobios, and the evidence could be reinterpreted as demonstrating a general evolution from oxygen-deficient to oxygen-rich habitats.

Recently I have put forward evidence, variously based on distribution, taxono-my, physiology and biochemistry, that the first Metazoa were thiobiotic and that this is also plesiomorphic in the Gnathostomulida, Platyhelminthes and Aschelmin-thes (Boaden 1975, Maguire & Boaden 1975). Even though the thiobios occurs in the oldest metazoan habitat on earth (and elsewhere since there are redu-cing atmospheres on other planets?) we must expect the "biological laws" to be, and to have been, as applicable here as in any other biome. What evidence is there that the same kind of mechanisms govern thiobiotic meiofauna in space and time as govern other biota? What evidence should be sought for? How might such evidence be applied to determine relationships between the thiobios and other modes of life? This paper suggests some answers to these three questions.

B. Speciation and local distribution

I. Vertical distribution and space

Gause's hypothesis states that no two species can occupy identical niches. This means that each niche must be separated from any other by at least one temporal or spatial feature. At present we are a long way from understanding niche specificity in the metazoan thiobios. According to Fenchel & Riedl (1970) "Patterns of distribution largely, but not exclusively, correspond to those of chemophysical gradients." Experience by my co-worker Cathy Maguire on several local beaches shows that this is a rather idealized statement. At Ballymaconell Beach, Bangor (54° 40' N, 5° 38' W) for example, there is often no clearly defined RPD layer and there may be several patches of black sand below any part of the beach and above

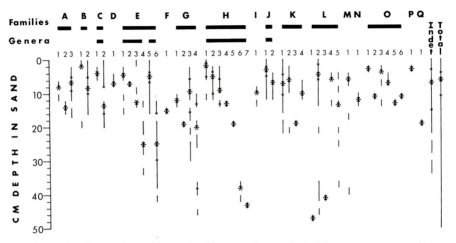

Fig. 1. Depth distribution of Nematoda. Numbers from a single 2.2 cm. diameter core (determined in 2 cm. depth subsections) from slightly above LWN, transect A, Firemore Beach, Wester Ross Scotland, 22nd July 1971 from unpublished data by R.M. Warwick and H.M. Platt. Genera in order of De Coninck 1965. Horizontal bars join species in common families and genera. Large asterisks at depth above which 50 % of each population occurred, small asterisks show 25 and 75 % levels in species with more than 10 individuals.

A1 *Ascolaimus elongatus* (Butschli) (2 individuals),

A2 *Axonolaimus orcombensis* Warwick (2),

A3 *Odontophora setosa* (Allgen) (18),

B1 *Leptolaimus* sp (2),

B2 *Stephanolaimus elegans* Ditlevsen (78),

C1 *Bathylaimus* sp. (1),

C2 *B. capacosus* Hopper (17),

D1 *Eumorpholaimus* sp. (3),

E1 *Theristus mirabilis* (Stekhoven & De Coninck) (7),

E2 *Theristus* sp. (2),

E3 *T. gelana* Warwick & Platt (5),

E4 *T.* aff *setifer* Gerlach (7),

E5 *Xyala striata* Cobb (106),

E6 *Xyala longicaudata* Ward (30),

F1 *Southernia zosterae* Allgen (1),

G1 *Chromaspirina inglisi* Warwick (2),

G2 *Metachromodora scotlandica* Warwick & Platt (1),

G3 *Sigmophora rufum* Cobb sensu Gerlach (33),

G4 *Leptonemella aphanotheca* Gerlach (12),

H1 *Microlaimus parhonestus* Gerlach (20),

H2 *M. teutonicus* Riemann (29),

H3 *M. acinaces* Warwick & Platt (10),

H4 *M. ostracion* Stekhoven (2),

H5 *Microlaimus* sp. (1),

H6 *M. honestus* De Man (2),

H7 *M. marinus* (Schulz) (1),

I1 *Dasynemella albaensis* Warwick & Platt (2),

J1 *Monoposthia mirabilis* Schulz (36),

J2 *M. costata* (Bastian) (5),

K1 *Hypodontolaimus schuurmans – stekhoveni* Gerlach (95),

K2 *Neochromadora tecta* Gerlach (9),

K3 *Chromadorissa* sp. (1),

K4 *Spiliphera hirsuta* Gerlach (8),

L1 *Choniolaimus tautraensis* (Allgen) (1),

L2 *Pomponema reducta* Warwick (179),

L3 *Neotonchus filiformis* Warwick (1),

L4 *Paracanthonchus caecus* (Bastian) (2),

L5 *Paracyatholaimus occultus* Gerlach (7),

M1 *Gammanema conicauda* Gerlach (6),

N1 *Halalaimus* aff. *longisetosus* Hopper (2),

O1 *Enoploides brunettii* Gerlach (1),

O2 *Mesacanthoides* aff. *caput-medusae* (Ditlevsen) (2),

O3 *Enoplolaimus propinquus* De man (6),

O4 *Mesacanthion africanthiforme* Warwick (3),

O5 *Gairleanema anagremilae* Warwick & Platt (2),

O6 *Rhabdodemania imer* Warwick & Platt (1),

P1 *Viscosia* sp. (nr. 1 Riemann 1966) (2),

Q1 *Desmoscolex* sp. (1), Undetermined (52), Total (819).

the main black layer. However in general it may be stated that particular species characterize particular layers in the sand (Fig. 1).

Four main types of vertical distribution may often be defined (Fig 2). These are for

I   Typical aerobic or yellow sand forms
II   Typical RPD or grey sand forms
III   Typical anaerobic or black sand forms
IV   Forms occurring mainly in the surface layers but extending into the black layer.

Types II and III comprise the bulk of thiobiotic species, but IV will contain some opportunistic species.

A kite diagram for various other metazoan taxa including some gnathostomulids and nematodes is presented in Fenchel & Riedl (1970). Ott & Schiemer (1973), in an important paper on nematode respiration, also give further details and references on anaerobic distribution of nematodes. The Ballymaconell results for kalyptorhynchid turbellarians are in close agreement with the excellent paper by Hoxhold (1974) on population structure and dynamics in the beach at Sylt. Such results

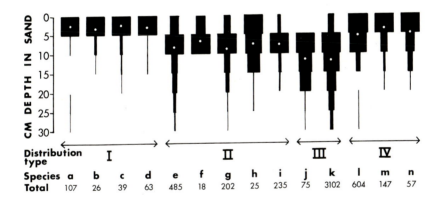

Fig. 2.   Depth distribution of various Gastrotricha and Turbellaria. Totalled distribution from three 3.2 cm. diameter cores taken each month June 1973 to August 1974 from just below MTL Ballymaconell Beach, Bangor, Co. Down, Northern Ireland from unpublished data by C. Maguire. Circles show depth above which 50 % of each species population occurred in the cores.

a   *Dactylopodola baltica* Remane,
b   *Neoschizorhynchus brevipharynggus* Schilke,
c   *Proschizorhynchus triductibus* Schilke,
d   *Schizochilus marcusi* Boaden,
e   *Neodasys chaetonotoideus* Remane,
f   *Neoschizorhynchus longipharynggus* Schilke,
g   *Paratomella* sp. (Fig. 5),

h   *Proschizorhynchus bivaginatus* Schilke,
i   *Pseudoschizorhynchoides ruber* Schilke,
j   *Neoschizorhynchus parvorostro* Ax & Heller
k   *Thiodasys sterreri* Boaden,
l   *Cephalodasys turbanelloides* (Boaden)
m   *Diascorhynchus rubrus* Boaden
n   *Schizochilus choriurus* Boaden

clearly demonstrate habitat partitioning associated with depth in the sand. This is of particular evolutionary interest where closely related species are concerned, for example of *Neoschizorhynchus*, of *Schizochilus* and of *Proschizorhynchus*, (Fig. 2) and *Microlaimus* and *Theristus* (Fig. 1) since it may result from sympatric speciation.

Most abiotic and biotic factors will change with depth in sand but unfortunately I have no idea which of these (such as type and abundance of food, presence of competitors, oxygen availability, temperature, etc.) are the main cause(s) of partitioning. This is hardly surprising since much more information about environmental parameters is required for all meiofaunal species before the factors likely to be limiting in any particular set of conditions can be predicted. However it may be stated as a general rule for benthic meiofauna that resource gradients which are vertical to the sediment surface impose much narrower space limitations than horizontal gradients.

Since, as competition theory predicts, niche breadth broadens with increased intraspecific competition but narrows with increased interspecific competition (O'Connor et al 1975), it should be instructive to compare the vertical distribution of a species where it occurs in biota with low and with high species diversity. In the latter case it will be much more restricted in distribution and therefore of more value as a "bioprobe" (although we need to know much more about response time) – a concept recently independently arrived at by both Hummon (1975) and Wieser (1975).

II. Vertical distribution and time

Local spatial distribution of the thiobios will of course vary with time. In beaches we have some knowledge of short-term migrations (Fig. 3) with upward movement associated with times of high activity in the aerobic biota. Such changes are correlated with diurnal, tidal and biological factors (Boaden & Platt 1973). Fenchel & Riedl (1970) state "The time pattern of the population dynamics shows progressive as well as rhythmical changes, as do the chemophysical parameters, analogous to the ones known for oxidized sands . . .: long-term, seasonal, semidiurnal, and tidal changes", but give no further information.

Vertical distribution of the gastrotrich *Thiodasys sterreri* Boaden over a one year period is shown in Fig. 4; the distribution changes appear to be associated with the temperature and oxygen regimes. Again according to Fenchel & Riedl (1970) there are seasonal changes in the depth of the RPD layer "since higher temperature causes higher position of RPD, while mixing by winter storms lowers RPD-level, whereas, at the same time input of organic matters causes it to rise again." Müller & Ax (1971) describe the distribution of *Gnathostomula paradoxa* Ax in relation to the RPD horizon over a one year period in the beach at Sylt; it "typically occurs in the upper 8 to 10 cm of the sulphide layer." For the same beach Hoxhold (1974) maintains that temperature is the determining factor in seasonal kalyptorhynchid population migration.

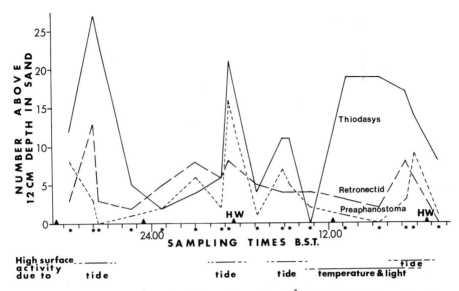

Fig. 3. Short term population distribution changes in three typical thiobiotic species. Numbers in six 1.6 cm. diameter cores from MTL $\frac{1}{2}$ HWN South Bay, Co. Down, Northern Ireland. For further details of sampling regime and physical parameters see Boaden & Platt (1973).

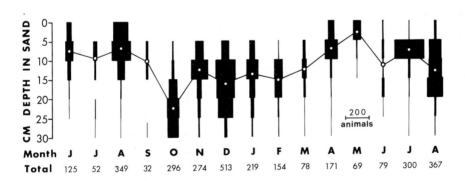

Fig. 4. Depth distribution of *Thiodasys sterreri*. Numbers in three 3.2 cm. diameter cores taken each month from June 1973 to August 1974 from Ballymaconell Beach, Bangor, Co. Down, Northern Ireland from unpublished data by C. Maguire. Circles show depth above which 50 % population in the cores occurred.

C. Speciation and red colouration

Among the meiofaunal taxa which may prove particularly useful as "bioprobes" in helping determine spatial and temporal changes in the environment are various species which are conspicuous by being coloured. Schilke (1970) for example lists several coloured Kalyptorhynchia but states the function of this feature, if any, is unknown.

During studies of beaches in Northern Ireland, North Wales and Scotland I have encountered many red turbellarians including representatives of the genera *Paratomella* and *Pseudaphanostoma* (Acoela), *Diascorhynchus, Karkinorhynchus* and *Pseudoschizorhynchoides* (Kalyptorhynchia) and *Subulagera* (Typhloplanoida). These species are all typical of the RPD layer or areas where there is variable and often low oxygen availability; they have a type II or IV vertical distribution pattern. Red gnathostomulids are also found in similar situations — for example see Sterrer (1968) for *Haplognathia ruberrima* (Sterrer), *H. rubromaculata* (Sterrer) and *H. rosea* (Sterrer); Kirsteuer (1969) for *Pterognathia grandis* Kirsteuer. Schiemer (1973) has shown by cartesian diver experiments, apparently using well aerated water, that *H.* aff. *ruberrima* has a very low oxygen uptake rate.

Since red colouration is particularly associated with thiobiotic conditions, is not usually associated with the gut and occurs in fairly widely differing taxa, this character may be cited as an example of convergent evolution and assumed to have some adaptive significance. I have a rather subjective opinion (or fancy) that the red species of a least *Diascorhynchus, Pseudoschizorhynchoides* and *Subulagera,* and also *H. ruberrima,* are more intensely coloured when recovered from, or left in, well oxygenated conditions than when found in true thiobiotic conditions. It is therefore tempting to assume that the colour is due to respiratory pigment.

I. Function

Preliminary absorption spectra for *Diascorhynchus rubrus* and two undescribed species (*Paratomella* sp. and *Subulagera* sp.) are presented in Fig. 5. These were simply prepared by placing squash preparations of about 140, 60 and 30 animals respectively across the beam in a Bausch & Lomb 700 spectrophotometer, then reading absorbance against a microscope/seawater/coverslip blank at 5 nm wavelength intervals. Specimens had been stored in watch glasses of normally aerobic seawater for between 12 – 24h after collection and extraction but the oxygen availability once under the coverslip was quite unknown. The spectra clearly demonstrate the presence of haem (γ Soret band at about 415 nm). I must confess the fact that the presence of other bands which might confirm the pigment as haemoglobin (e. g. oxidized β at 520 – 535 and α at 550 –565; reduced single band at 550 nm) remains rather fanciful at present.

Fig. 5.   Preliminary absorption spectra of three typical red thiobiotic Turbellaria (undescribed *Paratomella* and *Subulagera* sp., *Diascorhynchus rubrus*). A, within 15 min. of preparation; B, after 45 min.; *Subulagera,* upper within 20 min., lower another 20 min. after drawing deoxygenated water under coverslip. Insets show male organ and pore in *Paratomella*, male stylet in *Subulagera*. For further details see text.

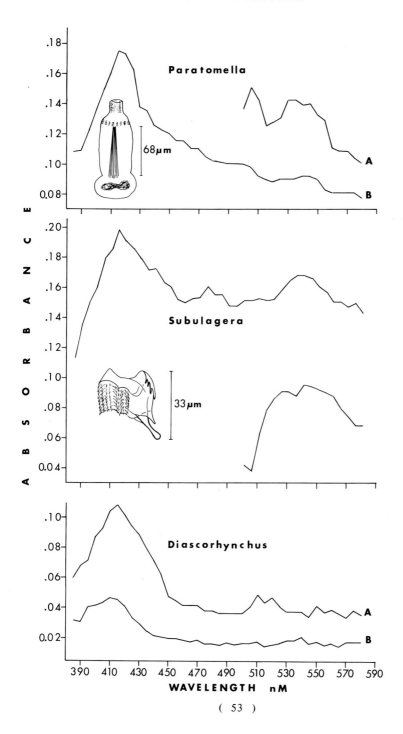

However, even if an oxygen carrying function were proved the pigment might be being used, not for respiration, but as an oxygen retention site serving to prevent oxygen interfering with anaerobic metabolism when the animals temporarily encountered oxygen rich conditions. The protective retention function would be probable if thiobiosis was plesiomorphic and the normal respiratory function probable if thiobiosis was apomorphic in the genera concerned.

II. Evolution of red species

Thiobiosis is a habitat symplesiomorph of the Gnathostomulida. It is interesting that detailed analysis of morphology in *Haplognathia* (*Pterognathia*) species has led Sterrer (1968) to consider *H. ruberrima* as remaining closest to the ancestral condition. *H. rosea, H. rubromaculata* and the yellowish *H. lunulifera* are the next most closely related species. The red members of this genus are therefore prime candidates for the protection theory; Schiemer's (1973) low "respiration rate" may therefore really represent an attempt to maintain essential anaerobic functions.

The question of thiobiosis as original or derived in the turbellarian genera with red species is much more difficult. Little phylogenetic evidence is available in two of the three genera for which preliminary pigment analysis is presented since only single species of *Paratomella* and *Subulagera* have so far been described. Both new red species are very similar to the known forms so it seems likely they are sister species and should conform to the "deviation rule" (Hennig 1966). Perhaps *P. unichaeta*

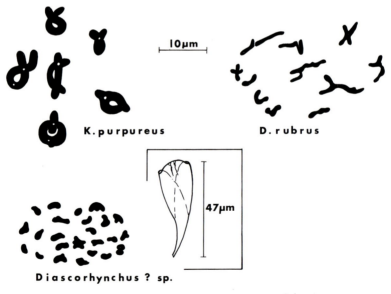

Fig. 6. Meiotic figures from *Karkinorhynchus* and *Diascorhynchus* species, indicating possible polyploidy.

Dörjes is more apomorphic ("chaeta" and parotomy) than the red *Paratomella* but the red *Subulagera* more apomorphic (complexity of male armament) than *S. mucronata* Ehlers.

There seems more hope of a future answer to the question in the Karkinorhynchidae and Diascorhynchidae especially in the species-rich type genera. Two red forms, *Diascorhynchus rubrus* and *Karkinorhynchus purpureus* Schilke, occasionally occur in the same sand cores, the deeper coloured *K. purpureus* deeper in the sediment than *D. rubrus*. Meiotic figures (Fig. 6) from the testis region are readily obtainable in aceto-carmine squash preparations, following the procedure of Austin (1959), though this has not proved an easy technique in some other kalyptorhynchids. The large chromosomes and obvious chiasmata, particularly in *K. purpureus,* offer good material for genetic studies. The indication of polyploidy, which is thought to be rather rare in sexually reproducing animals, also warrants further karyological investigation of the closely related (Schilke 1970) Karkinorhynchidae and Diascorhynchidae. Polyploidy offers obvious opportunities for sympatric speciation since it acts as an effective genetic isolating mechanism and would help explain the fairly common occurrence of congeneric species in meiofaunal samples, for example see Figs 1 & 2 and Boaden, Wieser & Sterrer in Sterrer (1966).

D. Speciation and geographic distribution

I. Transatlantic Distribution

Little is known of the geographical distribution of the thiobios. Sterrer (1973) has commented on the Atlantic distribution of gnathostomulid genera, in particular *Haplognathia,* and concluded that there has been little speciation since continental drift established present oceans. Thus *H. rosacea* Sterrer and *H. rosea* are "trans-allopatric" sister-species having evolved on either side of the widening Atlantic from a common mother-species. It seems likely that a similar situation holds in the thiobiotic turbellarian family Retronectidae, for as Sterrer & Rieger (1975) state "the European and American coasts of the Atlantic yield different (though often closely related) species."

In the thiobiotic Gastrotricha the genus *Thiodasys* is known to occur on both sides of the Atlantic (personal observation) and possibly in the Pacific (Schmidt 1975, Boaden 1975). The extraordinary new genus, a very elongated form without lateral tubules and with unusual reproductive features, being described by Gagne (personal communication) from the Massachusetts coast is also represented by at least two species in Northern Ireland where it appears to replace *Thiodasys* in finer sand beaches.

II. Lebensformmerkmale

These thiobiotic gastrotrichs are characterized by sparse ciliation, great length and poor adhesive power; sperm production, structure and transfer also often show unusual features. These trends are paralleled in the Gnathostomulida and Retronectidae and can therefore be cited as specific characters of the metazoan thiobios (Boaden 1974, Fenchel & Riedl 1970).

III. Dispersion

It is well known that the means of dispersal available to interstitial fauna are apparently very limited and certainly do not seem able to explain transoceanic distribution (Sterrer 1973). Short range transport, for example by adhesion to sand grains is obviously possible (Boaden 1968, Rieger & Ott 1971); if this did not occur much more allopatric speciation or subspeciation should be apparent intertidally since the population would be split into numerous small isolated reproductive communities. Evidence for allopatric speciation in interstitial fauna is discussed by Sterrer (1973) but is mainly concerned with "aerobic" species. The possibility of predominantly anaerobic species surviving transport by any of the means (other than plate tectonics) which have been invoked to explain distribution is remote if not impossible since they involve transport through an aerobic medium.

E. Speciation and habitat stability

It is also well known that uniformity of physical conditions such as salinity and temperature increases with depth in sand (Bruce 1928). There should, therefore, be even less abiotic "evolutionary pressure" in the thiobios than in other interstitial biota; new adaptations may mainly be initiated by biotic factors. The fringes of a dense population may be forced into areas of sub-optimal resource, adaptation and isolation then producing sister-species. Similarly a competing species may occupy the centre of another species resource gradient, thereby separating it into two populations leading to sister-species production. In either case gene flow along the resource gradient may already be slow or static because of poor distributive power (e. g. Wieser 1960).

Such mechanisms help explain the usually high diversity of stable biotopes as predicted by the stability-time hypothesis of Sanders & Hessler (1969). Similarly Bretsky & Lorenz (1970) believe that taxa from stable environments are characterized by low genetic variability with resulting low polymorphism, narrow niche breadth, provincial distribution and a tendency to speciate. However such hypotheses have been hotly contested by Boucot (1975) whose hypothesis is that evolutionary rate is inversely related to worldwide population size rather than environmental stability. All present available evidence for thiobiotic taxa is that they are both widely distributed and inhabit a very stable invironment. Are these hypotheses incompatible when applied to species diversity in the thiobios? Probably not: high evolutionary rate leads to high diversity in any particular habitat only if there is also a lower extinction or habitat emigration rate; high diversity can be achieved equally well by slow evolutionary rates over long time periods as by high rates over short periods.

Resource poverty

In general the metazoan thiobios is not nearly as diverse as the overlying "aerobios." This certainly seems to contradict the stability-time hypothesis but Sanders & Hessler (1969) provide an exemption clause by excluding from the category of physically stable habitats not only environments with great unpredictability but also those with great extremes of any one factor.

Two factors which are at an obvious extreme for present day thiobios are the lack of resources for primary production and the lack of resources for respiration. These all tend to be used in the upper sand layers; it is interesting to recall the upward migration of the thiobiotic Metazoa at times of high aerobic activity (see earlier). The presence of O_2, SO_4, NO_3 etc. in the upper sediment is necessary for the thiobios which is dependent on these substances, though perhaps in a transformed state e. g. $O_2 \rightarrow CO_2$, and on an organic input to survive (Fenchel 1969).

The apparent absence of thiobiotic Metazoa from anaerobic parts of the Baltic (Elmgren, personal communication) can be accounted for by the lack of any oxidative source sufficient for anaerobic metabolism of these forms even at the sediment surface.

However, regardless of the reasons for low diversity in an apparently stable environment and contrary to my snap judgement on *Pterognathia* (Sterrer 1966), it must be accepted that speciation "is very slow, in some cases quasi-stagnant . . . with species ages of 50 to 200 million years or more" (Sterrer 1973), if not in all interstitial fauna, in the sand thiobios at least.

F. The thiobios and early metazoan evolution

Further back in time the thiobios was not so restricted in evolutionary opportunity. From the evidence for the origin of life in reducing conditions (Oparin 1957), for sulphide-system Lebensort-Typen in Monera, Protista, Fungi and Animalia (Fenchel & Riedl 1970) and for the anaerobic origin of the gnathostomulids, platyhelminths and aschelminths (Boaden 1975) it may be concluded that thiobiosis is plesiomorphic for all the earths life, for 4 of the 5 living kingdoms and for the earliest metazoan phyla.

I. The original metazoan habitat

In my opinion the original metazoans were anaerobic, holobenthic, interstitial thiobionts feeding by epidermal absorption. Briefly:

anaerobic because this type of metabolism is the most widespread amongst life and whereas aerobic forms also always have anaerobic metabolic pathways the reverse is not true;

holobenthic because a planktonic phase could not exist in open water where insufficient food was available and ultraviolet radiation was intolerable;

interstitial because small size would have precluded burrowing existence and food (dissolved and particulate organic matter) would be concentrated in sediments; *absorptive feeding* because of the widespread occurrence of external absorptive epithelia in marine animals and the high levels of dissolved organic matter in redu-cing conditions.

This model — the Thiozoon — conflicts with several current zoological theories. It suggests, for example, that internal fertilization was the original condition in Metazoa whereas Franzén's (1956) primitive sperm type is associated with external fertilization. It contradicts the theory that the pelagobenthic life cycle is plesio-morphic in the Metazoa (Jägersten 1972) and is in partial conflict with Noodt (1974) who suggests most interstitial forms were derived via the epibenthos. Its most fundamental disagreement, however, is with the present generally held view that "eucaryotic organisms are fundamentally aerobic" and have always been associated with aerobiosis which is often given importance in discussing early evolution (Raff & Mahler 1972, Schopf 1970).

Reaney (1974) has concluded that large anaerobic heterotrophs were the main evolutionary line prior to the atmospheric oxygen level reaching the Pasteur Point but I do not agree with his associated conclusion that prokaryotes arose by D.N.A. reduction of such forms. If anything, the common occurrence of intimate symbio-sis of prokaryotes with protists, nematodes (Fenchel & Riedl 1970) and some turbellarians (personal observation) in the thiobios supports the theory of even more ancient prokaryote-eukaryote organelle transformation proposed by Margulis (1970), Sagan (1967) and others.

The present lack of fossil evidence cannot be held as a strong argument against extensive existence of metazoans in the Mid-Precambrian until a much more exten-sive search has been made; in any case it is only the development of predominantly aerobic metabolism which allows invertebrates to secrete calcareous hard parts which are the most usual fossil evidence (Rhoads & Morse 1971). Glaessner (1969) reports trace fossils from worm-like organisms dating from 1×10^9 years and it would presumably have taken a considerable time for the structure and size neces-sary to produce such traces to have evolved. Apparently by the late Precambrian some metazoans had evolved to arthropodean complexity (Glaessner & Wade 1971). For the reasons presented earlier in this paper and elsewhere (Boaden 1975, Magui-re & Boaden 1975) I believe many of the biological events set out in the Precambri-an timetable (Schopf 1975) must be considerably backdated.

II. Habitat evolution of the early metazoa

The first major evolution of the metazoan thiobios took place over 2×10^9 years ago and led to the establishment of the gnathostomulid, platyhelminth (+ cate-nulid) and aschelminth evolutionary lines.

The basic metazoan feeding strategy had been to live in — and remains, to obtain by digestion — food in dissolved concentration sufficient to transfer across membra-nes and transport to its utilization site to yield a net gain in energy. As animals (and

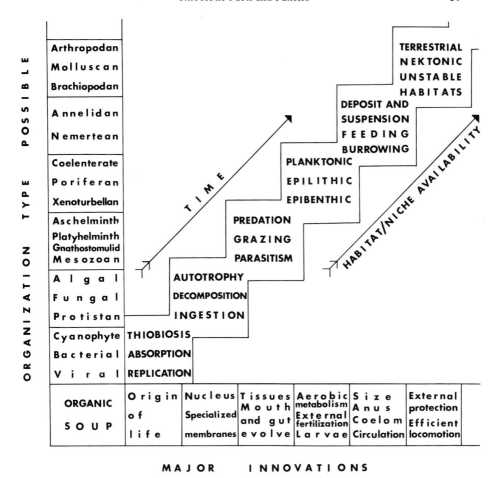

Fig. 7. A scheme to relate habitat, niche and the major innovative evolution of organisms in the Precambrian Era.

plants) evolved the supply of organic material gradually gave out. Species were forced to shift their food acquisition to simpler precursors (cf. biochemical synthesis, Horowitz 1945) or to seek other habitats or feeding mechanisms (Fig. 7).

In the plant evolutionary line, in some monerans and in some protistans such precursor and mechanism changes led to increased autotrophy. Other monerans, protistans and early fungi specialized in external decomposition of biologically produced material.

The metazoans predominantly evolved new feeding mechanisms; some specialized in grazing organic matter from sediment particles (gnathostomulid and aschelminth evolutionary lines?), others in predation of other organisms (platyhelminth and later aschelminth lines?). Either activity required the evolution of a mouth. A few forms (Mesozoa?) together with some monerans and protistans became parasitic (in a similar manner to some earlier protoorganisms which had evaded extinction by becoming viruses?). These early evolutionary lines were all established in the benthos.

Anaerobic and microaerobic conditions would have co-existed for a considerable time as oxygen production by autotrophs gradually emptied the ancient $FeCO_3$ reservoir (Garrels & Perry 1974). As oxygen levels rose there would be increased shielding from ultraviolet radiation and epibenthic forms could have survived at depths of less than 10 m (Lowenstam 1974) (evolution of different pigments in algae as other light wavelengths available?). The period from the Mid — to Late Precambrian as dissolved oxygen levels rose toward 1 ml/l must have been one of the most critical in early metazoan evolution. Adaptation of metabolism by reversing anaerobic Krebs cycle sequences (Hochachka & Somero 1972, Maguire & Boaden 1975) or by respiring oxygen rather than nitrate (Hall 1971) openned the way to numerous new habitats.

Perhaps some protistans were the first to exploit this opportunity some becoming colonial and adopting epilithic, epiphytic or planktonic existence (poriferan and coelenterate lines?).

At about the same time external fertilization and production of larvae were introduced and there was "experimentation" with oxygen-carrying pigments. The advent of aerobic metabolism and of an anus allowed more complex structures to develop in other forms which at first remained benthic (nemertean and annelid lines?).

The active filtration of seawater (suspension feeding), exploitation of the burrowing habitat available in mud, and the actual ingestion of sediment were possibilities which could only be fully exploited following this evolution of size and more efficient metabolism. (Perhaps the enigmatic *Xenoturbella bocki* Westblad survives from an early "attempt" at such adaptation?).

Life in shallow water benthos must have continued to present problems due to abiotic factors such as abrasion and sediment instability and normal biotic pressures such as predation. However with the evolution of aerobic metabolism the deposition of protective calcareous structures had become possible (arthropod and molluscan lines?). The size and spare metabolic capacity of such forms also led to further opportunities for parasitism by "lower" forms.

By now time for the exploitation of space by the thiobios had run out and it was only by adopting apomorphic habitats that major evolutionary opportunities were still open. There were now aerobic biota interposed between the thiobios and the sites of primary organic production. The gnathostomulids had failed to make the transition to aerobic habitats. The gastrotrichs were successful in aerobic sands but were excluded from other marine habitats by competition and/or lack of adaptive flexibility. Some of the nematodes and platyhelminths were even more successful in leaving both the thiobiotic and interstitial habitats, but apart from the evolution of parasitism (which in part represented a continuation or return to anaerobic conditions) they remained largely confined to habitats with intimate substratum contact.

G. Conclusion

The golden age of the thiobios was over one to two thousand million years ago. Evidence of evolutionary mechanisms in the thiobios is obscured by this enormous time factor and the ensuing ultra-conservative nature of its taxa. Further information can probably only be derived by a synthesis of new data from many fields of biology which should include detailed phylogenetic, morphological, karyological, ultrastructural, biochemical and palaeo-ecological studies.

Fact and fancy

The answer to the last of my three introductory questions, if it cannot be found in the text, is that fancy must be proven by the production of facts and facts by the production of fancies. In more formal terms we can only procede by the scientific method of formulating and testing hypotheses. My thiobiotic fancies may be regarded as fact only if they produce fresh insight, help explain the previously inexplicable and provide a firm basis for new information and inference.

Acknowledgments

I particularly wish to thank Cathy Maguire, Howard Platt and Richard Warwick for permission to include previously unpublished data. Part of the work was supported by a United Kingdom Natural Environment Research Council grant.

Bibliography

Austin, A.P.: Iron-alum aceto-carmine staining for chromosomes and other anatomical features of Rhodophyceae. Stain Technol. 34 (2), 69–75 (1959).

Ax, P.: Die Gnathostomulida, eine rätselhafte Wurmgruppe aus dem Meeressand. Abh. math. naturw. Kl. Akad. Wiss. Mainz, 8, 534–560 (1956).

Baas Becking, L.G.M.: Studies on the sulphur bacteria. Ann. Bot., 39, 613–650 (1925).

Boaden, P.J.S.: Water movement – a dominant factor in interstitial ecology. Sarsia 34, 125–136 (1968).

–: Three new thiobiotic gastrotrichs. Cah. Biol. mar. 15, 367–378 (1974).

–: Anaerobiosis, meiofauna and early metazoan evolution. Zool. Scripta. 4, 21–24 (1975).

– & H.M. Platt: Daily migration patterns in an intertidal meiobenthic community. Thalassia jugosl. 7 (1), 1–12 (1973).

Boucot, A.: Evolution and Extinction Rate Controls. Elsevier, Amsterdam. 427 pp. (1975).

Bretsky, P.W. & D.M. Lorenz: Adaptive response to environmental stability: a unifying concept in paleoecology. Proc. N. Am. Paleontol. Conv. F, 522–550 (1970).

Bruce, J.R.: Physical factors on the sandy beach. Part 1. Tidal, climatic and edaphic. J. mar. biol. Ass. U.K. 15, 535–552 (1928).

Fenchel, T.: The ecology of marine microbenthos IV. Ophelia 6, 1–182 (1969).

– & R.J. Riedl: The sulfide system: a new biotic community underneath the oxidized layer of marine sand bottoms. Mar. Biol. 7, 225–268 (1970).

Franzén, Å.: On spermiogenesis, morphology of the spermatozoan and biology of fertilization among invertebrates. Zool. Bidr. Upps. **31**, 355–480 (1956).

Garrels, R.M. & E.A. Perry Jr.: Cycling of carbon, sulfur and oxygen through geologic time. In The Sea **5** (ed. Goldberg, E.D.), 303–336, Wiley – Interscience, New York. 895 pp. (1974).

Glaessner, M.F.: Trace fossils from the Precambrian and basal Cambrian. Lethaia **3**, 369–393 (1969).

– & M. Wade: *Praecambridium* – a primitive arthropod. Lethaia **4**, 71–77 (1973).

Goldhaber, M.B. & I.R. Kaplan: The sulfur cycle. In The Sea **5** (ed. Goldberg, E.D.) 569–655, Wiley – Interscience, New York. 895 pp. (1974).

Hall, J.B.: Evolution of the prokaryotes. J. theor. Biol. **30**, 429–454 (1971).

Hennig, W.: Phylogenetic Systematics. University of Illinois Press, Urbana. 263 pp. (1966).

Hochachka, P.W. & G.N. Somero: Strategies of Biochemical Adaptation. Saunders. Philadelphia, 358 pp. (1972).

Horowitz, N.: On the evolution of biochemical synthesis. Proc. Nation. Acad. Sci. U.S.A. **31** (6), 153–157 (1945).

Hoxhold, S.: Zur Populationsstruktur und Abundanzdynamik interstitieller Kalyptorhynchia (Turbellaria, Neorhabdocoela). Mikrofauna Meeresboden **41**, 1–134 (1974).

Hummon, W.D.: Habitat suitability and the ideal free distribution of Gastrotricha in a cyclic environment. In Proc. 9th Europ. mar. biol. Symp. (ed. H. Barnes) 495–525, Aberdeen University Press, Aberdeen. 838 pp. (1975).

Jägersten, G.: Evolution of the Metazoan Life Cycle. Academic Press, London. 282 pp (1972).

Kirsteuer, E.: On some species of Gnathostomulida from Bimini, Bahamas. Am. Mus. Novit. **2356**, 1–21 (1969).

Lowenstam, H.A.: Impact of life on chemical and physical processes. In The Sea **5** (ed. Goldberg, E.D.) 715–796, Wiley – Interscience, New York. 895 pp. (1974).

Maguire, C. & P.J.S. Boaden: Energy and evolution in the thiobios: an extrapolation from the marine gastrotrich *Thiodasys sterreri*. Cah. Biol. **16**, 635–646 (1975).

Margulis, L.: The Origin of Eukaryotic Cells. Yale University Press, New Haven. 349 pp. (1970).

Meglitsch, P.A.: On the nature of the species. Syst. Zool. **3**, 49–64 (1954).

Müller, U. & P. Ax: Gnathostomulida von der Nordseeinsel Sylt mit Beobachtungen zur Lebensweise und Entwicklung von *Gnathostomula paradoxa* Ax. Mikrofauna Meeresboden **9**, 1–41 (1971).

Noodt, W.: Anpassung an interstitielle Bedingungen: Ein Faktor in der Evolution Höherer Taxa der Crustacea? Faun. – ökol. Mitt. **4**, 455–452 (1974).

O'Connor, R.J., P.J.S. Boaden & R. Seed: Niche breadth in Bryozoa as a test of competition theory. Nature, Lond. **256**, 307–309 (1975).

Oparin, A.I.: The Origin of Life on the Earth. Academic Press, New York. 495 pp. (1957).

Ott, J. & F. Schiemer: Respiration and anaerobiosis of free living nematodes from marine and limnic sediments. Neth. J. Sea. Res. **7**, 233–243 (1973).

Raff, R.A. & H.R. Mahler: The non symbiotic origin of mitochondria. Science, N.Y. **177**, 575–582 (1972).

Reanney, D.C.: On the origin of prokaryotes. J. theor. Biol. **48**, 361–371 (1974).

Rhoads, D.C. & J.W. Morse: Evolutionary and ecological significance of oxygen deficient marine basins. Lethaia **4**, 413–428 (1971).

Riedl, R.J.: Probleme und Methoden der Erforschung des litoralen Benthos. Zool. Anz., suppl. **26**, 505–567 (1963).

–: Biologie der Meereshöhlen. Paul Parey, Hamburg. 636 pp. (1966).

Rieger, R. & J. Ott: Gezeitenbedingte Wanderungen von Turbellarien und Nematoden eines nordadriatischen Sandstrandes. Vie Milieu **22**, 425–447 (1971).

Sagan, L.: On the origin of mitosing cells. J. theor. Biol. **14**, 225–274 (1967).

Sanders, H.L. & R.R. Hessler: Ecology of deep sea benthos. Science N.Y. **163**, 1419–1424 (1969).

Schiemer, F.: Respiration rates of two species of gnathostomulids. Oecologia (Berl.) **13**, 403–406 (1973).

Schilke, K.: Zur Morphologie und Phylogenie der Schizorhynchia (Turbellaria, Kalyptorhynchia). Z. Morph. Tiere **67**, 118–171 (1970).

Schmidt, P.: Interstitielle Fauna von Galapagos. IV. Gastrotricha. Mikrofauna Meeresboden **26**, 1–76 (1975).

Schopf, J.W.: Precambrian micro-organisms and evolutionary events prior to the origin of vascular plants. Biol. Rev. **45**, 319–352 (1970).

–: The age of microscopic life. Endeavour **34**, 51–58 (1975).

Sterrer, W.: Neue Gnathostomulida. Veröff. Inst. Meeresforsch. Bremerhaven **2**, 201–207 (1966).

–: Beiträge zur Kenntnis der Gnathostomulida I. Anatomie und Morphologie des Genus *Pterognathia* Sterrer. Ark. Zool. Ser 2., **22**, 1–125 (1968).

–: Systematics and evolution within the Gnathostomulida. Syst. Zool. **21**, 151–173 (1972).

–: Plate tectonics as a mechanism for dispersal and speciation in interstitial sand fauna. Neth. J. Sea Res. **7**, 200–222 (1973).

– & R. Rieger: Retronectidae – a new cosmopolitan marine family of Catenulida (Turbellaria). In Biology of the Turbellaria (ed. Riser, N. W. & Morse, M. P.) 63–92, McGraw-Hill, New York. 530 pp. (1974).

Wieser, W.: Problems of species formation in the benthic microfauna of the deep sea. In Perspectives in Marine Biology (ed. Buzzati-Traverso, A.A.) 513–518. University of California Press. Berkeley. 621 pp. (1960).

–: The meiofauna as a tool in the study of sediment heterogeneity: ecophysiological aspects. A review. Cah. Biol. mar. **16**, 647–670 (1975).

– , J. Ott, F. Schiemer & E. Gnaiger: An ecophysiological study of some meiofauna species inhabiting a sandy beach at Bermuda. Mar. Biol. **26**, 235–248 (1974).

Mikrofauna Meeresboden	61	65—88	1977

W. Sterrer & P. Ax (Eds.). The Meiofauna Species in Time and Space. Workshop Symposium, Bermuda Biological Station, 1975

The "Primitive" Interstitial Ciliates: Their Ecology, Nuclear Uniquenesses, and Postulated Place in the Evolution and Systematics of the Phylum Ciliophora[1]

by

John O. Corliss and Eike Hartwig

Department of Zoology, University of Maryland, College Park, Maryland, USA and Zoological Institute and Museum, University of Hamburg, Hamburg, West Germany

Abstract

Among the interstitial ciliates of the meiofaunal community are two allegedly "primitive" groups of special interest from an evolutionary point of view: the few homokaryotic forms; and, much more numerous, the species with diploid, non-dividing macronuclei. In the first group, represented solely by four species of *Stephanopogon* (not truly interstitial), there are two or more nuclei, each with a large RNA-rich nucleolus and each capable of division by mitosis. In the several genera (*Geleia, Kentrophoros, Loxodes, Trachelocerca,* etc.) of the other group (obligatorily-interstitial forms), a heterokaryotic nuclear complement is present, but the macronucleus is totally unlike the polyploid-polygenomic type so characteristic of the great majority of ciliate species, both facultatively psammobiotic forms and the thousands found in other biotopes. It is diploid and non-dividing, depending on the micronucleus for periodic replacement. Though containing nucleoli, it is small and often numerous; sometimes it is part of a loosely-bound nuclear complex.

Ecologically, these particular psammophilic ciliates appear to have a cosmopolitan distribution, being "endemic" only in the sense of being restricted to (or, for *Stephanopogon,* near to) interstitial habitats, fresh-water (*Loxodes*) or marine. They show marked euryhalinity and are eurythermic. Their global distribution leaves unanswered the problem of methods of dispersal. Continental drift may well have played a role, conveniently (if slowly) by transportation of the substratum itself.

Because of lack of fossil material, evidence for the primitiveness of the nuclear condition in these ciliates has had to be indirect, but it is compelling. The alternative would seem to require repeated, identical, independent origin either of the polyploid condition from the diploid (once nuclear dualism was established) or, by some kind of regression or degeneration, the diploid from the polyploid. The problem is discussed in detail. The impact on the systematics of the "lower" ciliates is considerable. Defense is offered of the separateness of the orders which have recently been erected, one — the Primociliatida — to contain the homokaryotic forms, and the other — the Karyorelictida — to embrace the group of species whose macronuclei are "karyological relicts." Because of their nuclear uniquenesses, these two groups are considered to be contemporary representatives — "living fossils" — of once-ancient "eociliate" groups which, in due time, gave rise to all of the more recently evolved groups still with us today.

[1] Contribution No. 690 from the Bermuda Biological Station for Research, Inc.

A. Introductory remarks

Forming a part of the great and still incompletely explored interstitial meiofaunal community are several hundred species of ciliated protozoa (Dragesco 1974).
Although some of these fascinating forms were first described nearly a century and
a half ago, by C.G. Ehrenberg, the "endemic" or obligatorily-psammophilic species
have generally been considered to be merely "ordinary gymnostomes," have been
buried in long-existing families comprising this large assemblage of "naked-
mouthed" ciliates among the so-called "lower holotrichs," and subsequently, in
effect, have been forgotten.

The revival of interest in these particular sand-dwellers, perhaps paralleling that
for certain of the metazoan groups sharing the biotope (e. g., see historical facts
and references in Ax 1966; Hulings & Gray 1971; and Swedmark 1964), commenced about 25 years ago with the stimulating works of Bock (1952a,b, 1953),
Fauré-Fremiet (1950, 1951, 1954, 1961), and Remane (1952), followed shortly by
the trail-blazing taxonomic-ecological monographs of Dragesco (1960, 1963a,b,
1965) and the first papers of the important continuing series by Raikov (1957a,b,
1958, 1959a,b, 1960, 1962, 1963a–d). Since those times, there has been a great
bloom of publications[1] – ecological, taxonomic, cytological, karyological, and
ultrastructural in nature – on the long-neglected interstitial ciliates, involving such
workers as Agamaliev, Ammermann, Bick, Borror, Burkovsky, Canella, Corliss,
Dragesco, Fenchel, Fjeld, Ganapati, Goulder, Grell, Hartwig, Jankowski, Jones,
Kattar, Kovaleva, Lee, Lepşi, Njiné, Nobili, Nouzarède, Petran, Poljansky, de
Puytorac, Raikov, Rao, Torch, and Tuffrau and best documented by reference to
the extensive bibliographies accompanying the recent papers or reviews by Corliss
(1974b, 1975c), Fenchel (1969), Hartwig (1973b), and Raikov (1969, 1972a).

Here we wish to concentrate on matters related to the evolutionary nature of
certain of these forms, hoping to throw light on their possible role in the very
origins of the phylum. First, something needs to be said about ciliates in general,
their occurrence through time and space, which of their characteristics may be considered "primitive" or "ancestral" as opposed to "derived," etc. With respect to
the psammobiotic species themselves, attention is focused on their peculiar nuclear
and other biological characteristics, their geographical distribution, adaptability to
their environment, and possible means of dispersal. We present the evolutionary-
phylogenetic alternatives in interpretation of such features, particularly of the
macronuclear characters. And, finally, briefly considered are the taxonomic implications of the conclusions which we believe to have been justifiably drawn on the
basis of data available to date.

---

[1] At the same time, curiously enough, there has occurred a general explosion in the overall
ciliatological literature, especially in areas of ultrastructural research (see Corliss 1974a, and
contrast with Corliss 1961). Many of the resulting new data have played a part in bringing
about new schemes of classification which, in turn, have focused fresh attention on the high-
level systematics of the particular groups under consideration in the present paper.

B. Ciliates through time

Many of the evolutionary "conclusions" drawn by biologists are no more than tacit assumptions, and perhaps some of the most agreed-upon ideas are supported by the least amount of "hard" data; yet it is not out of order to continue to accept – with proper caution – certain "natural" inferences until or unless we have good evidence of their invalidity. One such general conclusion or assumption is that the unicellular or protist organisms preceded the multicellular metazoa in the evolutionary scale. We know of no good reason to quarrel with this basic premise. But two rather frustrating facts become clear if we look at the geological time chart: (1) the protozoan progenitors of the earliest metazoa must have been in existence in the Precambrian era; but (2) the fossil record for the protozoa in such times is either non-existent (understandably, dealing primarily with soft-bodied organisms) or of little evolutionary significance (beyond limited intragroup usage, such as with the foraminiferans).

No one really knows where the ciliated protozoa stand in all of this; again, the reasonable assumption is that they, as protists, are also very ancient forms, from an evolutionary point of view. No serious attempt has been made to date to derive them from metazoa or mesozoa or the like, although some zoologists have hinted that this might, in truth, have been their origin (see discussion in Corliss 1972).

We do now have good fossil evidence that ciliates were certainly present more than 400 million years ago, early in the Ordovician period of the Paleozoic era. But such fossils are of the loricae of what we consider to be "advanced" ciliates, members of the order of oligotrichs, the tintinnines (Tappan & Loeblich 1968, 1973). A few folliculinid fossils, another loricate and another presumably highly-evolved ciliate group, are also known, but of a considerably younger age (Deflandre & Deunff 1957). They occur in rocks of Mesozoic age, as do also many tintinnids, 100–200 million years ago. But none of the groups which are being considered in the present paper as relatively primitive left a fossil record. (We may well be justified in considering them to be "living fossils," but there is no way to date such material!).

A number of protozoologists have enjoyed estimating the minimal age of parasitic or symbiotic forms from knowledge of the known ages of their hosts. For ciliates, for example, Mohr (1966) has suggested that the chonotrichs – a group of "lower" ciliates but surely highly specialized within its own line (see Jankowski 1973) – must have been around as epibionts or symphorionts on crustaceans for some 225 million years. But we really do not know when chonotrichs adopted their present phoretic mode of life, nor do we know what they might have been doing before the "right" crustacean groups came along. Furthermore, 225 million years is not very long ago for pre-metazoan protists, since, by then, even all of the great non-aquatic vertebrate groups (except for the mammals) were already flourishing. (One might mention, in passing, that the earth at this time was composed of but

one ocean and one continent, according to the emerging theory of "plate tectonics": continental drift — at least the most recent one — had not yet commenced.)

Nevertheless, it is not unreasonable to conclude, with or without the scant direct
fossil information mentioned above, that the ciliated protozoa represent an assemblage of great antiquity, with roots very likely well back into the Precambrian, at
least a billion or two years ago.

But we are dealing with a large, widely distributed, and taxonomically diverse
group when we speak of the phylum Ciliophora (Corliss 1974a, 1975a); and the
major question remains as to the relative ancientness of organisms comprising its
several orders and classes recognizable today (leaving completely to one side the
possibility that many ciliate groups of old have become extinct along the way).
We have to think about this evolutionary problem even if we cannot answer it posi-
tively. If we apply some of the appropriate and almost universally accepted canons
or precepts of phylogenetically-based systematic biology, as they have been so long
and presumably successfully employed for similar problems with various vertebrate
and invertebrate groups (see Dougherty et al. 1963; Hennig 1966; Mayr 1969;
Remane 1956; Rensch 1959; Simpson 1961; and many references within these
works), we cannot avoid concluding that some ciliate groups are older than others,
that some must have preceded others in a direct ancestor-descendant relationship,
and that some of the extant groups which we know today have probably changed
less (i.e., evolved more slowly) than others over the aeons of time that all of the
groups have presumably been in existence. Thus it is permissable to declare — even
without direct evidence — that some (groups of) species currently with us resemble
their earliest ancestors more closely (e.g., in morphology) than do certain other
contemporary (groups of) species. Thus, little-changed, "ancestral-looking" forms
may (be continuing to) resemble some very early ciliate group, the same group
which may have served as the source of, or given rise, in due time, to, the other
present-day groups of more complex or more "highly-evolved-looking" forms. This
convenient assumption, like some concepts of biogeography and homology, does
not require what for many groups of organisms is the impossible, viz., the existence
of fossilized "intermediate" forms.

Perhaps we have been belaboring the obvious. But we now come to a problem
made more difficult because pertinent data are available! In attempting to determine what group of ciliates "came first" (i.e., trying to decide which present-day
forms most closely still seem to resemble or are "most representative of" presumed
ancestral groups which served as progenitors for most of our contemporary groups),
we have to be selective; and, to be selective judiciously, we should state and justify
the bases chosen for our selectivity and explain the rationale behind our decisions.
This is not an easy task, and controversies are arising today in the growing fields of
comparative ciliate systematics and phylogenetics over the relative importance or
significance of various character-states or sets of characters.

C. Search for primitive characters

First we need — briefly — to reconsider what the ciliates are: what separates them from other protists and what distinguishes them from each other as groups within the single assemblage identified as the phylum Ciliophora.

Half a dozen traits set the organisms, as a whole, apart from other protozoa, algae, etc. (Corliss 1961, 1972, 1975a): manifestation of nuclear dualism[1] (with rare but important exceptions, to which I shall return later); possession of simple cilia or compound ciliary organelles; presence of a subpellicular infraciliature; exhibition of a perkinetal or, better, homothetogenic mode of binary fission; absence of "true" syngamy; and presence, very commonly, of a cytostome, a permanent oral opening often associated, further, with a more or less well-defined larger cavity (e. g., a vestibulum or a buccal cavity). There are additional characters, but they are not exclusively unique. Nevertheless, with the major exception of the curious and highly specialized opalinid flagellates, the following features (all related to the micronucleus only) are also diagnostic of ciliates and thus deserve mention: exhibition of diploidy; occurrence of acentric mitoses, with an intranuclear spindle and no dissolution of the nuclear envelope; and demonstration of so-called gametic (rather than sporic or zygotic) reduction during meiosis.

With respect to differentiation within the phylum, there are dozens of characters which show significant variation between and among the various higher-level taxa, although only a certain few "key" characters are generally used in textbooks. These "key" characters sometimes may be quite superficial in nature and not of basic importance in evolutionary or phylogenetic considerations. Traits which are useful in comparative studies may be found not only among morphological features, including ultrastructural characters, but also among physiological, behavioral, morphogenetic, ecological, and genetic properties of the organisms in their entire life histories. But we may recognize four broad sources of data as being the most important in supplying "constellations" of characters of highest utility not only in "alpha-level" and comparative taxonomic work but also in research (in areas of ciliate evolution and phylogeny) which involves looking for clues to a better understanding of the phylum from a historical perspective.

These sources of primitive characters — at this time still essentially restricted to descriptive morphological parameters, although fresh pleas have recently been made concerning the great potential value of experimental physiological approaches (Hutner et al. 1972; Hutner & Corliss 1976), and ecological considerations are receiving increasing attention as well (Bick 1972; Borror 1968, 1972, 1973; Corliss 1973a; Dragesco 1974; Fenchel 1969; Hartwig 1973b) — may be considered succinctly as follows, with minimal explanation:

[1] Independently evolved among certain of the sarcodinid foraminiferans (see Grell 1962, 1973; Raikov 1969), although there the "somatic" nucleus is unable to divide and also differs in other ways from the macronucleus typical of the majority of ciliates.

1. **The infraciliature.** The somatic infraciliature, composed essentially of kineto-
somes or basal bodies plus associated microtubular and microfibrillar structures,
appears to be a stable and universal attribute of ciliates, with variation among
"sub-parts" of it which makes it eminently appropriate as a source of characters
of comparative value. Today we have the tool of transmission electron microscopy
to add to such approaches as the still-indispensable methods of silver impregnation
(Corliss 1973b). The fundamental importance of the infraciliature and of cortical
structures in general for research in ciliate systematics and evolution cannot be
over-emphasized.

2. **The oral area.** Here we are beginning to appreciate the richness of data from
four different but interrelated aspects: the presence or absence or the kind of
"depression" in the oral area of monostomic ciliates; the nature of the associated
cytopharyngeal apparatus, simple or complex; the kind of oral ciliature, when
present; and the nature of the oral or buccal infraciliature and nearby cytoplasmic
inclusions.

3. **The mode of stomatogenesis.** Morphogenetic phenomena of all kinds in
ciliates are of systematic interest, but the mode of new mouth-formation represents an ideal example of a dynamic feature in the life cycle research on which
may cast considerable light on the phylogenetics of the organisms under comparative study (Corliss 1967, 1968). The four types of stomatogenesis recently
recognized (Corliss 1973d) appear to characterize levels of organization in steps
of increasing evolutionary complexity among ciliate groups; thus the advisability,
if not necessity, of understanding the process as it occurs in any ciliate being
thoroughly investigated should be made clear.

4. **The macronucleus.** Nuclear characteristics in general have been largely ignored
in ciliate systematics and phylogenetics (Corliss 1975c). Stages in the hypothetical
evolution of the macronucleus are particularly significant now that we have begun
to realize that some present-day species are karyologically "relict" forms: we return
to this topic in detail below. And even the common polyploid-polygenomic
macronucleus of the bulk of ciliate groups exhibits a diversity not always taken
sufficiently into consideration in past work.

If many of the characters derivable from studies in the areas briefly outlined
above appear to be equally qualified to serve as "the most reliable" in research on
the evolution of ciliates — that is, if they all seem capable of fitting the general
description of "good" characters: stable, conservative, widespread, universal,
constant, consistent, independent (and not highly variable, highly specialized, non-
homologous, or redundant) — then we are confronted with a further problem. In
cases of apparent conflict, how does or should one weight or favor one character
over another from among such "choice" data? This problem is partially taken care
of by keeping in mind the occurrence of mosaic evolution in ciliates, as in other
organisms, and by application of what may be called the "constellation of characters hypothesis," the latter tenet preventing exclusion from a group on the basis of

a single odd trait or odd organism. But a degree of judgment is still inevitable (this is perhaps why taxonomy is not wholly reducible to computerized management!). We need also to ask two other questions regarding potentially contradictory characters, in situations in which one character, or set of characters, may suggest that a given organism or group is relatively "primitive" and "ancestral" to other ciliates, and another set supports designating it as "highly-evolved" or "secondarily-derived": which character could more likely have become what it appears to be today by regressive evolution over a long period of time; and/or which character might be the result of convergent evolution and thus non-homologous with respect to morphologically similar-looking characters in other groups?

Corliss (1974c) has reviewed a number of these problems with respect to the protozoa, and we shall attempt to apply such general guidelines in the specific cases to be considered in the remainder of this paper.

D. Ecology of psammophilic forms

For the purposes of the present discussion, the psammophilic ciliate fauna may be considered to consist, roughly, of three major kinds or groups: (1) a very small, relatively rare contingent of species possibly not truly interstitial at all (but members of Deroux's, 1970, "microbiotecton" or "slime-film"), forms belonging to a single genus and markedly distinguishable by their possession of but one kind of nucleus; (2) a much larger group, widely distributed and obligatorily interstitial, with species often of large size or elongate form and now known to be identifiable by their universal exhibition of diploid, non-dividing macronuclei; and (3) any other ciliates without such nuclear characters, which may be long-term transient or fairly permanent facultative inhabitants of (on or in) the interstitial biotope but with close relatives widely distributed elsewhere and with many (but not all) of their species often further recognizable by their possession of a clearly more complex buccal ciliature in their oral area.

A number of the morphological and physiological-biological characteristics well known for members of various invertebrate groups sharing the interstitial biotope (e.g., see Ax 1966; Hartmann 1973; Laubier 1967; Swedmark 1964; and references therein; and other "Workshop Symposium" papers in this volume) are found also in the ciliates of all three of the kinds mentioned above (Dragesco 1962). The most conspicuous of these is the elongate and/or flattened form of the body, often drawn out into a thread-like "tail" at the posterior end (certainly a product of repeated convergence, as Ax, 1963, has pointed out). Also the ciliature may come to be reduced and restricted to the ventral surface, showing a strong tendency to become thigmotactic as well. As a result of such morphological alterations, psammophilic ciliates typically show an "interstitial" mode of locomotion, creeping agilely over and around the surfaces of the sand grains. No resistant cysts are produced by these forms. Accompanying such traits are commonly two others: a marked fragility of the body and a predominantly predatory feeding behavior. All of these striking characteristics appear to be adaptations to the psammobiotic way of life, whether the organisms are obligatory "endemics" of the biotope or are living there only on a facultative basis.

Interstitial ciliates are globally distributed, with "cosmopolitan" genera and even species: we can draw this conclusion from studies already made, although our detailed knowledge is still highly incomplete (Rao 1974). For example, Agamaliev (1966, 1967, 1968, 1971, 1972) has reported that the species composition from the shores of the Caspian Sea is very similar to that from other seas and from the Atlantic Ocean. Others have obtained comparable results: Borror (1963), with Gulf of Mexico material; Burkovsky (1970a, b, 1971) and Raikov (1962), with the White Sea; Hartwig (1973a–c, 1974), with the North Sea; Kattar (1970), with the Brazilian coast; Lepşi (1962) and Petran (1963, 1967, 1968), with the Black Sea; Raikov (1963d; see also Raikov & Kovaleva 1968), with the Sea of Japan; Rao & Ganapati (1968), with Indian Ocean material; and one of us (E.H.), in a preliminary survey, with Bermudian species. Agamaliev showed further that, despite the low salinity of the Caspian, the psammophilic ciliate fauna there is essentially marine (over 90 % of the species in his studies were of marine origin). Thus the "endemism" involved – if we may use the word in such a context – is world-wide and related to the biotope itself, not to a given geographical area.

The ability of these organisms to adapt to varying chemico-physical attributes of their habitats is interesting. Salinity and temperature appear to have an insignificant influence on their geographic distribution (Fenchel 1969); and the same species as those known elsewhere have been found under extreme climatic conditions (Fenchel & Lee 1972; Lee & Fenchel 1972). *Loxodes* is found solely in fresh-water sands (Dragesco 1966, 1970; Dragesco & Njiné 1971; and others). On the basis of such data, one can thus classify the psammophilic ciliates as eurytherm; and they also show marked euryhalinity. Availability of food seems to be the most important factor in determining their "local" abundance in any given interstitial habitat (Fenchel 1969).

The question of dispersal remains to be given brief consideration; but any answers suggested at this time can only be speculative in the absence of precise data on the subject. Sterrer (1973) has reviewed some contemporary means of distribution for the interstitial meiofauna in general, which would include the ciliates under discussion here. Methods cited for marine microorganisms or planktonic protists in general cannot, of course, be applied to sand-dwelling forms: they cannot be moved through the air; they do not have resistant cysts; their position and their lack of planktonic larval forms precludes migration via surface currents; etc. The possibility of transport by turbidity currents into the depths of the ocean and along the abyssal bottom cannot be answered because such little information on meiobenthic ciliates from the deep sea is available (Uhlig 1973), but it is not appealing. Dispersal by open-water means (e.g., surface oceanic currents) is conceivable only for those species which might occur in the plankton as well as in the sand lacunar system, and this would not include the two special groups of benthic forms under consideration here. And, at that, such a method would probably work only for short-distance transport, as in the case of island colonization. Rafting on drifting objects is possible; small amounts of sand on a floating substratum could represent a "mini-biotope" for psammobiotic species, though ideal conditions for this might obtain rather rarely. Scheltema (1973) has recently published a very interesting report on such intercontinental transport of the loricate heterotrich *Folliculina* (not a sand ciliate) attached to the shells of gastropod veliger larvae. Then again, maybe migration "around the edges" would be possible and significant over a long period of time?

But the known global distribution of identical genera and even species of interstitial forms is not, in our opinion, satisfactorily explained by any of the means described above. Perhaps one must agree with Sterrer's (1973) provocative suggestion that plate tectonics may somehow have been involved as a major factor in accounting for the present-day distributional patterns of the meiofauna. It is an intriguing idea, that the organisms comprising the meiofauna, vagile as individuals but sessile with respect to their biotope, may have become widely distributed primarily by the mobility of the substratum itself!

It has very recently been postulated that members of the first two groups of psammophilic (or near-psammophilic) forms referred to in preceding paragraphs — the essentially obligatory "endemics" — represent today the (evolved) remnants of the most primitive of the ciliated protozoa, in fact, the ancestral "eociliates" from which all other groups eventually arose (Corliss 1975a—c). We wish now to consider in some detail the possible evidence available to support such a hypothesis and to discuss, very briefly, its implications for the systematics of these forms (i.e., how best to reflect it in the scheme of classification at suprafamilial levels), in keeping with the overall ideas, pertinent evolutionary principles, etc. which we have presented in preceding pages as background information.

E. The homokaryotic species

Nearly 100 years ago, Entz (1884) first described and assigned to a new genus — *Stephanopogon* — a small marine benthic species of ciliate with a slit-like apical oral aperture, adorned with one or two finger-like protoplasmic protuberances but no cilia, and a dorsoventrally-flattened body with somatic ciliation confined principally to the ventral surface. Possessing two or more nuclei and a single posteriorly-located contractile vacuole, it apparently fed on small flagellates and diatoms of various sizes. Placed in the "lower holotrichs" among the gymnostomes, it was occasionally observed by subsequent workers with little addition of new information about it, until the startling observations of Lwoff (1923, 1936). This keen observer, scarcely 20 years of age at the time of his first observations on *Stephanopogon*, insisted that the species he was studying possessed but one kind of nucleus and that, as a truly homokaryotic ciliate, it was of great phylogenetic significance.

Although it never had fewer than two nuclei, they were all or both of the same type, divided by mitosis, etc. Protozoologists were loath to accept such iconoclastic information, perhaps overinfluenced by Metcalf's (1923, 1940) concept of "proto-ciliates" which he considered the parasitic opalinids, also homokaryotic forms, to be; they also suspected that the young Frenchman might have overlooked some smaller "micronuclei." Many years later, the views of Lwoff were fully vindicated by Dragesco (1963a) and Raikov (1969), and, in less intense studies, by Borror (1965, 1973), Fenchel (1968), Jones (1974), and Jones & Owen (1974).

Without going into the details of the life cycle of this fascinating group of species — four of which are now known — we should mention that the organisms show some interesting peculiarities. These include a stage of hypertrophic growth without cytokinesis and a stage of multiple fission within what may be called a reproductive cyst. During the growth stage, the number of nuclei increases by mitosis, but such divisions are often asynchronous. A large trophont may contain as many as 12–16 nuclei. In the multiple fission process, each presumptive tomite ends up with but a single nucleus; however, this divides before excystment so that binucleate forms, only, leave the cyst. No sexual phenomena have been described for *Stephanopogon*; one might speculate — along with Raikov (1969) — that they are non-existent in these ciliates.

Each nucleus has a large central nucleolus of RNA with finely granular chromatin (DNA) located peripherally. During nuclear division, the nucleolus constricts into two, and what appear to be chromosomes become visible in the equatorial region of the spindle; subsequently, the two "new" nucleoli or endosomes are well separated and the chromosomes move toward the poles in two groups, as in a typical mitosis. There are no centrioles involved and the nuclear envelope does not break down, as is also true of all other ciliates.

Is *Stephanopogon* primitively homokaryotic or was it originally heterokaryotic with subsequent loss of one of the kinds of nuclei? In other words, do its nuclei represent retention of an ancestral primitive condition or an evolutionary regression?

The question, no matter how phrased, needs to be answered rather than ignored. Largely from the perceptive observations of Raikov (1969), one may conclude that the former assumption is more likely the correct one, for the following reasons.

(1) The *Stephanopogon* nucleus does not have the characteristics of a macronucleus, as we know that organelle from studying it in other ciliates. Thus it is unreasonable to postulate loss of the micronucleus, either totally by drastic means or as is known for strains or races of several other ciliates, including the popular *Tetrahymena pyriformis* (Corliss 1973c). It is also not likely that micronuclei have been overlooked, as some workers have suggested. This is contravened in two ways: by the precision of the observations of such competent cytologists as Dragesco, Lwoff, and Raikov and by the fact that the nucleus which is present is not a macronucleus.

(2) The *Stephanopogon* nucleus does not have all the characteristics of a bona-fide ciliate micronucleus, either. Its possession of a prominent RNA-containing nucleolus, strongly reminiscent of the condition known in nuclei of numerous sarcodinid and flagellate groups, is the principal argument against considering it to be a micronucleus. Regular ciliate micronuclei have no nucleoli. It would be convenient to cite also the absence of a meiotic process; but this cannot (yet, at least) be used, since many other species of ciliates as well still remain to be caught in the act of exhibiting sexuality, though conjugation is widespread among all other groups within the phylum in general. It must be diligently sought in *Stephanopogon*.

(3) An additional reason, more in support of the overall primitiveness of the organism itself than of its nuclear condition, is *Stephanopogon*'s manifestation of non-specialized, non-highly-evolved characters in various of its other, non-nuclear, features. But one must recall that regressive characters can mimic primitive characters very closely — there are myriads of examples of this among metazoa and protozoa (e.g., forms adapted to an endoparasitic life-style) — so "simplicity" may have been secondarily acquired in (or nearby) the interstitial biotope, a point to which we return below. Nevertheless, the gymnostome-like characters of *Stephanopogon* (e.g., in its infraciliature) seem genuine enough; and one would expect to find the earliest or "dawn" forms among such relatively "simple" or non-highly-evolved groups of ciliates. (At the same time, the contemporary *Stephanopogon* may exhibit a number of characters "evolved" in their own way over millions of generations in genetic isolation.)

Unlike the situation in the opalinids, clearly related to (other) flagellates in a number of fundamental ways (mode of fission, production of gametes, exhibition of true syngamy, etc.), species of *Stephanopogon* are bonafide ciliates in all ways except in their unique exhibition of homokaryotic nuclei. They fit beautifully into what, following Raikov's (1969) hypothesis, may be called stage 1 in the origin and evolution of the ciliate macronucleus, viz., no macronucleus (and no micronucleus, as such, either). The possible significance of their occurrence as marine, bottom-dwelling organisms is returned to later; but, as pointed out elsewhere (Corliss 1975c), such a location may be considered to support — or, at least, certainly not contradict — their phylogenetic primitiveness. "Living fossils" they may truly be.

F. The diploid-macronucleate species

Well over a hundred species of ciliates, mostly belonging to the seven genera *Geleia, Kentrophoros, Loxodes, Remanella, Trachelocerca, Trachelonema*, and *Tracheloraphis*, appear to possess in common the unique trait of having only diploid rather than polyploid macronuclei. A larger number of such forms could be known if ciliatologists were to become genuinely interested in finding and describing them, for they abound (as Dragesco, 1974, has recently reminded us) in the truly interstitial biotope, predominantly at the edges of the seas and oceans but also in fresh-water sands (the home of *Loxodes*). For scores of years these ciliates have puzzled protozoologists, despite insightful work by such early researchers as Bütschli (1876). The exact structural nature of their nuclei and, particularly, the functions of these nuclei in the organism's life cycle have been elucidated only in relatively recent years, due mainly to the precise observations of workers such as Dragesco, Fauré-Fremiet, and Raikov. Interpretation of such findings as representing informa-tion of possibly notable phylogenetic significance has occurred even more recently, principally embodied in the ideas of Raikov (conveniently summarized in Raikov, 1969, endorsed and extended by Corliss, 1975c).

Both the structure and behavior of the nuclei or nuclear sets in these obligato-rily-psammophilic ciliates — heterokaryotic but only diploid-macronucleate — have to be considered here, albeit briefly (see Raikov, 1969, for a longer and beautifully-illustrated account). For our purposes, the differences-in-detail which exist among species of the several genera listed above (and others not mentioned at all) do not need to be stressed, important though they may be from cytological and cytogene-tic points of view.

There is invariably more than one macronucleus present. The minimal size of the nuclear complement, two macronuclei plus one micronucleus, occurs in certain species of the fresh-water genus *Loxodes* and its marine counterpart *Remanella*. But there may be several dozen of each kind of nucleus in many other species (in these same two or in different genera), sometimes scattered as isolated pairs or small sets or sometimes gathered into larger nuclear "complexes" in which the mi-cronuclei are typically in the center and the macronuclei are located peripherally. The entire cluster, including macronuclear anlagen, is sometimes surrounded by a special and rather fragile cytoplasmic membrane (as in some *Tracheloraphis* species: see Raikov & Dragesco 1970).

The micronuclei are essentially those typical of all ciliates — compact, homoge-neous, Feulgen-positive, rich in DNA, lacking in RNA, etc. — and thus need concern us no longer here. The macronuclei, however, show several unusual features structu-rally. They are small (seldom more than $10-15$ μm in diameter), spherical to bean-shaped, vesicular, poor in DNA (the amount seems to be approximately equivalent to that found in the still smaller accompanying micronuclei, therefore these macro-nuclei are considered to be diploid), and rich in RNA (localized in a huge central nucleolus or in smaller, scattered nucleoli or both). In many species of genera be-longing to the large family Trachelocercidae there are curious "protein crystalloid" bodies in the karyolymph. The whole cytological picture, now being confirmed and extended by ultrastructural observations, is quite unlike that known for the typical polyploid ciliate macronucleus (e.g., see Kovaleva 1974; Kovaleva & Raikov 1974; de Puytorac & Njiné 1971; Raikov 1972b, c, 1973a, 1974, 1975; Raikov & Dra-gesco 1970; and compare with findings, cited in Raikov 1969, and Raikov & Am-mermann 1974, on the typical ciliate macronucleus).

Furthermore, these small, unusual macronuclei of the psammobiotic genera mentioned above never divide. During cytokinesis of the ciliate, they are simply randomly distributed between the proter and the opisthe; they would thus ultimately disappear were it not for their periodic replacement by the transformation of some micronuclear products forming macronuclear anlagen following a regular mitotic division of a micronucleus. This atypical situation was first noted by Bütschli (1876), in *Loxodes*, and thoroughly explored in *Kentrophoros* (called "*Centrophorella*") and other forms by Fauré-Fremiet (1954) who, perhaps unwisely, termed the process "endomixis[1]." The peculiarity is now known to be common to all ciliates with diploid macronuclei discovered to date. This kind of "nuclear reorganization" is, of course, correlated with cytokinesis; yet it is important to note that the processes may not be synchronous. Especially in cases of species with a highly multinucleate condition, the replacement of macronuclei by micronuclei may be completely asynchronous with fission of the cell; that is, many of the transformations may occur — and, again, at various times — during the interdivisional stage of the life cycle. In forms with few nuclei, however, the nuclear processes generally occur in phase with cytokinesis.

There is no endomitotic duplication of chromosomal material in the macronuclear anlagen, nor does any DNA synthesis occur in these developing macronuclei (Raikov et al. 1963). The Feulgen-negative nucleoli, containing RNA, appear early and grow rapidly; in some species they coalesce to form the large centrally located nucleolus or endosome so characteristic of many of these psammophilic ciliates. The life of an "old" diploid macronucleus, incidentally, is probably no more than half a dozen cell generations, based on observations of occurrence of the pycnotic condition in a certain percentage of the macronuclei in species of *Loxodes* (Fauré-Fremiet 1954; Raikov 1959a). Thus the production of macronuclei from micronuclei described above is not at all identical to the situation obtaining in conjugation among polyploid-macronucleate ciliates, beyond the point of basic involvement of micronuclei (although even here it is, in conjugation, the synkaryon — a nucleus of micronuclear origin but not a micronucleus per se — which produces both micro- and macronuclear anlagen). Unlike *Stephanopogon*, the diploid-macronucleate ciliates are capable of (i.e., have been seen) undergoing conjugation. The phenomenon has been described in species of several genera (see Raikov's 1972a review); there are some peculiarities to it, and, of course, no polyploid macronuclei are generated by the process in these ciliates.

Other, non-nuclear, characteristics of this second and much larger group of sand-associated ciliates demand attention here, even if only briefly. As mentioned in a preceding section of this paper, the body form is typically flattened; and many species are elongate, even vermiform, reaching a length of greater than three millime-

---

[1] This same name was used by Woodruff many years earlier for the ill-fated "unique" process in *Paramecium* which Diller, Sonneborn, and others showed was actually no more and no less than the now well-known sexual phenomenon of autogamy.

ters. Many also are highly contractile, and, as pointed out before, have thigmotactic ciliature confined to the ventral surface and are incredibly fragile outside their micro-environment. Mucocysts, toxicysts, and contractile vacuoles (in one form or another) are present; and the feeding habit is often a predaceous, omnivorous one. There are additional cytoplasmic specializations too detailed for mention here. The oral area is apical or subapical and ventral, slit-like on the concave surface of the body in the latter case. Unlike the situation in *Stephanopogon*, there may be special (non-somatic) ciliature near the mouth. Its exact nature when complex (e.g., as it is in the perplexing *Geleia*: see preliminary reports by Nouzarède 1975, and de Puytorac et al. 1973) remains to be completely elucidated, but it presents a phylogenetic and taxonomic problem returned to below. With the principal exception of species of *Loxodes*, all of these diploid-macronucleate forms are found in marine or brackish mud, sands, and sediments as obligatory components of the vast and multitudinous interstitial micro- and meiofauna. They have no close congeners living elsewhere.

G. Evolutionary alternatives

The diploid-macronucleate ciliates seem to fit very neatly into stage 2 in the hypothesized evolution of the macronucleus, viz., primary differentiation of the genetically identical nuclei characteristic of the homokaryote (exemplified contemporaneously by *Stephanopogon*) into separate (though both only diploid) somatic (macro-) and generative (micro-) nuclei (see Corliss 1975 c, and Raikov 1969 for details). But do their various characteristics truly support only the one conclusion, that they, too, should be thought of as "living fossils," primitive forms, phylogenetically ancient species (or surviving representatives of such a group), taxonomically and evolutionarily "relict" groups, even as their nuclei appear to be "karyological relicts"? Along with Raikov (1973b) and Corliss (1975c), we believe so; but other options must be searched out and properly treated.

First, we ought to make what may seem to be a brief digression but is really most relevant to the question: mention must be made of Raikov's (1969) third step or major stage in his proposed theory of macronuclear evolution. It is called "polyploidization of the macronucleus." This refers to the effective amplification of the genomic, DNA-rich material of the ciliate's somatic nucleus, which results in providing templates for synthesis of much greater amounts of RNA, etc. This is the condition of the macronucleus in the overwhelming majority of present-day ciliate species, with all its attendant properties: large and compact, nucleolus-rich, autonomous, "amitotically" dividing, genome-segregating, controller of the cell's phenotype, often capable of regeneration, and independent and long-living (except for the inevitability of its regular replacement following occurrence of sexual phenomena). Raikov lists no specific stages beyond this, but we know that there are further refinements and specializations (e.g., reorganization-band formation) in the structure and comportment of macronuclei in various specialized taxonomic groups: but these do not concern us here.

The question posed above may be put this way: if primitiveness is not considered as an acceptable explanation of the occurrence of these diploid macronuclei, how else may they have come into existence? And if the forms exhibiting such apparent "karyological relicts" are not direct descendants or remnants of some very early group in the evolution of ciliates, what else might they be?

It is true, theoretically, that the diploid state of the macronucleus could represent a regressive character. Under selective pressures, perhaps related directly to adaptation to the interstitial habitat (which we admit has undoubtedly brought about uniformity and "simplicity" in other features held in common by psammophilic ciliates: see a preceding section of this paper), a polyploid condition might have been "reduced" to the diploid condition. This is essentially the explanation endorsed some years ago by Fauré-Fremiet (1954, 1961; and also see 1970). Taxonomically, it allowed the assignment of various diploid-macronucleate species to (what would be today) quite different orders and suborders. Such allocations were based on presumed similarities in, for example, characters of the somatic or oral ciliature with that typical of (other) members of such separate taxa. Some ciliatologists apparently still follow this view. Since species possessing both kinds of macronuclei would thus monophyletically be bound together in the same several taxa, required would be the independently-repeated occurrence — within each taxon — of the reduction or "regression" or "degeneration" of the macronucleus from the polyploid to the diploid state, if the consequences of the Faurean view are fully followed through.

We believe that there are several compelling reasons why the above-described alternative explanation is untenable, with the information we have available today.

First, it seems unlikely that stages 2 and 3 in the proposed model of evolution of the ciliate macronucleus could be reversed, especially repeatedly and independently in separate higher-level taxa. The same holds true if the argument is suggested that perhaps stage 2 evolved into 3 repeatedly and independently, instead of the reverse. The steps in the process must have been altogether too complex and too complicated — structurally, genetically, behaviorally, molecularly — to "go backwards" once established (or "forward" repeatedly by identical pathways), as Raikov (1969) has stressed. The micro-chemical data recently accumulated militate against having the pieces of such intricate and surely delicately balanced molecular patterns falling into identical place many times by chance, no matter — it seems to us — what the selective pressure might have been.

Secondly, it is also unlikely that the marine interstitial habitat could be responsible alone for causing such a complex process of "degeneration," even though it does contain allegedly secondarily-simplified meiofauna from among other groups of organisms living there (Swedmark 1964). For that matter, if the habitat supports so well secondarily-primitive forms why would it not support equally well primarily-primitive forms with similar needs? But, in the case of ciliates, several other reasons are even more directly pertinent.

(1) The third group of psammobiotic forms mentioned near the beginning of this paper — a taxonomic mixture of facultatively "endemic" species — also thrives well in the same biotope, with apparently no signs of effect on their universally-possessed polyploid-polygenomic macronuclei. The nuclei seen in members of this group are structurally and functionally identical with those of their look-alike (surely closely-related) congeners often living in far different kinds of habitats. Yet, even though these "higher" ciliates may be only long-visiting "transients," some of their other characters have, interestingly enough, apparently been influenced by the interstitial environment — with elongation of body and development of thigmotactic responses serving as concrete examples. Morphological-physiological adaptations of these types have occurred in the facultative as well as the obligatory members of the meiofaunal community: why should the macronucleus of long-residing facultative forms be "immune" to such influence?

(2) *Loxodes*, with solely fresh-water species, also shows the diploid-macronucleate, while living far from the marine interstitial biotope. Yet it is strikingly similar in both nuclei and other characteristics to its marine psammophilic counterpart, *Remanella* — so similar, in fact, that some protozoologists place species of both genera into a single genus. A close common diploid-macronucleate ancestry for both of these now widely separated "relict" forms seems indicated, rather than convergence by coincidental and identical macronuclear "degeneration" under the influence of at least partially dissimilar habitats.

(3) A third point might be made: would one expect major regressive characters to occur among free-swimming, non-parasitic, macrophagous, predaceous protozoa?

The vexatious problem of possible infraciliary differences of importance (e.g., with respect to the oral area) requires our attention still, although the question might well be postponed until more information of a comparative nature is made available. Yet there are such structural differences among the diploid-macronucleate forms; and also some of the more complex oral organelles present in certain species resemble conditions found in some groups of the so-called "higher" ciliates. Two explanations of this may be offered.

(1) First, the common ancestor of some of our present-day interstitial ciliates and of other orally similarly-appearing species may have already exhibited such complexities in the mouth area: that is, the phylogenetic separation of prepolyploid and all polyploid-macronucleate groups may not have occurred as early as one might otherwise have supposed. Thus, by parallel evolution, these particular homologous ciliary structures may have been allowed to retain their basic resemblance, assuming a low selection pressure for any change (a reasonable enough assumption). And, by mosaic evolution, not only did the nuclear picture change (being retained solely in the sand-dwellers) but additional (infraciliary and other) characteristics might have done so as well, still leaving oral structures (for example) strikingly similar.

(2) Or, secondly, an entirely different set of changes might have been involved. Perhaps the features of the oral areas in question and which now seem so similar in divergent groups are not homologous at all. Maybe they have been the result of convergent evolution, occurring separately in the several groups under consideration. Physiological-environmental selective pressures, working over long periods of time, could certainly have brought about strong resemblances in the feeding apparatus without necessitating close ancestor-descendant relationship of the implicated groups. In fact, the repeated occurrence of similarity in food-gathering structures seem more likely than repeated regression – independently identical – in a highly-structured and obviously highly-successful (macro)nucleus carrying the genic material vital to the long-time survival of the race.

H. Taxonomic implications

If "relict" nuclei are borne by (morphologically) "relict" species, then these relict forms may well deserve separation into higher-level "relict" taxa, keeping in mind that our contemporary "endemic" sand species may have evolved (mosaically) in other characteristics – though probably slowly – while retaining the early stages in macronuclear evolution. The interstitial biotope, especially when marine, seems to have favored preservation of these "living fossils," providing a haven for the primitive forms once they invaded it, with a subsequent "slowdown" in speciation (Fenchel & Riedl 1970). But such a fauna might have served as one source of progenitors for later groups, groups whose members explored and settled in new and different habitats and, evolving themselves, gave rise in due time to the richly-diverse assemblage comprising the bulk of the phylum today.

The fact that *Stephanopogon* resembles other "lower" gymnostome ciliates except for its striking homokaryotic condition supports Raikov's conclusion (see Poljansky & Raikov 1961; Raikov 1957b, 1963a, 1969) that nuclear dualism arose within the ciliates, that is, after an "eociliate" (Corliss 1975b, c) had already been fully developed from some sort of (unknown) flagellate ancestry. Yet it must have been (or, today, represents the group which must have been) early on the scene indeed, since all other 7,200 species of ciliates have both kinds of nuclei, with the macronucleus highly polyploid in its composition (with the important exception of the obligatorily-interstitial group with its diploid-macronucleate condition). Therefore, the nuclear complement of ciliates represents a conservative structure, since the vast majority of these organisms manifest tremendous diversity in their ciliary and infraciliary characters (Corliss 1974a, 1975a) while exhibiting the single, established polyploid-polygenomic type of macronucleus.

Thus, applying the precepts of systematic biology mentioned in an early section of this paper, the nuclear condition is highly deserving of recognition, taxonomically as well as phylogenetically. On the basis of the facts presented and the rationale developed and defended on preceding pages, one is surely justified in placing the two groups of benthic and mainly psammobiotic ciliates discussed here in taxa separated from other groups at quite a high level. The matter needs to receive only brief consideration: for details, see Corliss (1974b, 1975a, c).

Stephanopogon is left among the gymnostomes (note, though, that that category has now grown to a subclass); but it is placed in an order, the Primociliatida[1], of its own (Corliss 1974b, 1975a). Phylogenetically, it is certainly a most significant group and deserves the distinctiveness of such a separation, despite its otherwise re-latively-unpretentious appearance.

For the much larger group, the truly "endemic" interstitial forms, those with the diploid, non-dividing macronuclei, Corliss (1975a) has erected another order, the Karyorelictida; it is also placed in the subclass Gymnostomata of the first class (the Kinetofragminophora) in the phylum. It may be debatable whether or not these karyorelictids warrant a higher-level separate taxon than an order, and also whether or not they belong altogether in the same single taxon. But Corliss (1975a) has not even erected suborders — though some families seem to be very dissimilar (cf. the Trachelocercidae and the Geleiidae, for example) — because of lack of crucial data. Unfortunately, there is as yet practically no solid information on modes of stoma-togenesis, little on the ultrastructural makeup of the cortex, etc.

By their nuclear characteristics alone, the karyorelictids are distinct from other kinetofragminophoran orders — and similar to each other. Were they to be split apart (e.g., at the family-level), on the basis of other characteristics (completely ignoring the macronucleus), and placed separately within other gymnostome or trichostome orders (in the same or even different ciliate subclasses), one would be forced, as mentioned on a preceding page but worthy of repetition, to endorse one or the other of two unlikely notions: that the evolution of the macronucleus from the diploid to the polyploid state (with other attendant characteristics) occurred repeatedly and identically in different high-level taxonomic (i.e., evolutionary) lines; or, even worse, that the complex polyploid macronucleus so well established in the "higher" ciliates was reduced or degenerated repeatedly and identically to the diploid condition within these separate taxa (lines), for whatever reasons. No other explanation is possible[2], even if some ciliatologists are loath to admit it. An important point is that it is much more than a matter simply of differential taxono-mic weighting of nuclear versus infraciliary characters. And the problem is only in-tensified if orders within the two "higher" classes, the Oligohymenophora and the Polyhymenophora (see Corliss 1974a, 1975a; and de Puytorac et al. 1974) are for-ced into involvement.

Future refinements may be anticipated in this difficult area of systematic ciliato-logy. But we believe that it would be a backward ("regressive"?) step to deny taxo-

---

[1] The name "Protociliatida" would have been ideal, but, unfortunately, it has been preempt-ed by Metcalf's (mis) use of it for the opalinids.

[2] One can visualize the first transition, from step 2 to 3, as possibly happening independent-ly a couple of times, perhaps in parallel; but more than that seems inconceivable, as discussed in earlier sections of this paper.

nomically the distinctness of these phylogenetically-significant groups, ciliates eco-logically held together by their long-time common association in or in the vicinity of the relatively stable and certainly unique interstitial biotope.

Acknowledgements

The present paper was prepared jointly during our stay at the Bermuda Biologi-cal Station, and we wish to acknowledge the library and other facilities so kindly put at our disposal by the Director, Dr. Wolfgang E. Sterrer. It is also a pleasure to dedicate the work to the memory of the late Dr. Bertil Swedmark, inspiring leader in meiofaunal research. We are grateful for the assistance of a National Science Foundation grant GB-41172 (to J. O. C.), and an Exxon Corporation Fellowship and a grant from the Deutsche Forschungsgemeinschaft, Ha 979/1; the latter sup-ports allowed one of us (E.H.) to work at the Station in Bermuda for several weeks in the fall of 1975.

Bibliography

Agamaliev, F.G.: New species of psammobiotic ciliates of the western coast of the Caspian Sea. Acta Protozool. **4**, 169–184 (in Russian with English summary) (1966).

–: Faune des ciliés mésopsammiques de la côte ouest de la mer Caspienne. Cah. Biol. Mar. **8**, 359–402 (1967).

–: Materials on morphology of some psammophilic ciliates of the Caspian Sea. Acta Protozool. **6**, 225–244 (in Russian with English summary) (1968).

–: Complements to the fauna of psammophilic ciliates of the western coast of the Caspian Sea. Acta Protozool. **8**, 379–404 (in Russian with English summary) (1971).

–: Ciliates from microbenthos of the islands of Apšeronskij and Bakinskij archipelagos of the Caspian Sea. Acta Protozool. **10**, 1–27 (in Russian with English summary) (1972).

Ax, P.: Die Ausbildung eines Schwanzfadens in der interstitiellen Sandfauna und die Verwert-barkeit von Lebensformcharakteren für die Verwandtschaftsforschung. Zool. Anz. **171**, 51–71 (1963).

–: Die Bedeutung der interstitiellen Sandfauna für allgemeine Probleme der Systematik, Öko-logie und Biologie. Veröff. Inst. Meeresforsch. Bremerhaven, Sonderbd. **2**, 15–66 (1966).

Bick, H.: Ciliated protozoa: an illustrated guide to the species used as biological indicators in freshwater biology. World Health Organization, Geneva. 198 pp. (1972).

Bock, K.J.: Über einige holo- und spirotriche Ciliaten aus den marinen Sandgebieten der Kieler Bucht. Zool. Anz. **149**, 107–115 (1952a).

–: Zur Ökologie der Ciliaten des marinen Sandgrundes der Kieler Bucht I. Kieler Meeresforsch. **9**, 77–89 (1952b).

–: Zur Ökologie der Ciliaten des marinen Sandgrundes der Kieler Bucht II. Kieler Meeresforsch. **9**, 252–256 (1953).

Borror, A.C.: Morphology and ecology of the benthic ciliated protozoa of Alligator Harbor, Florida. Arch. Protistenk. **106**, 465–534 (1963).

–: New and little-known tidal marsh ciliates. Trans. Amer. Micros. Soc. **84**, 550–565 (1965).

–: Ecology of interstitial ciliates. Trans. Amer. Micros. Soc. **87**, 233–243 (1968).

–: Tidal marsh ciliates (Protozoa): morphology, ecology, systematics. Acta Protozool. **10**, 29–71 (1972).

–: Marine flora and fauna of the northeastern United States. Protozoa: Ciliophora. NOAA Tech. Rep. NMFS CIRC. No. 378. 62 pp. (1973).

Burkovsky, I.V.: The ciliates of the mesopsammon of the Kandalaksha Gulf (White Sea). I.
      Acta Protozool. 7, 475–489 (in Russian with English summary) (1970a).
—: The ciliates of the mesopsammon of the Kandalaksha Gulf (White Sea). II. Acta Protozool.
      8, 47–65 (in Russian with English summary) (1970b).
—: Ecology of psammophilous ciliates in the White Sea. Zool. Zh. 50, 1285–1302 (in Russian
      with English summary) (1971).
Bütschli, O.: Studien über die ersten Entwicklungsvorgänge der Eizelle, der Zellteilung und die
      Conjugation der Infusorien. Abh. Senckenberg. Naturforsch. Ges. 10, 213–452 (1876).
Corliss, J.O.: The ciliated protozoa: characterization, classification, and guide to the literature.
      Pergamon Press, London and New York. 310 pp. (1961).
—: An aspect of morphogenesis in the ciliate Protozoa, J. Protozool. 14, 1–8 (1967).
—: The value of ontogenetic data in reconstructing protozoan phylogenies. Trans. Amer.
      Micros. Soc. 87, 1–20 (1968).
—: The ciliate Protozoa and other organisms: some unresolved questions of major phylogenetic
      significance. Amer. Zool. 12, 739–753 (1972).
—: Protozoan ecology: a note on its current status. Amer. Zool. 13, 145–148 (1973a).
—: Protozoa. In Gray, P., ed., Encyclopedia of microscopy and microtechnique, Van Nostrand
      Reinhold, New York, pp. 483–484 (1973b).
—: History, taxonomy, ecology, and evolution of species of *Tetrahymena*. In Elliott, A.M., ed.,
      Biology of *Tetrahymena*, Dowden, Hutchinson, and Ross, Stroudsburg, PA, pp. 1–55
      (1973c).
—: Evolutionary trends in patterns of stomatogenesis in the ciliate Protozoa. (Abstr.) J. Proto-
      zool. 20, 506 (1973d).
—: The changing world of ciliate systematics: historical analysis of past efforts and a newly pro-
      posed phylogenetic scheme of classification for the protistan phylum Ciliophora. Syst. Zool.
      23, 91–138 (1974a).
—: Remarks on the composition of the large ciliate class Kinetofragmophora de Puytorac et al.,
      1974, and recognition of several new taxa therein, with emphasis on the primitive order Pri-
      mociliatida n. ord. J. Protozool. 21, 207–220 (1974b).
—: Time for evolutionary biologists to take more interest in protozoan phylogenetics? Taxon
      23, 497–522 (1974c).
—: Taxonomic characterization of the suprafamilial groups in revision of recently proposed
      schemes of classification for the phylum Ciliophora. Trans. Amer. Micros. Soc. 94, 224–267
      (1975a).
—: Which extant ciliates represent the most primitive group in the phylum? (Abstr.) J. Proto-
      zool. 22, 34A (1975b).
—: Nuclear characteristics and phylogeny in the protistan phylum Ciliophora. BioSystems 7,
      338–349 (1975c).
Deflandre, G. & J. Deunff: Sur la présence de ciliés fossiles de la famille des Folliculinidae dans
      un silex du Gabon. Compt. Rend. Acad. Sci. 244, 3090–3093 (1957).
Deroux, G.: La serié "chlamydonellienne" chez les Chlamydodontidae (holotriches, Cyrtopho-
      rina Fauré-Fremiet). Protistologica 6, 155–182 (1970).
Dougherty, E.C., Z.N. Brown, E.D. Hanson & W.D. Hartman, eds.: The lower metazoa, compa-
      rative biology and phylogeny. University of California Press, Berkeley. 478 pp. (1963).
Dragesco, J.: Ciliés mésopsammiques littoraux. Systématique, morphologie, écologie. Trav.
      Stat. Biol. Roscoff (N. S.) 12, 1–356 (1960).
—: On the biology of sand-dwelling ciliates. Sci. Progress 50, 353–363 (1962).
—: Compléments à la connaissance des ciliés mésopsammiques de Roscoff. I. Holotriches. Cah.
      Biol. Mar. 4, 91–119 (1963a).
—: Compléments à la connaissance des ciliés mésopsammiques de Roscoff. II. Hétérotriches. III.
      Hypotriches. Cah. Biol. Mar. 4, 251–275 (1963b).

−: Ciliés mésopsammiques d'Afrique Noire. Cah. Biol. Mar. **6**, 357−399 (1965).

−: Quelques ciliés libres du Gabon. Biol. Gabon. **2**, 91−117 (1966).

−: Ciliés libres du Cameroun. Ann. Fac. Sci. Yaoundé, Yaoundé. 141 pp. (1970).

−: Ecologie des protistes marins. In Puytorac, P. de J. Grain, eds., Actualités protozoologiques, Vol. 1, University of Clermont, France, pp. 219−228. (1974).

− & T. Njiné: Compléments à la connaissance des ciliés libres du Cameroun. Ann. Fac. Sci. Cameroun No. **7−8**, 97−140 (1971).

Entz, G., Sr.: Über Infusorien des Golfes von Neapel. Mitt. Zool. Stat. Neapel **5**, 289−444 (1884).

Fauré-Fremiet, E.: Ecologie des ciliés psammophiles littoraux. Bull. Biol. Fr. Belg. **84**, 335−375 (1950).

−: The marine sand-dwelling ciliates of Cape Cod. Biol. Bull. **100**, 59−70 (1951).

−: Réorganisation du type endomixique chez les Loxodidae et chez les *Centrophorella*. J. Protozool. **1**, 20−27 (1954).

−: Quelques considérations sur les ciliés mésopsammiques à propos d'un récent travail de J. Dragesco. Cah. Biol. Mar. **2**, 177−186 (1961).

−: A propos de la note de M. Thomas Njiné sur le cilié *Loxodes magnus*. Compt. Rend. Acad. Sci. **270**, 523−524 (1970).

Fenchel, T.: The ecology of marine microbenthos. II. The food of marine benthic ciliates. Ophelia **5**, 73−121 (1968).

−: The ecology of marine microbenthos. IV. Structure and function of the benthic ecosystem, its chemical and physical factors and the microfauna communities with special reference to the ciliated protozoa. Ophelia **6**, 1−182 (1969).

− & C.C. Lee: Studies on ciliates associated with sea ice from Antarctica. I. The nature of the fauna. Arch. Protistenk. **114**, 231−236 (1972).

−: R.J. Riedl: The sulphide system: a new biotic community underneath the oxidized layer of marine sand bottoms. Mar. Biol. (Berlin) **7**, 255−268 (1970).

Grell, K.G.: Morphologie und Fortpflanzung der Protozoen (einschließlich Entwicklungsphysiologie und Genetik). Fortschr. Zool. **14**, 1−85 (1962).

−: Protozoology. Springer-Verlag, Berlin, Heidelberg and New York. 554 pp. (1973).

Hartmann, G.: Zum gegenwärtigen Stand der Erforschung der Ostracoden interstitieller Systeme. Ann. Spéléol. **28**, 417−426 (1973).

Hartwig, E.: Die Ciliaten des Gezeiten-Sandstrandes der Nordseeinsel Sylt. I. Systematik. Mikrofauna Meeresboden **18**, 387−453 (1973a).

−: Die Ciliaten des Gezeiten-Sandstrandes der Nordseeinsel Sylt. II. Ökologie. Mikrofauna Meeresboden **21**, 3−171 (1973b).

−: Die Nahrung der Wimpertiere des Sandlückensystems. Mikrokosmos **62**, 329−336 (1973c).

−: Verzeichnis der im Bereich der deutschen Meeresküste angetroffenen interstitiellen Ciliaten. Mitt. Hamburg. Zool. Mus. Inst. **71**, 7−21 (1974).

Hennig, W.: Phylogenetic systematics. University of Illinois Press, Urbana. 263 pp. (1966).

Hulings, N.C. & J.S. Gray, eds.: A manual for the study of meiofauna. Smithsonian Contr. Zool. **78**, 84 pp. (1971).

Hutner, S.H., H. Baker, O. Frank & D. Cox: Nutrition and metabolism in protozoa. In Fiennes, R.N., ed., Biology of nutrition, Pergamon Press, London and New York, pp. 85−177 (1972).

− & J.O. Corliss: Search for clues to the evolutionary meaning of ciliate phylogeny. J. Protozool. **23**, 48−56 (1976).

Jankowski, A.W.: [Fauna of the USSR: Infusoria subclass Chonotricha.] Vol. 2, No.1. Akad. Nauk SSSR, Nauka, Leningrad. 355 pp. (in Russian) (1973).

Jones, E.E.: The Protozoa of Mobile Bay, Alabama. University of South Alabama Monograph. Vol. **1**, 113 pp. (1974).

Jones, E.E. & G. Owen III: New species of protozoa from Mobile Bay, Alabama. J. Mar. Sci. Alabama 2, 41–56 (1974).

Kattar, M.R.: Estudo dos protozoarios ciliados psamofilos do litoral Brasileiro. Bol. Zool. Mar. (São Paulo), N. S. 27, 123–306 (1970).

Kovaleva, V.G.: The fine structure of ciliary and cortical organoids and some structures of the ectoplasm and endoplasm of *Trachelonema sulcata* (Ciliata, Holotricha). Tsitologia 16, 217–233 (in Russian with English summary) (1974).

–: & I.B. Raikov: Ultrastructure de l'appareil nucléaire de *Trachelonema sulcata* Kovaleva, cilié holotriche gymnostome à macronoyaux diploïdes. Protistologica 9 (year 1973), 471–480 (1974).

Laubier, L.: Adaptations chez les annélides polychètes interstitielles. Ann. Biol. 6, 1–15 (1967).

Lee, C.C. & T. Fenchel: Studies on ciliates associated with sea ice from Antarctica. II. Temperature responses and tolerances in ciliates from Antarctica, temperate and tropical habitats. Arch. Protistenk. 114, 237–244 (1972).

Lepşi, J.: Über einige insbesondere psammobionte Ciliaten vom rumänischen Schwarzmeer-Ufer. Zool. Anz. 168, 460–465 (1962).

Lwoff, A.: Sur un infusoire cilié homocaryote à vie libre. Son importance taxonomique. Compt. Rend. Acad. Sci. 177, 910–912 (1923).

–: Le cycle nucléaire de *Stephanopogon mesnili* Lw. (cilié homocaryote). Arch. Zool. Exp. Gén. 78, 117–132 (1936).

Mayr, E.: Principles of systematic zoology. McGraw-Hill, New York. 428 pp. (1969).

Metcalf, M.M.: The opalinid ciliate infusorians. Bull. U.S. Nat. Mus. 120, 1–484 (1923).

–: Further studies on the opalinid ciliate infusorians and their hosts. Proc. U.S. Nat. Mus. 87, 465–635 (1940).

Mohr, J.L.: On the age of the ciliate group Chonotricha. In Barnes, H., ed., Some contemporary studies in marine science, G. Allen and Unwin, London, pp. 535–543 (1966).

Nouzarède, M.: Sur un nouveau genre de protozoaires ciliés géants mésopsammiques appartenant à la famille des Geleiidae Kahl. Compt. Rend. Acad. Sci. 280, 625–628 (1975).

Petran, A.: Contributii la cunoasterea microfaunei de ciliate psamofile din Marea Neagra (litoralul romănesc). Stud. Cercet. Biol. (Anim) 15, 187–197 (in Roumanian) (1963).

–: Cercetări asupra faunei de ciliate psamobionte la plajele din sudul litoralului romănesc al Mării Negre. Ecol. Mar. 2, 169–191. (in Roumanian) (1967).

–: Sur l'écologie des ciliés psammobiontes de la Mer Noire (littoral Roumain). Rev. Roum. Biol. Zool. 13, 441–446 (1968).

Poljansky, G.I. & I.B. Raikov: Nature et origine du dualisme nucléaire chez les infusoires ciliés. Bull. Soc. Zool. Fr. 86, 402–411 (1961).

Puytorac, P. de, A. Batisse, J. Bohatier, J.O. Corliss, G. Deroux, P. Didier, J. Dragesco, G. Fryd-Versavel, J. Grain, C.-A. Grolière, R. Hovasse, F. Iftode, M. Laval, M. Roque, A. Savoie & M. Tuffrau: Proposition d'une classification du phylum Ciliophora Doflein, 1901 (réunion de systématique, Clermont-Ferrand). Compt. Rend. Acad. Sci. 278, 2799–2802 (1974).

– & T. Njiné: Sur l'ultrastructure des *Loxodes* (ciliés holotriches). Protistologica 6 (year 1970), 427–444 (1971).

– I.B. Raikov & M. Nouzarède: Particularités des ultrastructures corticale et buccale du cilié marin *Geleia nigriceps* Kahl. Compt. Rend. Soc. Biol. 167, 982–985 (1973).

Raikov, I.B.: Nuclear apparatus and its reorganization during the fission cycle in the infusoria *Trachelocerca margaritata* (Kahl) and *T. dogieli,* sp. n. (Holotricha). Zool. Zh. 36, 344–359 (in Russian with English summary) (1957a).

–: Reorganization of the nuclear apparatus in ciliates and the problem of the origin of their binuclearity. Vest. Leningrad Univ., No. **15,** 21–37 (in Russian with English summary) (1957b).

–: Der Formwechsel des Kernapparates einiger niederer Ciliaten. I. Die Gattung *Trachelocerca.* Arch. Protistenk. **103,** 129–192 (1958).

–: Der Formwechsel des Kernapparates einiger niederer Ciliaten. II. Die Gattung *Loxodes.* Arch. Protistenk. **104,** 1–42 (1959a).

–: Cytological and cytochemical peculiarities of the nuclear apparatus and division in the holotrichous ciliate *Geleia nigriceps* Kahl. Tsitologia **1,** 566–579 (in Russian with English summary) (1959b).

–: [The interstitial infusoria of the sandy littoral regions of the Dal'Nezelenetskaya Bay (eastern Muran).] Murmanskogo Morskogo Biol. **2,** 172–185 (in Russian) (1960).

–: Les ciliés mésopsammiques du littoral de la Mer Blanche (U.R.S.S.) avec une description de quelques espèces nouvelles ou peu connues. Cah. Biol. Mar. **3,** 325–361 (1962).

–: On the origin of nuclear dualism in ciliates. lst Int. Cong. Protozool., Prague, Aug. 1961, Progress in protozoology, pp. 253–258 (1963a).

–: The nuclear apparatus of *Remanella multinucleata* Kahl (Ciliata, Holotricha). Acta Biol. **14,** 221–229 (1963b).

–: The nuclear apparatus of the holotrichous ciliates *Geleia orbis* Fauré-Fremiet and *G. murmanica* Raikov. Acta Protozool. **1,** 21–29 (in Russian with English summary) (1963c).

–: Ciliates of the mesopsammon of the Ussuri Gulf (Japan Sea). Zool. Zh. **42,** 1753–1767 (in Russian with English summary) (1963d).

–: The macronucleus of ciliates. In Chen, T.-T., ed., Research in protozoology, Vol. 3, Pergamon Press, London and New York, pp. 1–128 (1969).
Nuclear phenomena during conjugation and autogamy in ciliates. In Chen, T.-T., ed., Research in protozoology, Vol. 4, Pergamon Press, London and New York, pp. 147–289 (1972a).

–: The nuclear apparatus of the psammophilic ciliate *Kentrophoros fistulosum* Fauré-Fremiet: structure, divisional reorganization and ultrastructure. Acta Protozool. **10,** 227–247 (1972b).

–: Ultrastructures macronucléaires et micronucléaires de *Tracheloraphis dogieli,* cilié marin à macronoyaux diploïdes. Compt. Rend. Soc. Biol. **166,** 608–613 (1972c).

–: Ultrastructures cytoplasmiques de *Kentrophoros latum* Raikov, cilié holotriche psammophile: organisation corticale, endoplasme et trichocystes. Ann. Stat. Biol. Besse-en-Chandesse, No. **6**–7 (year 1972), 21–54 (1973a).

–: Holotrichs with diploid never-dividing macronuclei: "karyological relicts" or regressed forms? (Abstr.) 4th Int. Cong. Protozool., Clermont-Ferrand, Sept. 1973, Progress in Protozoology, p. 340 (1973b).

–: Fine structure of the nuclear apparatus of a lower psammobiotic ciliate, *Tracheloraphis dogieli* Raikov. Acta Protozool. **13,** 85–96 (1974).

–: Cytoplasmic fine structure of the lower holotrichous ciliate *Tracheloraphis prenanti.* Tsitologia **17,** 739–747 (1975).

– & D. Ammermann: The macronucleus of ciliates: recent advances. In Puytorac, P. de & J. Grain, eds., Actualités protozoologiques, Vol. 1, University of Clermont, France, pp. 143–159 (1974).

– E.M. Cheissin & E.G. Buze: A photometric study of DNA content of macro- and micronuclei in *Paramecium caudatum, Nassula ornata,* and *Loxodes magnus.* Acta Protozool. **1,** 285–300 (1963).

– & J. Dragesco: Ultrastructure des noyaux et de quelques organites cytoplasmiques du cilié *Tracheloraphis caudatum* Dragesco et Raikov (Holotricha, Gymnostomatida). Protistologica **5** (year 1969), 193–208 (1970).

– & V.G. Kovaleva: Complements to the fauna of psammobiotic ciliates of the Japan Sea (Posjet Gulf). Acta Protozool. **6**, 309–333 (1968).

Rao, G.C.: On the geographical distribution of interstitial fauna of marine beach sand. Proc. Indian Nat. Sci. Acad. **38** (year 1972), 164–178 (1974).

– & P.N. Ganapati: The interstitial fauna inhabiting the beach sands of Waltair coast. Proc. Nat. Inst. Sci. India (B) **34**, 82–125 (1968).

Remane, A.: Die Besiedelung des Sandbodens im Meere und die Bedeutung der Lebensform- typen für die Ökologie. Verh. Deutsch. Zool. Ges. 1951, 327–359 (1952).

–: Die Grundlagen der natürlichen Systeme der vergleichenden Anatomie und der Phylogene- tik. 2nd ed. Geest und Portig, Leipzig. 400 pp. (1956).

Rensch, B.: Evolution above the species level. Columbia University Press, New York. 419 pp. (1959).

Scheltema, R. S.: Dispersal of the protozoan *Folliculina simplex* Dons (Ciliophora, Hetero- tricha) throughout the North Atlantic Ocean on the shells of gastropod veliger larvae. J. Mar. Res. **31**, 11–20 (1973).

Simpson, G.G.: Principles of animal taxonomy. Columbia University Press, New York. 247 pp. (1961).

Sterrer, W.: Plate tectonics as a mechanism for dispersal and speciation in interstitial sand fauna. Neth. J. Sea Res. **7**, 200–222 (1973).

Swedmark, B.: The interstitial fauna of marine sand. Biol. Rev. **39**, 1–42 (1964).

Tappan, H. & A.R. Loeblich, Jr.: Lorica composition of modern and fossil Tintinnida (ciliate Protozoa), systematics, geologic distribution, and some new tertiary taxa. J. Paleont. **42**, 1378–1394 (1968).

–: Evolution of the oceanic plankton. Earth-Sci. Rev. **9**, 207–240 (1973).

Uhlig, G.: Preliminary studies on meiobenthic ciliates from the deep sea. (Abstr.) 4th Int. Cong. Protozool., Clermont-Ferrand, Sept. 1973, Progress in protozoology, p. 417 (1973).

Addendum

The following references represent pertinent papers not cited above because of their ina- vailability when our manuscript was submitted to the Editors:

Corliss, J. O.: On lumpers and splitters of higher taxa in ciliate systematics. Trans. Amer. Micros. Soc. **95**, 430–442 (1976).

–: Annotated assignment of families and genera to the orders and classes currently comprising the Corlissian scheme of higher classification for the phylum Ciliophora. Trans. Amer. Micros. Soc. **96**, 104–140 (1977a).

–: Flagellates, opalinids, and the search for the most primitive ciliate and its progenitor. Ceylon J. Sci. (in press) (1977b).

–: A note on the abundance of ciliated protozoa, their morphological and ecological diversity, and their potential in biological and biomedical research. Proc. Zool. Soc., Calcutta (in press) (1977c).

Hartwig, E.: On the interstitial ciliate fauna of Bermuda. Cah. Biol. Mar. **18**, 113–126 (1977).

–: & J. G. Parker: On the systematics and ecology of interstitial ciliates from beaches of North Yorkshire (England). J. Mar. Biol. Assoc. U. K. (in press) (1977).

| Mikrofauna Meeresboden | 61 | 89—103 | 1977 |

W. Sterrer & P. Ax (Eds.). The Meiofauna Species in Time and Space. Workshop Symposium, Bermuda Biological Station, 1975

Means of Meiofauna Dispersal[1]

by

Sebastian A. Gerlach

Marine Biological Laboratory, Helsingør, Denmark*

Abstract

Meiofauna dispersal is possible by waterfowl, by floating substrates including sea ice, and by turbulent water, either in suspension or adhering to sediment particles. During severe storms sea water turbulence erodes sediments not only in shallow water, but also in the sublittoral region. In the past centuries meiofauna had a chance of dispersal in sand that served as ballast for sailing vessels. Direct evidence for meiofauna dispersal is scarce. In an appendix to the paper several authors report on meiofauna found on drifting coconuts and in plankton samples, and on other observations bearing on meiofauna dispersal.

A. Introduction

Only very few meiofauna species have pelagic larvae. In most groups juveniles develop within the sedimentary habitat, and their number is generally low (Swedmark 1964). Sterrer (1973) therefore has concluded that meiofauna might be the most sedentary of all marine faunas and exhibit a very low rate of dispersal. Cosmopolitan distribution patterns of meiofauna should therefore reflect the great age of meiofauna taxa in the geological past. Another conclusion, however, would be that meiofauna animals, although lacking pelagic larval stages, are themselves so small that even as adults they might disperse over large distances of open water, transported by floating materials or while suspended in the water, or attached to suspended sediment.

Along a coastline meiofauna could, of course, disperse within the benthal by active locomotion; locomotory rates of 0.3—0.4, sometimes 1 mm/sec have been observed for instance in interstitial gastrotrichs (Hummon 1975). However, the ability for active longshore locomotion is insufficient explanation for the occurrence of meiofauna animals in salt marsh and in brackish water biotopes, which represent rather small patches on the shore line, separated from each other by large

[1] Contribution No. 700 from the Bermuda Biological Station for Research, Inc.

* Present adress: Institut für Meeresforschung, Am Handelshafen 12, D 285 Bremerhaven, Federal Republic of Germany.

areas with other environmental qualities. Salt marsh sites and brackish water localities may develop and perish in short periods of time, according to changes of the
shore line. Nevertheless these biotopes are invariably inhabited by the appropriate
meiofauna which, therefore, must have reliable means of dispersal, independent
from active locomotion within the sediment.

In addition, there is evidence that rather young oceanic islands are inhabited by
a rich interstitial beach meiofauna, which must have crossed deep water; for
example, the Galapagos Islands exist only since less than 3 million years (Ax &
Schmidt 1973; Bailey 1976).

However, long distance transport of meiofauna does not have to be regular, but
may be occasional, for example during a hurricane. For this reason it is difficult to
directly observe dispersal of meiofauna animals, or to achieve experimental evidence bearing on this matter. The following considerations point out some possible
means of meiofauna dispersal and will hopefully stimulate the publication of casual
anecdotic observations which could provide more arguments. Some observations
made by several marine biologists are collected in the appendix to this paper.

B. Dispersal at the air-sea interface

I. Airborne animals

As Sterrer (1973) pointed out, aerial transport of meiofauna would require special adaptations like hard-shelled eggs or cysts, which generally are not found in
meiofauna. In addition, experimental evidence is scarce; Lang (in Boaden 1964) reported that brushings from the feet of mallards (*Anas platyrhynchos*) and other
ducks yielded copepods, nematodes and larval hydracarids. Purasjoki (1948) suggested transport by migrating terns (*Sterna hirundo* and *macrura*), when he found the
African ostracod *Pomatocypris (Cyprilla) humilis* in a rockpool on a Finlandian
skerry. This species, however, is well established on skerries of the Tvärminne area
and was found again later by Hagerman (1967). Though it still seems a possibility,
there is no proof for airborne transport of marine meiofauna; this has been established for terrestrial nematodes (Spaull 1973).

A similar argumentation is possible to explain the occurrence of about twenty
brackish water species of marine nematodes in inland saline biotopes of Central
Germany, about 150 km away from North Sea and Baltic shores. Some of these
saline biotopes are in direct contact with Permian salt structures in the under-
ground. Therefore they may have been colonized at any time by members of the
coastal brackish water meiofauna, transported for instance with waterfowl which
regularly visit saline lakes. During the past some hundred years, however, mining for
potassium salts became more and more important in the region between Halle and
Westfalia, and numerous deposits of waste sodium chloride were piled up. The
runoff provided brackish water biotopes: salt marshes and salt ponds, which are inhabited by brackish water nematodes (Paetzold 1958). Saline biotopes are found,
as a rule, not farther apart than 5–15 km; therefore Meyl (1955) suggested that

halophilic insects may provide transport for meiofauna from one locality to the other.

II. Rafting on drifting materials

Any drifting material which can carry some sediment or provides, like a coconut with its fibrous layer, an interstitial habitat itself, is a potential means of transport for intertidal meiofauna. It is well known among geologists that the quantities of rock and sediment transported by driftwood, and the distances travelled are considerable (Emery 1955), but there is, to my knowledge, no direct evidence that driftwood carried meiofauna. Reports on drifting coconuts are given in the appendix to this paper.

In the colder regions shore ice is formed every winter and may transport sediment. Waves wash sediment on top of ice cakes, or stranded ice blocks freeze on the sediment surface during low tide. When they float up they take sediment layers away; with drift ice large quantities of sediment may be transported over long distances. It has been established, too, that bivalves, for instance *Mytilus edulis,* are transported away from the German wadden sea with ice (Reineck 1976). Many meiofauna animals survive ice conditions, live specimens of several meiofauna groups have been found in totally frozen sand (Jansson 1968). Drift ice, therefore, may have a certain importance for meiofauna dispersal, and it certainly was available over larger geographical areas during the ice ages.

Algae drifting at the sea surface certainly provide a habitat for meiofauna, especially for species adapted to the phytal. The nematodes *Monhystera parva* and *Chromadorella filiformis* have been found on drifting *Sargassum* in the eastern Sargasso Sea, *Chromadora hentscheli* off the Brazilian coast. Though the specimens were smaller than animals from the shore, there were females carrying eggs and juveniles present, indicating reproduction while drifting (Micoletzky 1922). Among copepods there are two species (the cyclopoid *Macrochiron sargassi* and the harpacticoid *Harpacticus gurneyi*) apparently specialized to drifting *Sargassum*, while five more harpacticoid species from the Atlantic between 35 ° and 70 ° West are as well known from European, and partly from American and African shores. These species are continuously breeding on the drifting *Sargassum,* and they are preyed upon by flying fish (Yeatman 1962).

It should be mentioned that drifting sea weed occurs in the sublittoral region, too. This has been observed, in Kiel Bay, by members of „Sonderforschungsbereich 95" of Kiel University, who had to protect their instruments installed on the sea bed in 10–20 m depth against drifting algae. Furthermore, there is evidence that flint stones and other stones of about one kilogram weight are washed up on the beach because adhering *Laminaria* algae provide a large area on which the wave forces act. One finds, occasionally, sublittoral phytal-inhabiting nematodes like *Anticoma limalis, Thoracostoma trichodes* and *Symplocostoma longicolle* close to the water level in Kiel Bay (Gerlach 1954), and one can be sure that they were transported with sublittoral seaweeds.

III. Ballast of sailing vessels

Hessland (1954) attributed the term "Mya-Period" to the past three centuries when ocean going vessels provided special means of transportation for marine organisms. For quite a number of bivalves, crustacea, polychaetes and other groups including plants, it is documented that they extended their area of distribution by travelling in the water of ballast tanks, or while fouling the ship bottoms.

Ballast tanks are a rather recent invention which came largely into use with steel hulls. When wooden sailing vessels sail without cargo, they use, for ballast, any material that is heavy and cheap: stone and rock, earth, and sand. In 1880 at the port of Frederikstad, Norway, 10 000 tons of ballast were unloaded (Ouren 1974), more than 10 000 tons of Cretaceous flintstone have been transported, as ballast, from Europe to ports on the US east coast (Emery 1968), and the ballast islands in Georgia salt marshes testify that in 1700 a strict reglementation for ballast was necessary to keep the access to the ports intact (Reineck 1970). These examples may give some kind of idea as to the amount of material involved.

Freshly dredged sand or sand from the tidal zone was wet, heavy to handle and eventually lost part of its valuable weight when drying. Therefore other kinds of ballast were preferred. But in some places sand was the only type of ballast available. Lindroth (1957) reproduces a photograph showing barges taking sand and gravel north of Appledore, North Devon, England (fig. 25) and reports (p. 164) that "the vessels would be laid on Sandridge at Appledore and given a list, so that when the tide was out the crew could jump out and heave the sand up over the side with shovels."

In larger harbours there was an important trade with ballast sand located on certain quays, certainly not a first class environment for more delicate meiofauna animals, even if some species may be rather hardy (see appendix). However, it certainly happened that freshly dredged sand was loaded into a ship, and did not dry out in the ship's bilge. Therefore we must concede a good probability that between 1500 and 1900 some meiofauna animals were transported by sailing vessels, over the oceans, and over shorter distances as well.

The following historical report, while not of importance for marine meiofauna dispersal because the Thames at London is freshwater, tells about shipping procedures more than a century ago (Owen 1923): "There are reaches in the Lower Thames where the river flows over beds of golden gravel, and in the old days this gravel made excellent ballast for the Newcastle colliers. Having discharged their coal they probably moored at a spot where at low water they would lie high and dry on the gravel, and this gravel they shovelled up and hoisted aboard as a necessary stiffening for their run back to Newcastle. The process was going on continuously year in and year out, so that in the centuries no small part of the bed of the Thames must have been carried to Newcastle and there dumped overside. A huge bank was in fact thus created at the northern port."

C. Dispersal in the water column

Pelagic ways of life

Under certain circumstances, benthic macrofauna animals without special adaptations to a pelagic way of life have been observed in the free water: the lugworm *Arenicola marina* in the German wadden sea (Werner 1954), and 5 mm long *Macoma balthica* (bivalves) in the Dutch wadden sea, where they are transported over distances up to 10 km by the tidal currents (Beukema 1973). The mud snail *Hydrobia ulvae*, on British coasts, floats with the incoming tide by means of a mucous raft, feeding at the water surface (Newell 1962), or may float in rather dry condition for longer periods (Schwarz 1929). Finally it should be mentioned that benthic amphipods, for instance *Tritaeta gibbosa* which lives within the sponge *Halichondria*, at night in summer swims in the open water (Jones et al. 1973).

There are only a few observations to prove that meiofauna, too, under certain circumstances, leave the sediment. This habit is well established for some harpacticoid copepods which may be found in sediment traps well above the sea bed; the species in question live at the sediment-water interface and are not interstitial in their habits. However, Karling (1974) observed sometimes typical subsoil Turbellaria swimming freely and abundantly in the water of shore pools, and he suggests that this more or less temporary occurrence in the water may play a role in the dispersal of subsoil animals.

In a petri dish under the dissecting microscope, some members of the nematode family Monhysteridae display wriggling movements quite different from the normal locomotion. Might it be that such movements keep the animal in suspension? From the freshwater habitat, one should remember a cryptic observation by Schneider (1913) which needs confirmation: on one occasion in August 1912, he found many females of the nematode *Tripyla* sp. in the gut of a plankton feeding fish, *Coregonus*, which is believed never to feed upon benthos. Schneider postulated that ripe females of *Tripyla* swim up into the water column and spawn, so that their offspring may have a better chance of dispersal.

Sometimes nematodes have been found in plankton samples (Schulz 1961: Elbe Estuary, Germany; Nyqvist 1975: sediment traps in the Baltic); more observations are reported in the appendix to this paper. Finally it should be mentioned that the kinorhynch *Echinoderes levanderi* has often been found in plankton samples from the Gulf of Finland (Purasjoki 1945).

Sometimes such findings happened after stormy weather, and it is obvious that animals were suspended in the water, together with sediment. Sometimes, however, there is no indication of high turbulence. Further observations are neces-

sary before a statement can be given on the chance that occasionally meiofauna
may appear in the free water.

Sediment of different kind, suspended in containers well above the sea bed in
Kiel Bay, was rather quickly colonized by copepods, nematodes and other meiofauna (Sarntheim & Richter 1974, Scheibel 1974). Unfortunately it cannot be
excluded that for example nematodes crept up the suspending wires; their presence
is no proof for dispersal through the water column, even if this may have occurred.

D. Dispersal in suspension

Benthic Foraminifera were found 3–5 months after the start of sediment colonization experiments in Kiel Bay, mostly medium sized (0.2 mm) members of the
genus *Elphidium*, which subsequently built up rather dense populations in the artificial sediment. There are no pelagic dispersal stages known in benthic Foraminifera; Wefer & Richter (1977) therefore conclude that animals had been resuspen-
ded, together with sediment, by the strong bottom currents which occasionally
occur in certain areas of Kiel Bay. Species of *Elphidium* seem to be specially
capable to float in turbid water. They recolonize rapidly the deeper areas of the
Baltic, when life conditions improve after anoxic periods, and they have been
found, together with other benthic Foraminifera, in plankton net tows off Bodega
Head, California (Loose, 1970; additional information on benthic Foraminifera in
plankton samples: Murray, 1965).

Boaden (1964) considers that the main mode of dispersal in meiofauna is probably wave and current displacement of sand grains with adhering eggs and individuals, followed by deposition in a new locality. Anyone who has seen surf action
on a beach and knows about large scale movements which beach sand undergoes,
will admit that meiofauna could easily be transported in those regions of the shore
where turbulence is high enough to keep sand and meiofauna in suspension. 1–2 m
above a beach sediment, Boaden (1968) collected suspended sediment in the brea-
king waves; he found, together with 100 ml sand, 10 gastrotrichs of the genus
Turbanella, 90 cyclopoid copepods, 8 nematodes and 60 larval polychaetes. However, most meiofauna animals migrate downward in the deeper sediment layers
when the sand is disturbed by turbulence.

Erosion takes place in the sublittoral region, too. A wave of a given amplitude
provides decreasing energy levels in increasing water depths. Gienapp (1973) calculated that for a wave 6 m high (length 100 m, period 9 sec) acting upon the sea bed
in 20 m of water in Helgoland Bight, an orbital current of 2 m/sec can be expected
at the bottom which is sufficient to erode a fine sand sediment. Hickel (1969)
meanwhile, in September 1967 at 25 m depth northwest of Helgoland, had observed a 4–10 cm thick layer of freshly deposited sand, and concluded that this might
have been the result of the heavy storm that hit Helgoland Bight in February
1967, with wave heights above 8 m. We have evidence (Rachor & Gerlach 1977)

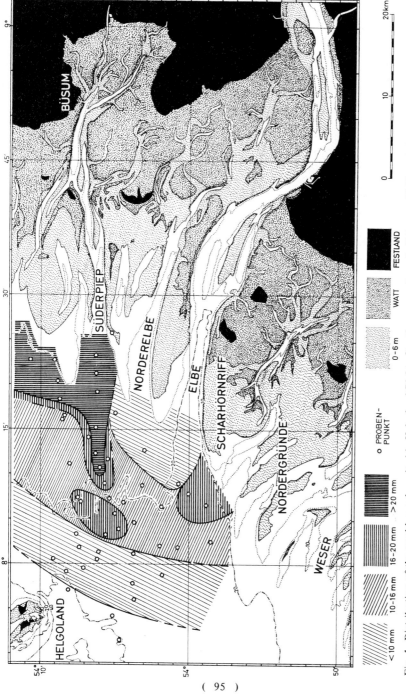

Fig. 1. Distribution of the sand layer deposited in Helgoland Bight after the severe storm in February 1967, and thickness of this layer (hatched area: deeper than 10 m). The sand was eroded from the shallower area in the west (dotted area, shallower than 6 m below Mean Spring Low). From Gadow & Reineck 1969.

that this storm rather drastically changed the composition of macrofauna in Helgoland Bight, carrying away small and delicate organisms; unfortunately, meiofauna
were not investigated before and after the impact of the storm.

Hydrographic conditions are responsible for the fact that during westerly storms
water currents in the northern part of Helgoland Bight are directed westwards, with
speeds up to 45 cm/sec (Gienapp & Tomczak 1968). In the area southwest of
Helgoland scientists from the Senckenberg Institute in Wilhelmshaven found layers
of sand interspersed within cores from silt and clay sediments at 40 m water depth.
Gadow & Reineck (1969) concluded that the heavy storm of February 1967 eroded
about 5 million tons of sediment from the shallows at both banks of the outer
Elbe Estuary (Fig. 1). This huge amount of sand was transported westward over
distances up to 50 km. When the turbulence of the water decreased, the sand was
deposited and formed a layer 10–20 mm thick which, of course, was then covered
by silt and clay which sedimentated later. In sediment cores from an area of silty
fine sand 10–12 m deep, east of Helgoland, layers of *Hydrobia*-shells are interspersed, which have been transported from the wadden sea during storms. From characteristic escape traces penetrating the clay sedimentated in top of the shell layer one
can conclude that many of the *Hydrobia*-snails were still alive when deposited (Reineck et al. 1968). Of course these snails are better protected in their shells than
meiofauna animals, but one might suppose that quite a number of living meiofauna
animals from the shallow areas close to the coast were transported offshore.

Marine engineers round the North Sea are frightened with the probability that
once within a century weather conditions will be such to create 20 m high waves
which will turn large areas of the North Sea into a surf region. Wave heights at open
ocean coasts may occasionally be even higher. Shelf sediments down to 200 m water depth are sometimes eroded. Not only for intertidal meiofauna, but for sublittoral animals, too, there is a good chance for transport together with suspended sediment.

E. Conclusion

Most meiofauna animals have no pelagic larval stages, nevertheless they may be
transported by birds, by floating materials, by ballast sand in sailing vessels, and
while suspended in the water or attached to suspended sediment during periods of
stormy weather. These means of dispersal are more regularly available in shallow
water, and the distances travelled in suspension will be short, in most cases. Occasionally, however, sublittoral and shelf sediments are eroded during severe storms.
Generally, benthic animals with pelagic larvae have a better year-to-year chance
than meiofauna to establish populations in other areas. Stormy weather conditions
of various grades, however, occur every several years, decades or only once within
a century. On the long run they provide for meiofauna animals a good chance to
become dispersed. Normally, pelagic larvae of benthic animals have a restricted life
span. Compared with such benthic animals, meiofauna should not be classified as

specially restricted in the capacity of dispersal. There is no reason to attribute more importance to the geographical distribution patterns of meiofauna than to the distribution of most other marine animals. Even transport by man cannot be excluded.

Acknowledgements

Thanks to the cooperation of many colleagues this paper can report on many more facts and arguments than I was aware of when I presented it at the workshop "The Meiofauna Species in Time and Space", Bermuda, September 1975. I want to thank all those who provided contributions for the appendix, and I am specially grateful for bibliographic hints and amendments of my manuscript to P. Boaden (Portaferry), B. Coull (Columbia), R. Higgins (Washington), T. Karling (Stockholm), H.-E. Reineck (Wilhelmshaven), F. Riemann (Bremerhaven) and to Handels-og Søfartsmuseet, Helsingør.

Appendix

1) Wolfgang Sterrer (Biological Station, St. George's West, Bermuda):Meiofauna dispersal by coconuts

On July 17, 1973, a ripe coconut was found drifting several miles off Bermuda. A cursory investigation of meiofauna that might be present in its fibrous coat was carried out by washing and shaking the nut in an isotonic solution of magnesium chloride. The following numbers of meiofauna animals have been found:

Turbellaria (Microstomidae)	1
Nematoda (J. Ott. det.)	
Oncholaimus dujardini De Man, 1876	9 ⎫
Syringolaimus striaticaudatus De Man, 1888	2 ⎪
Synonema braziliense Cobb, 1920	1 ⎬ Nematoda
Phanoderma sp.	1 ⎭
Polychaeta (2species)	4
Copepoda, incl. nauplii	present
Amphipoda (Gammaridea, 1 species)	13
Isopoda (G. Schultz det.)	
Bagatus bermudensis Richardson, 1902	6

All of the nematode species are true cosmopolitans, and the isopod is found wherever floating *Sargassum* is present.

Since coconuts only rarely mature in Bermuda, it is likely that drifting mature nuts originate further south in the Caribbean, i.e. at least in the Bahamas. Coconut palms often grow on beaches, and nuts fall into the surf where they roll in the sand before drifting away. An analysis of their passengers, therefore, may give valuable clues to the question of meiofauna dispersal.

2) Franz Riemann (Institut für Meeresforschung, Bremerhaven, Federal Republic of Germany): Nematodes from a drifting coconut at Bermuda.

On September 27, 1975, Miss Mary White, Bermuda Biological Station, found a ripe coconut at Whalebone Beach, Bermuda, which was demonstrated to the members of the meiofauna symposium. A storm hit the island on the preceding day, and it seems to be probable that the coconut came drifting from the Caribbean region (see the preceeding note by W. Sterrer).

The coconut was washed first in sea water, then in diluted formalin. The following meio-fauna animals have been found, together with some fragments of Rhodophyceae:

Polychaeta (one species of Syllinae)	2
Copepoda (Harpacticoidea, 2 species)	2
Nematoda	
Spilophorella paradoxa (De Man, 1888)	4
Euchromadora sp.	3
Cyatholaimus sp.	1
Anticoma sp.	1
Viscosia sp.	1

Spilophorella paradoxa is a cosmopolitan species with a very prominent spinneret (outlet of the adhesive caudal glands). It has been recorded from muddy sand, sand and algae, prefe-rably in the littoral zone of the European coasts, including the Mediterranean and the Black Sea, from the African coast of the Atlantic, from the Canary Islands, from the US east coast, from Bermuda itself, from the Caribbean, and from California, Japan, the Bay of Ben-gal and the Red Sea (Gerlach, S.A. and F. Riemann 1973/74: Veröff. Inst. Meeresforsch. Bremerh. Suppl. 4).
There is no proof that the fauna encountered came as passengers on the drifting coconut; the animals could have been picked up directly from the Bermuda beach where the coconut was found.

3) Olav Giere (Zoologisches Institut, Universität Hamburg, Federal Republic of Germany): Meiofauna in plankton samples from the Elbe-Estuary, Germany

In 1963 and 1965, during a zooplankton survey of the region between Glückstadt and Lightvessel Elbe 1, I found regularly and in all seasons 1–3, once 15 nematodes in 20 liter of water (station "Elbe 3", 31. August 1965, calm weather). These findings have been men-tioned shortly in Arch. Hydrobiol. Suppl. **31** (1968) on page 388 and 400–401. There have been found some harpacticoid copepods and Tardigrada, too, together with non pelagic Amphipoda and Isopoda.

4) Carlo Heip (Laboratorium voor Morfologie en Systematiek, Gent, Belgium): Nematodes in plankton samples from the North Sea

In the course of a large sampling programme, in many samples from the plankton of the southern bight of the North Sea nematodes were found, mostly in coastal stations, but also sometimes in quite deep water.

5) Wilfried Westheide (II. Zoologisches Institut, Universität Göttingen, Federal Republic of Germany): The polychaet *Microphthalmus sczelkowii* found in wadden sea plankton

In spring 1965 I took zooplankton samples at 2–4 m water depth outside the jetty of the harbour of List, Sylt Island, Germany. I found one specimen of *Microphthalmus sczel-kowii* Mecnikow, a species normally living in nearby intertidal and shallow subtidal coarse sand with detritus. Most probably the animal did not enter the water column voluntarily, but was eroded from the sediment. Preceding the sampling day was a period of severe storms, and the water was still rather agitated.

6) Bruce C. Coull (Belle W. Baruch Institute for Marine Biology and Coastal Research, Columbia, South Carolina, USA): Nematodes in the plankton of North Inlet Estuary, South Carolina
During a 20 month (Jan. 1974–Aug. 1975) zooplankton survey of the North Inlet Estuary,

South Carolina (33° 20'N; 79° 10'W) we irregularly collected nematodes in the plankton. We encountered nematodes in 18 samples during the time period with maximum numbers of 15 nematodes per m^3 of water.

7) Bernt Zeitzschel (Institut für Meereskunde, Universität Kiel, Federal Republic of Germany): Nematodes in plankton samples from Kiel Bay

Summarizing observations from our area of study, we can state that nematodes sometimes are found in plankton samples. Dr. Lenz recorded them sometimes in 3–5 m water depth outside of the jetty of Institut für Meereskunde, Kiel Harbour. Mr. Smetacek, working on about 250 samples of 50 ml water each, under the Utermöhl-microscope, found nematodes twice: on 10. November 1972, 2m above the sediment, 12 specimens, and on 4 December 1972, 7 m above the sediment, 2 specimens. The station is the "Hausgarten"-area at Boknis Eck, Kiel Bay, water depth to the bottom is 22 m. Finally Mrs. Schnack reports that she did not find nematodes in about 50 plankton samples (55 μm and 300 μm nets) from station Breitgrund (Kiel Bay), collected 1970–1972.

8) Reinhard M. Rieger (Department of Zoology, University of North Carolina, Chapel Hill, USA): Meiofauna in the plankton of Atlantic Beach, North Carolina

On the night of October 18, 1975 a 10 minute surface plankton tow was taken off a fishing pier at Atlantic Beach, North Carolina. Though it was calm when the tow was made, the previous night had been very rough. The water depth off the pier was 3 to 4 m. In looking through the silt and sediment collected in the net, I found 7 nematodes, 1 tardigrade (*Batillipes*), 2 juvenile acoels and 1 gastrotrich (*Tetranchyroderma*).

9) George Hagerman (Department of Zoology, University of North Carolina, Chapel Hill, USA): Dispersion of soft-bodied meiofauna in Bogue Sound, North Carolina

On October 17, 1975 a device for detecting the colonization of sterile sand by meiofauna was suspended from a pier off the Institute of Marine Sciences in Morehead City, North Carolina. The apparatus is similar to that used by Oviatt and Nixon; 1975 (Estuarine and Coastal Mar. Sc. 3: 201–217) with some modification. The most important difference is that the frame containing the sediment traps is suspended from above the water line, and not anchored to the bottom, thus eliminating the possibility of colonization via mooring lines. The frame is located 1,5 m from the bottom, which is mostly mud with some patches of sand, and is about a meter from the neighboring pilings. At low tide the frame hangs 70 cm below the water surface.

While there is little heavy wave action in this part of Bogue Sound, there are strong longshore currents which reverse with the tide, reaching speeds of 0,5 to 1 m/sec. Furthermore, there is a heavy sediment load with an average of 20–40 ml of silt being deposited in each trap per week.

The traps consist of funnels (15 cm diameter) and 8 oz (236 ml) catch bottles. Each catch bottle contains 100 ml of beach sand which has been autoclaved for 20 minutes at 120 °C and 20 psi (1.3 Atmospheres) and then dried in an oven at 230 °C for one hour. The bottles of sterile sand are put out in sets of three for periods of 1 week, 2 weeks and 3 weeks. Upon retrieval the animals are extracted using the magnesium-chloride technique. Aside of about 30 to 130 nematodes per bottle, the following soft bodied meiofauna have been found in the bottles:

Turbellaria

Acoela: juveniles of different species, adult Haploposthiidae
Macrostomida: *Microstomum* sp. and *Paromalostomum* sp.
Prolecithophora: juvenile Plagiostomidae
Proseriata: *Minona* sp.?
Thyphloplanoidea: *Messoplana falcata*
Kalyptorhynchia: two species of *Carcharodorhynchus* und an unknown species of Gnatho-
rhynchidae

Gastrotricha

Macrodasyoidea: *Urodasys* sp., *Macrodasys* sp., *Thaumastoderma* sp., *Acanthodasys* sp.
Chaetonotoidea: *Neodasys* sp., *Heterolepidoderma* sp. and a species of the *Chaetonotus
schulzei* group.
With a few exceptions, these represent isolated findings, one or two times. *Microstomum*
sp. occurs commonly (5 to 10 per set of bottles) as do the juvenile acoels. *Neodasys* also
occurs regularly, but less abundantly (1 to 3 per set of bottles). In addition, three or four
types of ciliates appear regularly in the bottles, but these have not been identified.
Based on the data collected so far, inshore dispersal through suspension of adults, juveniles
and eggs is clearly indicated for a number of interstitial forms. To what extent this mechanism is sufficient to allow genetic exchange between meiofauna populations, and the geographic range of such exchange remains to be seen.

10) Nathan W. Riser (Marine Science Institute, East Point, Nahant, Massachusetts, USA): Turbellaria in the plankton

Recently Dörjes and Karling (1975, Zoologica Scripta **4**: 175–189) confirmed the syno-
nymy of the acoelous turbellarian *Childia spinosa* Graff, 1911, found from benthic algae
and swimming free in the water at Woods Hole, Massachusetts, with *Childia groenlandica*
(Levinsen, 1879). I collected the material in July, 1973 in Pleasant Bay, Cape Cod, Massachusetts. Vast numbers of this species were obtained in the plankton near eel grass (*Zostera*) beds. I have never encountered the species in any of the interstitial samples taken
throughout the year and over several years in Pleasant Bay, or in other areas of the coast.
It is possible that the species is inhabiting algae, and frequently swims.
Several species of the turbellarian genus *Plagiostomum* also are frequently encountered
in the plankton at Nahant.

11) Carlo Heip (Laboratoria voor Morfologie en Systematiek, Gent, Belgium): Pelagic nauplii
in the harpacticoid copepod *Canuella perplexa*

In cultures the nauplius larvae of *Canuella perplexa* T. und A. Scott 1893 are pelagic; the
copepodites go down to the sediment. Therefore this harpacticoid copepod is an example
for a meiofauna animal with a pelagic larval stage.

12) Tor G. Karling (Swedish Museum of Natural History, Stockholm, Sweden): Sediment
transport by sea ice

Dispersal with ice floes is evidently important in regions where the sea periodically is
covered with ice. Sand and wreck will be transported wide distances when the ice is released
from the shore. I have observed this phenomenon in many places in the Baltic Sea and in

the Kristineberg area (Gullmarfjord). Heaps of sea weed are often loosened and transported by the ice, though not totally frozen.

13) Robert P. Higgins (Smithsonian Institution, Washington D.C., USA): Meiofauna survives in dry beach sand

Both Dr. Leland Pollock and I have found that beach sand can be transported great distances — wet or dry — stored for several months and when rewetted, certain meiofauna, especially tardigrades and some nematodes, revive. This phenomenon (cryptobiosis) has been reported for freshwater tardigrades, nematodes, rotifers, etc., but both of us were surprised to find it true for marine equivalents.

14) Robert P. Higgins (Smithsonian Institution, Washington D.C., USA): Sediment entangled in *Enteromorpha* algae fouling ships

Algae such as *Enteromorpha* growing on the hulls of ships may accumulate some sand at their point of attachment. Sand is entangled by a net of the algal filaments and this might well provide some vehicle for transport of meiofauna. The tardigrade *Echiniscoides sigismundi* M. Schultze 1865 is often found associated with barnacles which have a close association of sand and algae. This kind of fouling community, when attached to ships could account for some additional transport, especially if the hulls are cleaned in a foreign port.

15) Hartmut Goethe (Bernhard-Nocht-Institut für Schiffs- und Tropenkrankheiten, Abteilung für Schiffahrtsmedizin, Hamburg, Federal Republic of Germany): Reports of Cape Horniers on sand ballast

With several sailship captains organized in the Society of Cape Horniers I discussed the chance that meiofauna animals could have been transported by sand ballast. Captain Schnegelsberg and Captain Nielsen, the former inspector of Laeiss-Line, report that prior to the time when ballast was gained by dredging, sand ballast was taken from the shores close to the harbors, and transported by barges to the sailing vessels. Sand ballast was shipped from European harbors for example to Australia and to Iquique, Chile. In fact the sand was always wet down in the ship's hull, and at sea water temperature. Together with professor Ulli Schmidt, the former director of Institut für Seefischerei in Hamburg, the captains are of the opinion that life conditions for meiofauna could have been good during the ocean passage.

Bibliography

Ax, P. & P. Schmidt: Interstitielle Fauna von Galapagos I. Einführung. Mikrofauna Meeresboden **20**, 1–38 (1973).

Bailey, K.: Potassium-argon ages from the Galapagos Islands. Science (N. Y.) **192**, 465–467 (1976).

Beukema, J. J.: Migration and secondary spatfall of *Macoma balthica* (L.) in the western part of the Wadden Sea. Netherlands J. Zool. **23**, 356–357 (1973).

Boaden, P. J. S.: Grazing in the interstitial habitat: a review. In: D. J. Crisp (ed.): Grazing in terrestrial and marine environments. Brit. Ecol. Soc. Symp. **4**, 299–303 (1964).

—: Water movement — a dominant factor in interstitial ecology. Sarsia **34**, 125–136 (1968).

Emery, K. O.: Transportation of rocks by driftwood. J. Sediment. Petrol. **25**, 51–57 (1955).

—: Ballast overboard! Science **162**, 308–309 (1968).

Gadow, S. & H.-E. Reineck: Ablandiger Sandtransport bei Sturmfluten. Senckenbergiana marit. **1**, 63–78 (1969).

Gerlach, S.A.: Die freilebenden Nematoden der schleswig-holsteinischen Küsten. Schr. naturw. Ver. Schlesw.-Holstein 27, 44–69 (1954).

Gienapp, H.: Strömungen während der Sturmflut vom 2. November 1965 in der deutschen Bucht und ihre Bedeutung für den Sedimenttransport. Senckenbergiana marit. 5, 135–151 (1973).

– & G. Tomczak: Strömungsmessungen in der Deutschen Bucht bei Sturmfluten. Helgoländer wiss. Meeresunters. 17, 94–107 (1968).

Hagerman, L.: Ostracods from the Tvärminne area, Gulf of Finland. Comment. Biol. Soc. Sc. Fenn. 10 (3), 1–7 (1967).

Hessland, J.: On the quarternary Mya Period in Europe. Ark. Zool. 37 A (8), 1–51 (1945).

Hickel, W.: Sedimentbeschaffenheit und Bakteriengehalt im Sediment eines zukünftigen Verklappungsgebietes von Industrieabwässern nordwestlich Helgolands. Helgoländer wiss. Meeresunters. 19, 1–20 (1969).

Hummon, W.D.: Habitat suitability and the ideal free distribution of Gastrotricha in a cyclic environment. Proc. 9th Europ. mar. biol. Symp., 495–525 (1975).

Jansson, B.O.: Quantitative and experimental studies of the interstitial fauna in four Swedish sandy beaches. Ophelia 5, 1–71 (1968).

Jones, D.A., N. Peacock and O.F.M. Phillips: Studies on the migration of Tritaeta gibbosa, a subtidal benthic amphipod. Netherlands J. Sea Res. 7, 135–149 (1973).

Karling, T.G.: Turbellarian fauna of the Baltic proper. Identification, ecology and biogeography. Fauna Fennica 27, 1–101 (1974).

Lindroth, C.H.: The faunal connections between Europe and North America. Stockholm (Almqvist & Wiksell): 1–344 (1957).

Loose, T.L.: Turbulent transport of benthonic Foraminifera. Contr. Cushman Fdn foramin. Res. 21, 164–166 (1970).

Meyl, A.H.: Nematoden aus einer Salzwiese bei Artern. Zool. Anz. 154, 233–240 (1955).

Micoletzky, H.: Freilebende Nematoden aus den treibenden Tangen der Sargassosee. Mitt. Hamb. zool. Mus. Inst. 39, 1–11 (1922).

Murray, J.W.: Significance of benthonic foraminiferids in plankton samples. J. Palaeont. 42, 156–157 (1965).

Newell, R.: Behavioural aspects of the ecology of Peringia (= Hydrobia) ulvae (Pennant) (Gastropoda, Prosobranchia). Proc. Zool. Soc. London 138, 49–75 (1962).

Nyqvist, B. G.: Notes on the sedimentation of particulate organic matter in a Baltic coastal zone. Merentutkimuslait. Julk. 239, 338 (1975).

Ouren, T.: Om at skyde ballast. Wiwar 1/1974, 13–16 (1974).

Owen, D.: Thames ballast. The Mariner's Mirror 6, 123–124 (1923).

Paetzold, D.: Beiträge zur Nematodenfauna mitteldeutscher Salzstellen im Raum von Halle. Wiss. Z. Univ. Halle Math.-Nat. 8, 17–48 (1958).

Purasjoki, K.J.: Quantitative Untersuchungen über die Mikrofauna des Meeresbodens in der Umgebung der Zoologischen Station Tvärminne an der Südküste Finnlands. Comment. Biol. Soc. Sc. Fenn. 9 (14), 1–24 (1945).

–: Cyprilla humilis G.O. Sars, an interesting ostracod discovery from Finland. Comment. Biol. Soc. Sc. Fenn. 10 (3), 1–7 (1948).

Rachor, E. & S. A. Gerlach: Changes of macrobenthos in a sublittoral sand area of the German Bight, 1967–1975. Rapp. P.-v. Réun. Cons. perm. int. Explor. Mer, in press (1977).

Reineck, H.-E.: Die Ballast-Inseln. Natur und Museum 100, 105–110 (1970).

–: Dift ice action on tidal flats, North Sea. Rev. Géogr. Montr. 30, 197–200 (1976).

––, J. Dörjes, S. Gadow and G. Hertweck: Sedimentologie, Faunenzonierung und Faziesabfolge vor der Ostküste der inneren Deutschen Bucht. Senckenbergiana lethaea 49, 261–309 (1968).

Sarntheim, M. & W. Richter: Submerse experiments on benthic colonization of sediments in the western Baltic Sea I. Technical layout. Mar. Biol. **28**, 159–164 (1974).

Scheibel, W.: Submerse experiments on benthic colonization of sediments in the western Baltic Sea II. Meiofauna. Mar. Biol. **28**, 165–168 (1974).

Schneider, G.: Nematoden als Fischnahrung. Int. Rev. ges. Hydrobiol. Hydrogr. **6**, 489–490 (1913).

Schulz, H.: Qualitative und quantitative Plankton-Untersuchungen im Elbe-Ästuar. Arch. Hydrobiol. Suppl. **26**, 5–105 (1961).

Schwarz, A.: Die Ausbreitungsmöglichkeiten der Hydrobien (Die Bildung eines einheitlichen Hydrobien-Sediments). Natur und Museum **59**, 50–51 (1929).

Spaull, V. W.: Distribution of soil nematodes in the maritime Antarctic. Brit. Antarct. Surv. Bull **37**, 1–6 (1973).

Sterrer, W.: Plate tectonics as a mechanism for dispersal and speciation in interstitial sand fauna. Netherlands J. Sea Res. **7**, 200–222 (1973).

Swedmark, B.: The interstitial fauna of marine sand. Biol. Rev. **39**, 1–42 (1964).

Wefer, G. & W. Richter: Colonization of artificial substrates by Foraminifera. Kieler Meeresforsch., in press (1977).

Werner, B.: Eine Beobachtung über die Wanderung von *Arenicola marina* L. (Polychaeta sedentaria). Helgoländer wiss. Meeresunters. **5**, 93–102 (1954).

Yeatman, H.C.: The problem of dispersal of marine littoral copepods in the Atlantic Ocean, including some redescriptions of species. Crustaceana **4**, 253–272 (1962).

Mikrofauna Meeresboden	61	105–112	1977

W. Sterrer & P. Ax (Eds.). The Meiofauna Species in Time and Space. Workshop Symposium, Bermuda Biological Station, 1975

On the Evolution of Reproductive Potentials in a Brackish Water Meiobenthic Community

by

Carlo Heip

Department of Zoology, State University of Ghent, Ghent, Belgium

Abstract

The realized and intrinsic rates of increase of several meiobenthic populations are compared. The realized rate of increase is remarkably similar among four species of copepods, which leads to the conclusion that dominance is not a function of rate of increase or parameters influencing this rate. The result is investigated in the light of Vandermeer's (1975) β-competition. There is also a remarkable agreement between the values of the predator *Protohydra leuckarti* and its principal prey *Tachidius discipes*.

The intrinsic rates of natural increase vary more, but are a linear function of temperature in all species investigated. This is in agreement with the theory that natural selection acts against nonlinear dynamic characteristics (Patten 1975). The low intrinsic rate of increase of the nematode *Oncholaimus oxyuris* may be the result of a group selection process.

A. Introduction

In the last few years several mathematical models of complex ecosystems seemed to indicate that increased complexity makes for instability of the system (Gardner & Ashby 1970; May 1972, 1973). This is contrary to the common belief of most ecologists that ecosystems tend to be more stable the larger the number of interacting species they contain. May (1973) brings this apparent contradiction into the light of natural selection by observing that if complex natural systems are in fact stable, this can only be because the interactions occurring in the system, between the species, are highly non-random. It appears that the contradiction is not as keen as May believed, and both Roberts (1974) and De Angelis (1975), by either eliminating biologically meaningless negative densities or by constructing more plausible food web models, recently showed that increased complexity may in fact coincide with increased stability. Nevertheless, it remains obvious that the parameters describing interactions between species have highly non-random values which are the product of natural selection (Maynard-Smith 1974).

Finding it unreasonable to think of selection working upon an ecosystem as a whole, Maynard-Smith (1974) cites two ways in which selection might act to produce complex and stable ecosystems; these are genetic feedback and species exclusion. Actual ecosystems, complex or otherwise, are persistent only because species whose presence would be inconsistent with persistence have been excluded. Genetic feedback, a concept introduced by Pimentel (1961), is still a controversial topic and I will not go into this further, but the main idea is that if two species coexist for some time in an ecosystem, and if they interact, then they are likely to undergo genetic changes caused by this interaction: they will coevolve.

Natural selection operates by differential reproductive success. A convenient indicator of the reproductive potential of a population is the intrinsic rate of natural increase r_m, the rate of increase of a population unrestricted by density effects as described by the equation for exponential growth $dN/dt = r_m N$ (1). This intrinsic rate of increase generally has an inverse hyperbolic relation with generation time and a negative correlation with body size. In this paper I will present some calculations of both the intrinsic and the realized rate of increase of some meiobenthic populations from a brackish water pond in northern Belgium. These populations have been studied from 1968 onwards and most of the material concerning life-cycles has yet to be published.

B. Material and methods

Values of density are provided from fortnightly samples taken over a period of four years. Methods of sampling and extraction are as described by Heip (1973a). In calculations the running average of three samples is used instead of the middle value. Numerous calculations have shown that the value of the rate of increase is not changed substantially by this procedure.

Culture methods for the copepods have been described by Smol & Heip (1974). Both the nematode *Oncholaimus oxyuris* and the polyp *Protohydra leuckarti* were cultured at different temperatures using the nematode *Panagrellus silusiae* as food given in excess of the requirements. A more detailed account on this will be published shortly (Heip, Smol & Absillis, in preparation).

C. Results

It has been shown (Heip 1972, 1973b and in preparation) that all populations which have been investigated show one or more distinct peaks during the year, which can be described by an exponentially increasing and an exponentially decreasing function. These populations belong to the crustacean subclasses Ostracoda and Copepoda, and to the Hydrozoa. The realized rate of increase has been calculated for all peaks for all years, and the mean value for the different populations is given in Table 1. For further illustration the corresponding time needed to double population size, calculated as $t_2 = (\ln 2)/r$, is also given in Table 1.

The calculation of the intrinsic rate of natural increase r_m ideally should use life-tables and requires knowledge of the age-specific survival and fecundity. It is difficult to obtain this information from species with a short life cycle. Heip & Smol (1976b) therefore estimated r_m in copepods from:

$$r_m = \frac{1}{T_1} \ln pN_e \qquad (2)$$

Tab. 1. Mean realized rate of increase of all peaks over a four year period (r per day) and corresponding doubling time (t_2 in days)

Hydrozoa	r	t_2
Protohydra leuckarti Greeff, 1870	0.049	14.1
Copepoda		
Halicyclops magniceps (Lilljeborg, 1853)	0.042	16.5
Canuella perplexa T. & A. Scott, 1893	0.045	15.4
Tachidius discipes Giesbrecht, 1882	0.048	14.4
Paronychocamptus nanus (Sars, 1908)	0.044	15.8
Ostracoda		
Cyprideis torosa (Jones, 1850)	0.015	46.2
Loxoconcha elliptica Brady, 1868	0.030	23.1

in which T_1 is the time of first appearance of gravid females of the next generation, p is the percentage of females and N_e is the number of eggs per egg-sac. The value of r_m obtained in this way may not be entirely accurate but it is the best available in the absence of very detailed data which can be used to construct life tables.

Values of the intrinsic rate of natural increase of two harpacticoid copepods as a function of temperature are given in fig. 1. From this figure it is obvious that r_m is a nearly linear function of temperature. This result is robust against large changes in the values of the parameters used in the calculation of r_m. However, the parameters as used in eq. (2) themselves do not show this linear relationship with temperature.

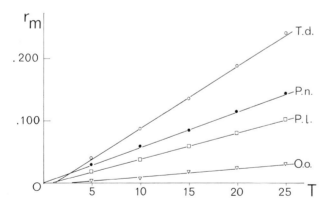

Fig. 1. Intrinsic rate of natural increase as a function of temperature in four meiobenthic populations: T.d. = *Tachidius discipes* (Copepoda); P.n. = *Paronychocamptus nanus* (Copepoda); P.l. = *Protohydra leuckarti* (Hydrozoa); O.o. = *Oncholaimus oxyuris* (Nematoda).

The intrinsic rate of natural increase of the nematode *Oncholaimus oxyuris* has been calculated in essentially the same way and is also shown in fig. 1. Again there appears to exist a linear relationship between r_m and temperature, in spite of the fact that the parameters used in the calculations are nonlinear functions of temperature (Heip, Smol & Absillis, in preparation).

The intrinsic rate of natural increase of the polyp *Protohydra leuckarti* was calculated from the observed division rate in laboratory experiments at three different temperatures (Table 2). The influence of temperature on the division rate is profound and can be described by $D = 254\ T^{-1.12}(3)$. As this species nearly always reproduces asexually the intrinsic rate r_m is given by $r_m = (\ln 2)/D = 0.69 / (254\ T^{-1.12})$. Selected values of r_m and the corresponding doubling time t_2 are also given in table 2. As in the other species, the relationship between r_m and temperature is nearly linear (fig. 1).

Tab. 2. *Protohydra leuckarti*: Division time, as observed and as calculated from eq. (3), intrinsic rate of natural increase and corresponding doubling time at five different temperatures (after Smol, unpublished).

Temperature (°C)	Division time observed	Division time calculated	r_m per day	t_2 days
5	–	41.9	0.017	40.8
10	–	19.3	0.036	19.3
15	12.5 ± 2.5	12.2	0.057	12.2
20	8.4 ± 2.3	8.9	0.078	8.9
25	7.1 ± 1.1	6.9	0.100	6.9

The intrinsic rate of natural increase of the four species investigated seems to be a linear function of temperature. When we consider r_m as such the regression of r_m against temperature yields:

Protohydra leuckarti: $r_m = 0.0042\ T - 0.0048$
Oncholaimus oxyuris: $r_m = 0.0013\ T - 0.0043$
Tachidius discipes: $r_m = 0.0100\ T - 0.0137$
Paronychocamptus nanus: $r_m = 0.0057\ T - 0.0006$

D. Discussion

The fact that the realized rates of increase of the four copepod species investigated are nearly the same is remarkable, more so because differences in the intrinsic rate of increase appear to be much larger. This shows that dominance in these species is not a function of the rate of increase, and this in turn may have consequences on the realism of current models describing competition, when in fact it could be demonstrated that these species are competitors. This is not the case. Nevertheless, it remains certain that dominance is a matter of how long the increase can be maintained and has something to do with tolerance to environmental conditions in the broadest sense, including other populations, and not with the rate of increase, hence not with fertility or generation time.

The same is also apparent from the realized rates of increase of the two ostracod species, where the species with the lowest rate of increase is in fact dominant and the other species occurs much more erratically with much larger fluctuations in density. This might be in agreement with current theory on r- and K-selection, but as the species with the larger r-value is also the smaller one there could be an influence of body size, too. However, in the copepod species body size does not seem to be too important, as the range here is larger than the difference between the two ostracods.

The intensity of interspecific competition is normally expressed as the change in the per capita growth rate of species i due to the presence of an individual of species j relative to the change in the per capita growth rate of species i due to the presence of an individual of species i itself, formally:

$$\frac{\delta(dN_i/N_idt) / \delta N_j}{\delta(dN_i/N_idt) / \delta N_i} = \frac{a_{ij}}{a_{ii}} = \alpha_{ij}$$

The interspecific effect of one species is expressed as a fraction of the intraspecific effect of the other species. Vandermeer (1975) made the perfectly valid proposition that, instead of comparing interspecific competitive ability a_{ij} to the intraspecific competitive ability of the other species a_{ii}, we could compare it with the intraspecific competitive ability of the species perpetrating the competition.

$$\frac{\delta(dN_i/N_idt) / \delta N_j}{\delta(dN_j/N_jdt) / \delta N_j} = \frac{a_{ij}}{a_{jj}} = \beta_{ij}$$

When competition is defined in this way it follows that, when two species are less and less distinguished in terms of the way that one individual depresses the survivorship and fecundity of other individuals, their intrinsic rates must become more and more equal. Ecologically equivalent species coexist only if their r's are relatively equal (Vandermeer 1975). Now we are dealing with realized rates; when these are not themselves the result of the competition it appears justifiable to explain their values as resulting from a selection process permitting coexistence, and use these values in describing the competition. Our data, therefore, seem to validate Vandermeer's approach.

The predator-prey relationship between *Protohydra leuckarti* and its main prey species *Tachidius discipes* has been studied by Heip & Smol (1976a). It is remarkable that the realized rates of increase of these two species are again nearly the same; moreover, there is a remarkable resemblance in the reaction of development time to temperature as expressed with the parameter b of the power equation D = aT^b. The value of b is -1.12 for *P. leuckarti* and -1.13 for *T. discipes*. The identical values seem to indicate an interaction during the process which brings them about. Once again, there is a much larger difference in the intrinsic rates of increase of predator and prey. However, the predator increases at a rate which is equal to its intrinsic rate ($r_m = 0.049$ per day taking the mean temperature of 13.2 °C into account), while the realized rate of increase of the prey is a result of the predation process, and thus a function of the realized rate of increase of the predator.

The ostracod *Cyprideis torosa* is a detritus feeder on top of the food chain. It has a relatively low realized rate of increase when compared with the other ostracod *Loxoconcha elliptica* and the copepod species. The nematode *Oncolaimus oxyuris* is a predator on other nematode species and also on top of a food chain. In this species the intrinsic rate of natural increase is very low. Both this species and *C. torosa* seem not to be restricted by either food or predation. It is possible that group selection might play a role in the lowering of reproductive potential. A plausible mechanism for this kind of selection has been described by Gilpin (1975). It has been shown (Rosenzweig & MacArthur 1963) that overexploitation in predator-prey communities arises because there is a selective pressure on the predator to improve its exploitation efficiency. If this is continued too far a point may be reached where the ecological steady state between predator and prey is destroyed and whereafter predator extinction may occur. The destabilization of predator-prey steady state leads to limit cycle oscillations which may effectively cause the predator's extinction by taking it to such low densities that extinction is biologically inescapable. In Gilpin's model, predator populations occupy a number of environmental patches in which overexploitation and extinction of predator populations that are too efficient may occur. Between patches there is migration and gene flow. The populations on each patch are finite, thus the frequency of a gene for cropping efficiency will be subject to genetic drift, a process that will tend to maintain intergroup variance. Genetic drift is necessary for group selection in ensembles of populations which show initially the same genetic structure. If the generation time is lowered the same genetic drift will occur each generation, but the amount of selection is decreased, so lowering the generation time is favorable for group selection. This is of obvious application to meiobenthic populations.

Gilpin (1975) showed that destabilization and predator extinction are very rapid phenomena that occur in ecological as opposed to evolutionary time. If this is true it appears improbable that coevolution would take place, in the sense that it is doubtful that the prey could induce a change in the parameters of the predator population. However, experiments of Pimentel et al. (1965) showed that changes in the competitive ability of a species can occur in a short time. This can be explained by observing that fitness is lowered in the presence of competition, therefore selection acts within competing species to lessen the effect of the competitor on them. This selection should be strongest on the losing competitor because it suffers the greatest depression of fitness. It seems possible that this selection could change the outcome of competition before extinction occurs (Park & Lloyd 1955). A similar mechanism can be conceived in predator-prey relations, and by Occam's razor we need not involve a group selection hypothesis. However, although it is possible that natural selection produces populations with a lowered intrinsic rate of natural increase (Hairston et al., 1970), there is another aspect of the biology of

O. oxyuris which is not easily explained by natural selection. This species, both in the laboratory and in the field, has a low percentage of females in the adult population, around 40 % when sexual maturity is attained and much lower later because males tend to live longer. This phenomenon is hard to explain because selection will always cause either equal sex-ratios or favor females (Emlen 1973); however, the advantages of this low percentage for the population are clear. It lowers the reproductive potential while permitting the species to maintain a relatively high density. Indeed, *O. oxyuris* is at times the dominating species in the nematode community (Heip, Smol & Absillis, in preparation) and such a system in which a top predator is dominant both in numbers and in biomass seems impossible unless a very low turnover of the predator is accompanied by a very rapid turnover of the prey. The latter might indeed be the case, given the very high reproductive potential of some nematode species (Tietjen and Lee 1977).

The fact that the intrinsic rates of increase are linear functions of temperature is in agreement with the hypothesis that linearization of ecosystems is evolutionary adaptative (Patten 1975) in the sense that natural selection acts against nonlinear dynamic characteristics. It seems that systems which express nonlinear dynamic characteristics are not reliable while linear systems are well behaved due to superposition, ecosystem complexity, homeostasis and hierarchical organization. Ecosystem behavior is in the large neighbourhood dynamics about equilibrium and thus linear. Patten (1975) observes furthermore that ecosystems are rich in feedback which again introduces linearization. All these mechanisms should make ecosystems nominally linear in their large-scale holistic dynamics. This might provide some justification for the calculation of linear regressions and offers some hope for future modelling of complex ecosystems.

Bibliography

Angelis, D.L. De: Stability and connectance in food web models. Ecology **56**, 238–243 (1975).

Emlen, J.M. Ecology: an evolutionary approach. Addison-Wesley Publishing Company. Reading Massachusetts. 493 pp. (1973).

Gardner, M.R. & W.R. Ashby: Connectance of large dynamical (cybernetic) systems: critical values of stability. Nature **228**: 784 (1970).

Gilpin, M.E.: Group selection in predator-prey communities. Monographs in population biology 9. Princeton University Press. 108 pp. (1975).

Hairston, N.G., D.W. Tinkle & H.M. Wilbur: Natural selection and the parameters of population growth. J. Wildl. Manag. **34**: 681–690 (1970).

Heip, C.: The reproductive potential of copepods in brackish water. Marine Biology **12**: 219–221 (1972).

–: Partitioning of a brackish water habitat by copepod species. Hydrobiologia **41**: 189–198 (1973a).

–: Een populatie-dynamische studie over de benthale Ostracoda en Copepoda van een brakwaterhabitat. Ph.D. Thesis State University of Ghent. 429 pp. (in Dutch) (1973b).

– & N. Smol: On the importance of *Protohydra leuckarti* as a predator of meiobenthic populations. Proc. 10th. Europ. mar. biol. Symp. Vol II: 285–296 G. Persoone & E. Jaspers, Editors. Universa Press, Wetteren, Belgium. (1976a).

— & N. Smol: Influence of temperature on the reproductive potential of two brackish water harpacticoids (Crustacea, Copepoda). Marine Biology **35**: 327–334 (1976b).

May, R.M.: What is the chance that a large complex system will be stable? Nature **237**: 413–414 (1972).

—: Stability and complexity in model ecosystems. Monographs in population biology 6 Princeton University Press. 235 pp. (1973).

Maynard-Smith, J.: Models in ecology. Cambridge University Press. 146 pp. (1974).

Park, T. & M. Lloyd: Natural selection and the outcome of competition. Amer. Natur. **96**: 235–240 (1955).

Patten, B.C.: Ecosystem linearization: an evolutionary design problem. Amer. Natur. **109**: 529–539 (1975).

Pimentel, D.: Animal population regulation by the genetic feedback mechanism. Amer. Natur. **95**: 65–79 (1961).

—, E.H. Feinberg, P.W. Wood & J.T. Hayes: Selection, spatial distribution, and the coexistence of competing fly species. Amer. Natur. **99**: 97–109 (1965).

Roberts, A.: The stability of a feasible random ecosystem. Nature **251**: 607–608 (1974).

Rosenzweig, M.L. & R.H. MacArthur: Graphical representation and stability conditions of predator-prey interactions. Amer. Natur. **97**: 209–223 (1963).

Smol, N. & C. Heip: The culturing of some harpacticoid copepods from brackish water: Biol. Jaarb. **42**: 159–169 (1974).

Tietjen, J.H. & J.J. Lee: Life histories of marine nematodes. Influence of temperature and salinity on the reproductive potential of *Chromadorina germanica* Bütschli: Mikrofauna Meeresboden **61**, 263–270 (1977).

Vandermeer, J.: Interspecific competition: a new approach to the classical theory. Science **188**: 253–255 (1975).

| Mikrofauna Meeresboden | 61 | 113–136 | 1977 |

W. Sterrer & P. Ax (Eds.). The Meiofauna Species in Time and Space. Workshop Symposium, Bermuda Biological Station, 1975

Introgressive Hybridization Between Two Intertidal Species of Tetranchyroderma (Gastrotricha, Thaumastodermatidae) with the Description of a New Species[1]

by

William D. Hummon

Department of Zoology and Microbiology, Ohio University, Athens, Ohio 45701 USA

Abstract

Tetranchyroderma papii Gerlach, 1953 is redescribed from specimens collected on Crane's Beach near Woods Hole, Massachusetts, and *T. enallosa* n. sp. is described from a specimen collected on nearby Wood Neck Beach. Specimens from several sampling sites on Wood Neck Beach are analyzed with respect to introgressive hybridization between these two parental species. After scoring each individual on a point basis, it was found that nearly "pure" *T. papii* occurred in the upper part of the *Tetranchyroderma* zone but gave way to specimens having a broad array of hybrid characters as one proceeded downbeach. The indication is that *T. enallosa* acted as the recurrent parent in the introgression process. "Pure" *T. enallosa* was never found in abundance, and evidence of introgression was present in the largest specimens found, which were nearly double the total length of the largest reproductively mature *T. papii*. Specimens from two other beaches in the northeastern U.S. on which both species were found also show signs of probable introgressive hybridization. The situation appears to represent a local breakdown of reproductive isolation between the two species and does not seem to represent a significant evolutionary pathway in these animals.

A. Introduction

It is not often that a field zoologist, working on the systematics and ecology of a specific group of micrometazoa such as the Gastrotricha, stumbles upon a phenomenon of the sort described and analyzed below. And it is probably less often that such a person has the good fortune to recognize the phenomenon, in this case one of introgressive hybridization, and has enough time to assess it in sufficient detail that it can be communicated to others. Thus, to the predominately vertebrate zoological literature on introgressive hybridization (see for instance reviews by Sibley 1961, Mayr 1963, Remington 1968 & Mayr 1970, and papers by Lewontin & Birch 1966 and Gorman et al. 1975) can be added a probable instance among the marine meiofauna.

[1]This research was supported in part by Grants 320 and 365 from the Ohio University Research Committee.

Wood Neck Beach (41° 35′ N lat., 70° 39′ W long.) lies 6 km north of the village of Woods Hole, Massachusetts, USA, on the barrier beach forming the western border of Sippiwisset Pond. The study transect was located 50 m south of the pond outlet, facing Buzzards Bay. Substratum consisted of fine sand (Mϕ 2.7 (155 μm), SDϕ 0.2) over the lower 18 m of the transect, from mean low water spring to mid 'tide level (profile slope of 3–4 %) and of mixed sand (Mϕ 1.3 (400 μm), SDϕ 1.1) and cobble grading to fine sand (Mϕ 2.2 (220 μm), SDϕ 0.5) over the upper 4 m of the transect, from mid to mean high tide level (profile slope 10–12 %). In this beach transect there occurred two species of *Tetranchyroderma*, *T. papii* Gerlach, 1963, and the *T.* sp. E of Hummon, 1974c (*T.* sp. D of Hummon 1969). *T. papii* occurred in the lower portion of the mixed sand and cobble region and tended to intergrade or hybridize with the second species in the fine sand lower portion of the beach. It is this pattern of introgressive hybridization and its description that forms the core of the present paper.

It is necessary, however, first to provide descriptive material on the two species of *Tetranchyroderma* involved. The description of *T. papii* is derived from animals collected on Crane's Beach (41° 32′ N lat., 70° 41′ W long.), located 0.5 km west of the village of Woods Hole on the thinnest portion of the proximal end of Penzance Point. The study transect, situated 0.1 x the beach length from a sea wall at the west end, facing Buzzards Bay, was described in its physical and dynamic aspects by Pollock & Hummon (1971). On this beach, only *T. papii* of the two species was found. The description of the second species of *Tetranchyroderma*, one new to science, is derived from animals collected on Wood Neck Beach.

B. Description of species

Family Thaumastodermatidae Remane, 1926, sensu Hummon, 1974
Genus Tetranchyroderma Remane, 1926
Tetranchyroderma papii Gerlach, 1953
Figs. 1d–f

Distribution. Mediterranean: Tyrrhenian sea – San Rossore (type locality) and Bagno Gorgoni near Pisa (Gerlach 1953, 1955; Luporini et al. 1971) and a beach midway between Gaeta and Sperlonga (Hummon present study), Italy; Gulf of Lions – Canet Plage near Perpignan (Delamare Deboutteville 1953, 1954) and several beaches near Sète and Carnon (Fize 1957, 1963), France; Adriatic Sea-San Cataldo near Lecce (De Zio & Grimaldi 1964), Italy; Aegean Sea-Loutsa Beach near Athens (Hummon present study), Greece; Raoud Plage and Amilcar near Tunis (Westheide 1972). Tunisia; and a series of beaches from Béni-Saf east to Port-aux-Poules (d'Hondt 1973), Algeria. North Atlantic: a series of beaches from the New Hampshire-Massachusetts border south and west onto Long Island, New York (Hummon 1967, 1968, 1969, 1974a, 1974b, 1975, present study), a beach at

Roosevelt Inlet near Lewes, Delaware (Hummon present study) and several beaches near Morehead City, North Carolina (E. E. Ruppert personal communication), U.S.A.

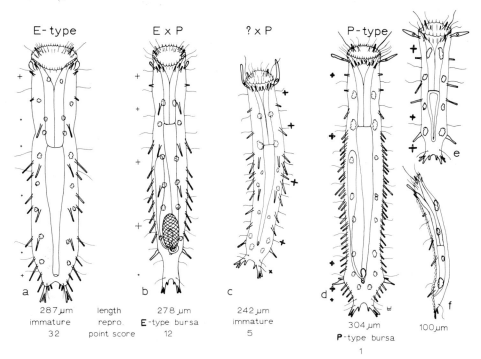

Fig. 1. Specimens of parental species from locations near Woods Hole, Massachusetts, *Tetranchyroderma enallosa* n. sp. (Fig. 1a) from Wood Neck Beach and *T. papii* Gerlach, 1953 (Fig. 1d adult, 1ef juvenile dorsal and lateral views) from Crane's Beach, along with one specimen of a *T. enallosa* x *T. papii* hybrid (Fig. 1b) from Wood Neck Beach and one specimen of an unknown parent x *T. papii* hybrid (Fig. 1c) from Loutsa Beach near Athens, Greece. Other data in the figure are: total length of body, reproductive condition and point score according to the scheme presented in Table II.

Description. Relaxed adults with mouth opening nearly as broad as trunk; body having two slight constrictions in the pharyngeal region, one at the level of the prominent lateral head tentacles and the other, the "neck", about two thirds back along the pharyngeal length; body reaches its maximal width in the anterior intestinal region; another lateral constriction may occur in the posterior intestinal region, particularly in North American specimens where it is often followed by paired, sculptured prominences bearing large, heavy lateral adhesive tubes, before narrowing once again to the base of the caudal feet. Dorsum of mouth overhangs the venter, forming an oral hood, and bears one or two small tentacles on either side; where two are present per side, they insert near one another with the medial one only 0.5–0.7 x the length of the lateral one (contra Gerlach 1953). Elongate tentacles insert lateral to the venter of the mouth opening and project anteriolaterally; as noted by d'Hondt (1973) they tend to be narrower at the base and apex than in the middle, with their maximal width perhaps one third out from the insertion; there is, however, a great deal of variability in the shape of these tentacles and they sometimes appear knobbed at their apices and are often somewhat asymmetric on one side relative to the other. Dorsal tentacles are absent from the trunk region.

Adhesive tubes are arranged in three groups. Anteriorly, close behind the ventral border of the mouth, are located 3–4 pairs, forming an arc with the medial-most pointing directly forward and the lateral-most pointing obliquely forward; a set of structures located in a transverse row even more closely appressed behind the ventral mouth surface appears to be cuticular supports rather than adhesive tubes, as held for members of this genus by Gerlach and others. Lateral series of adhesive tubes insert ventrolaterally and begin with 1–3 pairs directed laterally or posteriolaterally in pharyngeal region, followed by a host of posteriolaterally directed tubes located in the trunk region; arrangement of lateral trunk tubes is quite variable, ranging from more or less grouped to regular, though one must take care that a grouped appearance does not result from adhesive tubes of a more regular series being tucked up beneath the trunk where they can be seen only with difficulty; prominent tubes on either side occur adjacent to the anus, followed posteriorly by several additional tube pairs as the trunk narrows to the base of the caudal feet. Caudal feet appear to be a posterior continuation of the trunk; adhesive tubes of caudal feet typical for most species in the genus, consisting on either side of a pair of major tubes with a minor tube inserting dorsally between their fused bases, and bearing on the medial surface of each foot a fourth adhesive tube.

Short tactile cilia extend forward from the escalloping of the oral hood; dorsally on the hood there are several longer pairs of tactile cilia, one of which is laterally placed and more elongate and has been referred to by Luporini et al. (1971) as a pair of "flagelliform tentacles;" and additional sensory cilia are scattered along the lateral margins of the body, with two pairs being borne dorsally on the bases of the caudal feet. Locomotor cilia insert in transverse rows and cover nearly the entire ventral body surface.

The cuticular armature is comprised of robust pentancres, borne on dorsal and lateral body surfaces, with an occasional tetrancre interspersed among them; points of ancres are of subequal length. Pentancres vary somewhat in size from region to region of the body and from individual to individual, measuring 6–7 μm (occasionally 5–6 or 7–8 μm) in width across opposite points over much of the body but markedly decreasing in size as one proceeds forward onto the oral hood or backward onto the caudal feet; the forward portion of the oral hood is free of cuticular armature, but bears several fine longitudinal cuticular thickenings.

Refractile granules number seven to nine pairs and are mostly symmetrical from side to side in their dispersion; it is characteristic in adults I have seen of this species that the second pair from the front occurs at the level of the pharyngointestinal junction.

The pharyngeal to intestinal length ratio varies with age (see data in Table 1); pharynx narrows from the broad mouth to a fairly constant width for half of its length; pharyngeal pores are borne at the base of the pharynx, but are seen only with difficulty; intestine narrows gradually throughout its length and exits by means of a ventral anus.

The reproductive system consists of a single testis on the right side and a single ovary; testis seems to develop before a maturing ovum is apparent, though both may be present simultaneously; the bursal caudal organ is smaller than that figured by Luporini et al. (1971), but is of the same basic shape; youngest reproductively mature specimens were in the 250–274 μm length class; mean size of ovum when present (n = 19) was 66g x 22f (germinal vesicle 11g) μm, mean size when two ova were present (n = 4) were 66d x 26b(g.v. 11g) μm for the anteriormost of the two and 26o x 18.5h (g.v. 7h) μm for the posteriormost (letters refer to coefficients of variation in percent, as explained in Table 1, bottom); the posterior of two eggs being smaller than the anterior indicates a more rearward position for the ovary than that shown by Gerlach (1953) and Luporini et al. (1971).

A sequence of metric-meristic-eidostic portraits for specimens of *T. papii* from Crane's Beach are given in Table 1, as called for by Hummon (1971). Size and number of morphological characters are related to total body length of extended and relaxed specimens for each of several developmental length classes, ranging from juveniles to adults, and are supplemented by Figs. 1ef and 1d of juvenile and adult specimens respectively. Variability of characters within each length class is indicated by a letter, denoting coefficient of variation clustered to the nearest four percent. These data will be used later in the paper as an adjunct means of assessing the probability of genetic introgression from *T. papii* to the other species of *Tetranchyroderma* occurring on Wood Neck Beach.

Several anomalous specimens have appeared in the course of studying the Crane's Beach population of *T. papii* and will be commented upon briefly. One individual, 304 μm in length, bore no cleft between the caudal lobes, giving the impression of a rounded posterior end; its pentancres in the posterior half of the trunk were arranged in transverse rows widely spaced (11 μm) from one another compared with the more typical spacing (5.5 μm) between pentancres of a given row. Several individuals, lacking caudal feet altogether, were found in late July 1970 after a severe storm resulted in 6.5 cm of rain within a period of several hours at low tide.

In a final note, d'Hondt (1973) commented upon the similarity between the pentancrous *T. papii* Gerlach, 1953 and the tetranchyrous *T. bunti* Thane-Fenchel, 1970, and logically suggested that perhaps the two may ultimately be aligned in

William D. Hummon

Table 1. Body measurements in µm by length class for *Tetranchyroderma papii* Gerlach, 1953, from Crane's Beach, Woods Hole, Massachusettes. Lt, total length from anterior tip of head to posterior tips of caudal adhesive tubes; LPh, length of pharynx from anterior tip of head to pharyngointestinal junction; LIn, length of intestine, from pharyngointestinal junction to posterior tip of anus; LCFt, length of caudal foot, obliquely from medial junction of lobes to tip of furthest projecting adhesive tubes; LHdTn, length of lateral head tentacle from insertion to apex; WBc, max. width of head at level of the buccal cavity, WNk, min. width of body in posterior half of pharyngeal region; WTr, max. width of trunk in intestinal region; WFtBs, min. width of trunk at base of caudal foot, DTn, mean no. dorsal tentacles in trunk region; TbA, mean no. adhesive tubes per side ventrally behind mouth opening; TbL, mean no. adhesive tubes per side in lateral body series; TbCFt, mean no. adhesive tubes per side on and between caudal feet. All measurements were made on living, relaxed specimens.

No.	Midpoint of 25 µm length class	Lt	Lph	LIn	LCFt	LHdTn	WBc	WNk	WTr	WFtBs	Tb A	Tb L	Tb CFt
1	87 (75–99)	96 *	51	29	8	11	24	23	21	11	1	5	4
2	112 (100–124)	108a	58b	37a	10h	13.5j	30c	25c	26b	14c	1a	4a	4a
4	137 (125–149)	137b	69b	43c	12.5d	15.5g	31.5c	28c	30c	16d	1.7h	5j	4a
11	162 (150–174)	164b	73a	62b	14d	17b	34b	27.5b	30.5b	17.5d	1.8f	6.5i	4a
11	187 (175–199)	187a	75b	79d	16.5e	20c	37.5c	31c	33c	18c	2.1d	8.5f	4a
18	212 (200–224)	212a	84b	98b	17c	20.5b	40a	33b	37c	19.5b	2.4f	10.5a	4a
24	237 (225–249)	237a	89b	113b	18c	21b	41b	34.5b	40b	20b	2.8c	12.5e	4a
25	262 (250–274)	262a	95b	131b	17d	21.5c	43b	34.5b	44c	21b	3b	16.5e	4a
25	287 (275–299)	286a	100b	150b	19c	22.5c	46b	38.5b	49c	22b	3.1c	19.5d	4a
25	312 (300–324)	312a	107a	168b	20c	25b	47b	40b	54b	22.5c	3a	21.5c	4b
23	337 (325–350)	334a	109a	185a	20c	23.5c	48b	41c	54c	22.5b	3.1c	23c	4.1c
9	362 (350–374)	358a	114a	201b	21c	24.5b	50b	42b	55c	25c	3a	22e	4.1c

* Coefficient of variation 100 SD/x̄ in percent: a = 0 to 4, b = 4 to 8, c = 8 to 12, d = 12 to 16, e = 16 to 20, f = 20 to 24, g = 24 to 28, h = 28 to 32, i = 32 to 36, j = 36 to 40, k = 40 to 44, l = not used, m = 44 to 48, . . . z = 96 to 100.

the same species. However, based on my observations of both (Hummon 1974c) and those of E. E. Ruppert (personal communication) and Nixon (In press), the two are separate and distinct species.

Tetranchyroderma enallosa n. sp.
Fig. 1a

Of the several hundred specimens analysed during the course of this study (Summer, 1970), only a dozen were sufficiently near maturity, and possessing those characters which I believe to comprise *T. enallosa,* to be considered as belonging to this species. Unfortunately, none of these animals was preserved, so the species is described from the immature specimen drawn in Fig. 1a, without benefit of a Holotype specimen. The remaining specimens belonged to the *T.* spp. hybrid complex and showed signs of minute to predominate amounts of introgression from *T. papii.*

D e s c r i p t i o n. Body measurements for the specimen illustrated are as follows: (see Table 1 for key to symbols): Lt 287; LPh 112; LIn 146; LCFt 19; LHdTn 10, but see note below; WBc 53; WNk 44; WTr 54; WFtBs 30; DTn 4; TbA 1; TbL 11 and TbCFt 5. Body robust with mouth opening as broad as trunk, two slight constrictions at the level of the lateral head tentacles and a visible neck constriction in the posterior pharyngeal region; sides of trunk nearly parallel over most of the intestinal region, narrowing to form two broad posterior lobes upon which the caudal feet insert. Dorsum of mouth overhangs the venter, forming an oral hood, and bears one small tentacle on either side; short knobby tentacles insert lateral to the venter of the mouth opening and project anteriorly in a manner so closely appressed to the ventrolateral sides of the head that they appear to be absent. Four pairs of thin dorsal body tentacles, 15–16 μm in length, occur along the length of the body, one in the posterior pharyngeal region and three rather evenly spaced along the intestinal region.

Adhesive tubes are arranged in three groups. Anteriorly, close behind the ventral border of the mouth, is located a single pair of adhesive tubes lying medially to the insertion of the lateral head tentacles and pointing obliquely forward. Laterally is located a series of 11 adhesive tubes per side; two rather short pairs occur in the head region, one lying just behind the insertion of the lateral head tentacle and projecting obliquely forward beyond the profile of the body, and the other lying midway down the length of the pharynx and projecting obliquely rearward; the other nine pairs are of varying length and project posteriolaterally, being sporadically located behind the pharyngointestinal junction. Caudal feet appear to be distinct from and inserting on two broad posterior lobes of the trunk; adhesive tubes of the caudal feet are somewhat atypical for the genus, in that they consist, on either side, of but a pair of major tubes with a minor tube inserting dorsally between their fused bases; two pairs of adhesive tubes occur medially between the caudal feet, inserting not on the caudal feet themselves but rather on the posterior trunk lobes.

Short tactile cilia extend forward from the escalloping of the oral hood; dorsally on the hood there are several longer pairs of tactile cilia, one of which is laterally placed and more elongate than the others; and additional sensory cilia are scattered along the lateral margins of the body, with one pair being borne dorsally on the lobe-like bases of the caudal feet. Locomotor cilia insert in transverse rows and cover nearly the entire ventral body surface.

The cuticular armature is comprised of delicate pentancres, borne on dorsal and lateral body surfaces; points of ancres are of subequal length. Pentancres vary somewhat in size from region to region of the body, measuring 2–3 μm (occasionally 4–5 μm) in width across opposite points over much of the body but decreasing even further in size as one proceeds forward onto the oral hood or backward onto the caudal lobes; the forward position of the oral hood is free of cuticular armature, but bears several fine longitudinal cuticular thickenings.

Refractile granules number 9 pairs and are mostly symmetrical from side to side in their disposition; granules are nearly all small in size, with four pairs in the pharyngeal region and five pairs in the intestinal region; none occur directly adjacent to the pharyngointestinal junction.

The pharyngeal to intestinal length ratio is relatively large for this genus at 0.77 in an animal of 287 μm total length, and indicates both the immaturity of the specimen and the large size that reproductively mature adults can be expected to attain; pharynx narrows from the broad mouth to a fairly constant width for well over half of its length; pharyngeal pores are borne at the base of the pharynx, but are seen only with difficulty; intestine is broad anteriorly and narrow posteriorly with a fairly rapid decrease in width occuring about halfway down its length.

The reproductive system was not yet developed in the specimen being described. E t y m o l o g y. **Enallos** (Gk), changed or contrary.

D i s c u s s i o n. While keeping in mind all of the hazards that are entailed, the description of this species was based on a reproductively immature specimen because of the difficulty of obtaining assuredly non-hybrid material. Indications are that mature adults reach 600 μm in total length and that they can be recognized primarily by an increase in the number of dorsal body tentacles to perhaps eight or nine and by an increase in the number of adhesive tube pairs inserting on the posterior trunk lobes between the caudal feet to perhaps four pairs. None of the individuals I saw which exceeded 350 μm in total length and were reproductively mature was free of the evidence of hybridization.

The adult reproductive system, I believe, consists of a single testis located on the right side and a single ovary; the bursal caudal organ is of the type illustrated in Fig. 1b, large with rather complex internal and accessory structures and with highly visible cross-helical muscle striations.

The other species of *Tetranchyroderma* having dorsal trunk tentacles are *T. cirrophora* Levi, 1950; *T. vera* Wilke, 1954; *T. massiliense* Swedmark, 1956; *T. indica* Chandrasekhara Rao and Ganapati, 1958; and *T. hirtum* Luporini et al.,

1970. Of these, all but the latter have tetrancres, while *T. hirtum* differs from the presently described species in a number of characters; *T. hirtum* being reproductively mature at between 300 and 360 μm, having many more adhesive tubes than *T. enallosa* and the caudal feet of *T. hirtum* forming posterior extensions of the trunk.

Hybridization

In looking over some gastrotrich meiofauna from Wood Neck Beach during the late summer of 1968, I noticed that most of the *Tetranchyroderma* specimens were morphologically intermediate between *T. papii* and a second species. There was no single type of intermediate, but a broad range of intermediates, grading more or less from one species to the other. The opportunity to follow this up was not present until 1970, and during the early part of that summer, I sampled and analyzed a number of specimens from various parts of the study transect at Wood Neck Beach. Gradually a series of characters was sorted out which could serve to quantify the hybrid characteristics that appeared in most of the individuals analyzed. These were set up into a point scoring scale, after the manner of that presented by Anderson (1949). Then, that portion of the study transect in which *Tetranchyroderma* spp. occurred was subdivided into six sampling sites, horizontally equidistant from one another (Fig. 2). Each was sampled and the first 20 specimens (10 from the upper sampling site) whose total length lay between 250 and 350 μm were scored according to the scale of points shown at the head of Table II. The scale was set up so that "pure" *T. papii* theoretically would score

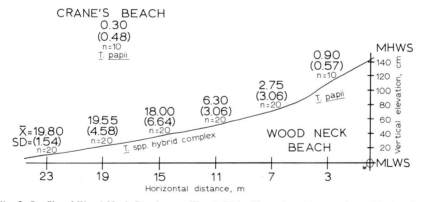

Fig. 2. Profile of Wood Neck Beach near Woods Hole, Massachusetts, together with sites from which samples were taken for the analysis of hybridization between *Tetranchyroderma enallosa* n. sp. and *T. papii* Gerlach, 1953. Number of specimens located between 250 and 350 μm total length examined, mean point score according to the scheme presented in Table II and standard deviation of point scores are given for each sampling site. Comparable data are given upper left for a sample of specimens taken from the reference population of *T. papii* at nearby Crane's Beach.

Table II. Index scale used to assess the amount of hybridization between P-type and E-type *Tetranchyroderma*, and scored values for 120 specimens.

Key to Point Scoring

Group I (max. range 0–6 pts)
General body configuration (see text):
Type P   0 pts   H (intermediate)   1 pt
Type E   2 pts
Percent body length back to max. neck constriction:
≤ 29   0 pts   30–34   1 pt
35 ≤   2 pts
Width of posterior foot base, μm:
≤ 24   0 pts   25–29   1 pt
30 ≤   2 pts

Group II (max. range 0–8 pts)
No. anterioventral adhesive tubes per side:
3–4   0 pts   1–2   2 pts
No. ventrolateral adhesive tubes per side:
15 ≤   0 pts   10–14   1 pt
≤ 9   2 pts
Excess No. medioposterior adh. tubes per side:
0   0 pts   1–2   2 pts
Posterior foot attachment type and length, μm:
P ≤ 24   0 pts   H (intermediate) 25 ≤   1 pt
E ≤ 24   2 pts

Group III (max. range 0–8 pts)
Shape and μm length of lateral head tentacles:
L (elongate)   16–24   0 pts   L 25 ≤   1 pt
C (club-shaped)   9–15   2 pts   C 4–8   3 pts
A (absent)   0–3   4 pts
No. dorsal trunk tentacles per side:
1 pt each

Group IV (max. range 0–6 pts)
Sum, diagonal μm widths of dorsal pentancres
25, 50 and 80 % back from the anterior tip
of the body:
19 ≤   0 pts   17–18   1 pt
15–16   2 pts   13–14   3 pts
11–12   4 pts   9–10   5 pts
7–8   6 pts

Group V (max. range 0–6 pts)
Type of bursal caudal organ
P   0 pts   H (intermediate)   2 pts
E   4 pts   I (immature)   6 pts

Table II (cont'd)

Specimen location and number		Total body length μm	General body config.: % length to max. neck constr./width of post. foot base, μm	No. ant.-ventral/No. ventrolat./No. excess No. medio-posterior adh. tubes per side	Posterior foot attachment type and length, μm	Shape and μm length of lateral head tentacles/No. dorsal trunk tentacles per side	Diagonal μm width of dorsal pentancres 25, 50 and 80 % back from ant. tip of body	Type of bursal caudal organ	Point total
Cr	1	324	P24/21	3/15/0	P20	L22/0	6/7/7	P	0
	2	298	P28/23	3/15/0	P21	L22/0	6/7/6	P	0
	3	295	p28/20	4/15/0	P23	L30/0	6/6/7	P	1
	4	252	P29/20	4/19/0	P12	L23/0	6/6/6	P	1
	5	295	P27/20	3/25/0	P20	L20/0	7/6/6	P	0
	6	281	P27/22	3/16/0	P16	L19/0	6/6/7	P	0
	7	260	P28/24	3/15/0	P20	L21/0	6/6/7	P	0
	8	270	P29/22	3/22/0	P20	L22/0	6/5/6	P	1
	9	332	P24/23	3/20/0	P20	L23/0	6/6/7	P	0
	10	305	P26/23	3/21/0	P21	L22/0	6/6/7	P	0
Wa	1	285	P29/20	3/23/0	P18	L22/0	6/6/7	P	0
	2	350	P25/22	3/24/0	P21	L21/0	6/6/6	P	1
	3	345	P24/20	3/22/0	P20	L23/0	5/6/7	P	1
	4	348	P25/22	3/22/0	P20	L21/0	6/5/6	P	1
	5	273	P23/21	3/19/0	P20	L23/0	6/6/6	P	1
	6	292	P27/22	3/22/0	P20	L24/0	6/5/7	P	1
	7	305	P27/22	3/20/0	P16	L22/0	6/7/7	P	0
	8	290	P27/21	3/18/0	P17	L22/0	6/5/5	P	2
	9	343	P23/20	3/21/0	P20	L22/0	6/6/6	P	1
	10	253	P29/19	3/19/0	P19	L21/0	5/6/6	P	1
Wb	1	284	P27/22	3/19/0	P19	L27/0	6/7/7	P	1
	2	265	P25/21	3/21/0	P20	L24/0	6/4/4	P	3

William D. Hummon

Table II (cont'd)

Specimen location and number	Total body length, μm	General body config.: % length to max. neck constr./ width of post. foot base, μm	No. ant.-ventral/No. ventrolat./No. excess No. medio-posterior adh. tubes per side	Posterior foot attachment type and length, μm	Shape and μm length of lateral head tentacles/ No. dorsal trunk tentacles per side	Diagonal μm width of dorsal pentancres 25, 50 and 80 % back from ant. tip of body	Type of bursal caudal organ	Point total
3	273	P29/20	3/21/0	P16	L20/0	6/5/6	P	1
4	325	P25/23	3/23/0	P22	L20/0	6/5/5	P	2
5	316	P26/24	3/23/0	P21	L22/0	6/6/6	P	1
6	302	P26/22	3/24/0	P20	L22/0	6/6/6	P	1
7	292	P25/20	3/15/0	P20	L20/0	5/6/6	P	1
8	294	P34/22	3/10/0	H28	L25/2	5/5/5	P	8
9	320	P24/20	3/17/0	P20	L21/0	6/5/6	P	1
10	250	P29/19	3/14/0	P21	L19/0	5/5/6	P	3
11	332	P24/22	3/18/0	P21	L23/0	5/6/6	P	1
12	262	P33/21	3/8/0	P22	L28/2	4/5/6	P	8
13	284	P28/22	3/16/0	P20	L22/0	6/6/7	P	0
14	267	P31/21	3/20/0	P21	L26/0	6/6/6	P	3
15	333	P26/23	3/22/0	P17	L22/0	5/6/6	P	1
16	287	P24/21	3/18/0	P20	L22/0	5/5/6	P	2
17	328	P27/25	3/22/0	P20	L20/0	6/6/6	P	2
18	323	P26/22	3/17/0	P21	L23/0	6/6/6	P	1
19	320	P24/22	2/24/0	P20	L23/0	6/6/6	P	3
20	284	H30/22	2/8/0	H29	L29/2	5/5/6	P	12
Wc 1	307	P32/20	3/16/0	P23	L29/1	6/6/5	P	4
2	278	P30/21	3/14/0	H29	L26/1	6/6/7	P	5
3	260	P33/21	3/9/0	H26	L27/2	5/4/5	P	10
4	285	P29/20	3/8/0	H27	L26/2	4/6/6	P	8

Table II (cont'd)

Specimen location and number	Total body length, μm	General body config.: % length to max. neck constr./ width of post. foot base, μm	No. ant.-ventral/No. ventrolat./ excess No. medio-posterior adh. tubes per side	Posterior foot attachment type and length, μm	Shape and μm length of lateral head tentacles/ No. dorsal trunk tentacles per side	Diagonal μm width of dorsal pentancres 25, 50 and 80 % back from ant. tip of body	Type of bursal caudal organ	Point total
5	327	P27/21	3/14/0	H26	L26/2	7/6/5	P	6
6	307	P29/21	3/8/0	H28	L27/1	5/6/6	P	6
7	322	P29/20	3/15/0	P22	L28/1	6/6/9	P	2
8	340	H26/16	2/8/0	H31	C8/1	4/5/6	E	16
9	292	P31/21	3/11/0	H29	L26/1	4/4/7	P	7
10	290	P32/22	3/11/0	H25	L23/2	5/6/5	P	7
11	348	P29/19	3/17/0	P22	L32/1	5/6/6	P	3
12	348	P30/21	4/16/0	H31	L24/2	6/7/8	P	4
13	270	P30/20	3/13/0	P24	L23/1	5/5/5	P	5
14	272	P34/22	3/11/0	P23	L25/2	4/5/5	P	7
15	286	P29/21	3/8/0	H29	L25/1	4/4/4	P	8
16	315	P28/21	3/11/0	H26	L23/2	5/5/6	P	6
17	318	P28/22	4/14/0	H29	L29/0	6/7/8	P	3
18	300	P31/20	4/17/0	H28	L29/2	5/5/5	P	7
19	325	P29/23	3/15/0	H29	L29/2	6/6/7	P	4
20	325	P30/22	3/9/0	H29	L31/2	5/5/7	P	8
Wd								
1	287	E34/30	3/10/1	E19	C9/4	3/2/4	I	27
2	278	E26/19	3/10/0	E24	C8/3	4/5/6	E	17
3	265	E36/27	2/8/0	E19	A/3	4/3/4	I	28
4	289	E23/22	3/14/0	E16	C8/3	5/6/7	E	16
5	260	E38/26	2/5/0	E16	A/3	3/2/5	I	29
6	254	E29/20	3/10/0	E18	C8/2	3/4/5	E	18

Table II (con'd)

Specimen location and number	Total body length, μm	General body config.: % length to max. neck constr./ width of post. foot base, μm	No. ant.-ventral/No. ventrolat./ excess No. medio-posterior adh. tubes per side	Posterior foot attachment type and length, μm	Shape and μm length of lateral head tentacles/ No. dorsal trunk tentacles per side	Diagonal μm width of dorsal pentancres 25, 50 and 80 % back from ant. tip of body	Type of bursal caudal organ	Point total
7	291	P27/19	3/16/0	P20	L28/1	5/6/8	P	2
8	310	E30/18	2/14/0	E20	C9/2	4/5/6	E	18
9	283	E27/18	2/9/0	E18	C8/3	5/5/6	E	20
10	288	E28/19	2/16/0	E19	C8/3	4/4/4	E	20
11	296	E24/27	2/13/0	H27	C8/3	4/6/6	E	19
12	256	E27/19	2/15/0	E20	C8/1	3/4/5	E	18
13	342	H30/20	3/6/0	H28	L27/1	6/6/7	H	9
14	265	E31/18	2/5/0	E17	C9/3	5/6/5	E	20
15	275	E29/18	2/10/0	E21	C8/3	5/5/6	E	19
16	279	E27/20	2/14/0	E20	C8/3	5/6/5	E	19
17	255	E29/23	2/12/0	E20	C8/3	6/7/7	E	17
18	285	E29/22	2/11/0	H26	C8/3	5/5/6	E	18
19	332	P27/22	3/16/0	H25	L27/2	5/6/6	P	5
20	306	E23/25	2/11/0	E18	C8/3	4/5/5	E	21
We 1	274	E27/20	2/7/0	E23	C8/3	4/5/5	E	21
2	294	E39/23	2/8/0	E14	A/3	3/3/3	I	28
3	302	E28/20	2/11/0	E21	C9/2	4/5/5	E	18
4	315	E28/20	2/13/0	E19	C9/3	5/3/4	E	20
5	339	E28/20	2/10/0	E21	C10/1	5/6/5	E	16
6	250	E38/22	2/8/0	E18	A/2	2/2/3	I	28
7	250	E28/20	2/5/0	E13	C8/3	5/5/5	E	20
8	258	E30/19	2/11/0	E17	C8/3	5/5/6	E	20

Table II (cont'd)

Specimen location and number	Total body length, μm	General body config.: % length to max. neck constr./ width of post. foot base, μm	No. ant.-ventral/No. ventrolat./ excess No. medio-posterior adh. tubes per side	Posterior foot attachment type and length, μm	Shape and μm length of lateral head tentacles/ No. dorsal trunk tentacles per side	Diagonal μm width of dorsal pentancres 25, 50 and 80% back from ant. tip of body	Type of bursal caudal organ	Point total
9	262	E29/21	2/8/0	E21	C7/3	4/6/5	E	20
10	268	E28/20	2/10/0	H25	C8/3	5/5/6	E	18
11	270	E29/21	2/10/0	E20	C7/3	5/5/5	E	19
12	300	E27/20	2/17/0	H26	C7/3	5/4/4	E	18
13	318	E22/19	2/14/0	E21	C7/3	5/7/5	E	18
14	274	E31/20	2/11/0	E21	C8/3	4/5/5	E	21
15	292	E25/18	3/19/0	P15	L20/0	5/5/5	P	5
16	276	E30/19	2/10/0	E19	C8/3	4/5/5	E	21
17	279	E30/21	2/9/0	E20	C8/3	6/6/6	E	20
18	270	E27/20	2/14/0	E23	C8/3	6/6/6	E	18
19	255	E29/28	2/8/0	E23	C8/3	4/4/4	E	23
20	250	E29/20	2/11/0	E20	C8/3	5/5/5	E	19
Wf 1	296	E27/20	2/13/0	E23	C6/3	5/6/5	E	19
2	310	E26/20	2/14/1	E18	C8/3	4/5/4	E	22
3	346	E27/20	2/5/0	E20	C8/3	4/6/5	E	20
4	286	E24/20	2/10/0	E20	C7/3	4/5/5	E	20
5	252	E32/18	2/10/0	E17	C7/3	5/4/4	E	20
6	303	E25/20	2/11/0	E20	C8/3	5/5/5	E	20
7	290	E25/18	2/11/0	E21	C8/3	4/5/4	E	19
8	290	E28/16	2/11/0	E16	C8/3	4/5/4	E	20
9	330	E26/20	2/14/0	E24	C9/3	4/4/4	E	21
10	296	E28/20	2/14/0	E20	C8/3	5/6/6	E	18

Table II (cont'd)

Specimen location and number	Total body length, μm	General body config.: % length to max. neck constr./ width of post. foot base, μm	No. ant.-ventral/No. ventrolat./excess No. medio-posterior adh. tubes per side	Posterior foot attachment type and length, μm	Shape and μm length of lateral head tentacles/ No. dorsal trunk tentacles per side	Diagonal μm width of dorsal pentancres 25, 50 and 80% back from ant. tip of body	Type of bursal caudal organ	Point total
11	252	E31/21	2/9/0	E22	C7/3	4/3/6	E	21
12	273	E26/19	2/13/0	E21	C7/3	4/4/4	E	21
13	250	E28/22	2/5/0	E20	C7/3	3/3/3	E	23
14	280	E29/14	2/9/0	E20	C8/3	4/4/4	E	22
15	338	E25/24	2/12/0	E20	C8/3	4/5/5	E	20
16	314	E27/18	2/11/0	E21	C10/3	4/5/6	E	18
17	302	E25/20	2/14/0	E22	C8/3	5/6/5	E	19
18	322	E26/24	2/15/0	H26	C8/3	6/5/6	E	17
19	350	E27/25	2/15/0	H31	C9/3	4/4/4	E	19
20	300	E26/21	2/12/0	E23	C8/2	4/5/6	E	18

zero points and "pure" *T. enallosa*, the second species, theoretically would score 34 points. These specimens were also measured over a metric-meristic series of the sort presented in Table I. The range 250–350 µm represented roughly the extremes of reproductively mature specimens of *T. papii* previously studied on Crane's Beach, where *T. enallosa* was not found.

Larger specimens were included in the metric-meristic series from Wood Neck Beach when encountered. Results, individual by individual, are given in Table II, along with those of the 10 individuals from the Crane's Beach reference population of *T. papii*. In Table II and thenceforth, Cr represents the specimens from Crane's Beach, Wa the 3m sampling site at Wood Neck, Wb that at 7m, Wc at 11m, Wd at 15m, We at 19m and Wf at 23m.

Character group I included a general subgroup based on body configuration and was determined on the basis of several characters which were either not easily quantifiable but presented a visual impression – eidostic characters – or which were not included in other groups or subgroups. Among the latter were the delicacy or robustness of the pentancres, the absence or presence of refractile granules at the pharyngointestinal junction, the size and number of refractile granules and the absence or presence of prominent lateral adhesive tubes just anterior to the base of the caudal feet. Other scoring groups and subgroups are self evident, though several should be commented upon. In the last subgroup of Group II it should be noted that the posterior foot length of hybrids was greater than that of either parent species; in the first subgroup of Group III the lateral head tentacle length of hybrids tended to be greater than that of the local *T. papii* population, approaching that of Mediterranean specimens from the Tyrrhenian Sea; and in Group V specimens which were sexually immature in this length range were considered to be "pure" *T. enallosa* with respect to this important character, since specimens of this parent species do not achieve reproductive maturity until they are larger than the limits set on specimens to be analyzed.

Table III provides summaries of mean (and standard deviation) hybridization index scores by location and morphological character group. The Crane's Beach reference population of *T. papii* scored 0–1 points, with scoring points coming from the lateral head tentacle length and aggregate pentancre size. Wood Neck site Wa (3m) specimens scored 0–2 points, with all scoring points resulting from variability in aggregate pentancre size. Site Wb (7m) specimens scored 0–12 points, showing sizable variability in all characters save that of the bursal caudal organ. Site Wc (11m) specimens scored 2–16 points, with sizable variability in all characters. The same was true of specimens from site Wd (15m) and We (19m), which scored 2–28 and 5–28 points respectively, while those from site Wf (23m) scored 17–23 points and showed variability in all characters save that of the bursal caudal organ. The aggregate total mean and standard deviation data are also shown in Fig. 2, where one can better visualize the pattern, or lack thereof. I conclude that specimens from site Wa (3m) were nearly as "pure" *T. papii* as those from the

Table III. Summaries of mean (and standard deviation) hybridization index scores by location and morphological character group (see Key to Point Scoring, Table II); localization of parent populations at Crane's Beach (Cr) and along neap high (Wa) to spring low (Wf) water transect at Wood Neck Beach.

	I	II	III	IV	V	Total	% of Specimens	
							"Pure" P-type	"Pure" E-type
Cr	0	0	0.10(0.32)	0.20(0.42)	0	0.30(0.48)	100	0
Wa	0	0	0	0.90(0.57)	0	0.90(0.57)	100	0
Wb	0.30(0.57)	0.60(1.27)	0.55(1.10)	1.30(0.73)	0	2.75(3.06)	65	0
Wc	0.55(0.51)	1.85(1.27)	2.27(0.79)	1.45(1.15)	0.20(0.89)	6.30(3.06)	5	0
Wd	2.45(1.36)	4.15(1.63)	5.25(1.48)	2.35(1.57)	3.80(1.58)	18.00(6.64)	5	15
We	2.45(0.69)	5.00(1.21)	5.45(1.54)	2.65(1.31)	4.00(1.12)	19.55(4.58)	0	10
Wf	2.15(0.37)	5.20(0.70)	5.80(0.41)	2.80(1.06)	4.00(0.00)	19.80(1.54)	0	0
Range, pts 0 to 6	0 to 6	0 to 8	0 to 8	0 to 6	0 to 6	0 to 34	0–2	27–34

Crane's Beach reference population ($\bar{x}$ = 0.90 compared with 0.30; SD = 0.57 compared with 0.48). Specimens from site Wb (7m) showed slight but definite signs of hybridization ($\bar{x}$ = 2.75; SD = 3.06), though they were still basically of *T. papii* stock. Moving to site Wc (11m) there was an increase in the mean value ($\bar{x}$ = 6.30) but not in the variability (SD = 3.06), whereas at site Wd (15m) there was a sharp increase in *T. enallosa* characters and in the variability from individual to individual ($\bar{x}$ = 18.00; SD = 6.64). Increase in mean point scores slowed over the last two sites, We (19m) and Wf (23m), while variability of point scores declined ($\bar{x}$ = 19.55 and 19.80; SD = 4.58 and 1.54 respectively).

The picture is one of nearly "pure" *T. papii* in the upper part of the beach with an increasing intrusion of *T. enallosa* characters as one progresses lower down the beach. Counting specimens which score 0–2 points as *T. papii* (Fig. 1d) and those which score 27–34 points as *T. enallosa* (Fig. 1a), the last two columns in Table III show a decrease in the percent of *T. papii* specimens as one moves downbeach, but surprisingly no consistent increase in *T. enallosa* specimens. (A hybrid scoring 12 points, but showing only one of a large variety of character combinations, is illustrated in Fig. 1b). The maximum percent of *T. enallosa* specimens occurred not at the bottom site but at the two sites above, and then as a distinct minority of specimens analyzed. Two additional and very important points need to be made. One is that mean and standard deviation point scores provide an inadequate picture of the broad array of hybrid character combinations that occurred in the midbeach area. A perusal of Table II and of the variability of point scores by character group in Table III will confirm this point. The second is that, taking the beach transect as a whole, hybrids were vastly dominant numerically to both parent species. The situation is one which fits all of the criteria of introgressive hybridization (Anderson 1949, 1962), with fertile hybrids being produced and *T. enallosa* being the recurrent parent species, and the subject of backcrossing cross-fertilization.

Evidence supporting this judgment is presented in Table IV. Comparing morphometric characters of Wood Neck Beach specimens with those of specimens from the reference population of *T. papii* at Crane's Beach indicates that Wood Neck Beach specimens were less *papii*-like with increasing size and increasing scoring point range. Using .10 probability as a cutoff, only those Wood Neck specimens scoring 0–2 points consistently showed morphometric characters of *T. papii*, regardless of size. Unfortunately, there were insufficient specimens of the other parent species, *T. enallosa*, to make such a comparison in reverse, but morphometric measurements made on six individuals exceeding 380 μm in total length are given in Table V, along with their probable scoring point range. Two of the six, including the longest specimen taken, would in my judgment not score in the 27–34 point range, indicating introgression of *papii*-like characters. And the other four specimens all show some minor but important *papii* features, generally associated with the number of anterior and lateral adhesive tubes.

Table IV. Mean deviations of morphometric characters in SD units for specimens of *Tetranchyroderma papii* Gerlach, 1953 and *Tetranchyroderma* spp. hybrid complexes, relative to comparable characters for specimens of *T. papii* from Crane's Beach, Woods Hole, Massachusetts (Table I). Mean deviation values are presented by location, length class and index of hybridization score. P represents the probality that the mean deviation of these measurements in SD units will be exceeded by the stated decimal fraction of such measurements in the reference population from Crane's Beach itself.

Location	Length class, μm	Scoring point range	No. specimens	No. measurements	Mean SD deviation per measurement	P
Wood Neck Beach, near Woods Hole, Massachusetts	100–249	(3–12)	10	110	1.64	P > .10
		(16–23)	9	99	2.34	.05 > P
		(27–34)	1	11	2.85	.02 > P
	250–274	0–2	3	33	0.70	P > .40
		3–12	7	77	1.22	P > .20
		16–23	18	198	1.99	.05 > P
		27–34	4	44	2.83	.01 > P
	275–299	0–2	8	88	0.77	P > .40
		3–12	8	88	1.57	P > .10
		16–23	16	176	2.11	.05 > P
		27–34	2	22	2.45	.05 > P
	300–324	0–2	6	66	1.03	P > .20
		3–12	6	66	2.44	.02 > P
		16–23	12	132	4.15	.001 > P
	325–349	0–2	8	88	0.93	P > .20
		3–12	7	77	2.10	.05 > P
		16–23	7	77	2.85	.01 > P
	350–380	(3–12)	3	33	2.35	.05 > P
Pisa, Italy *	268	1	1	11	1.14	P > .20
Gaeta, Italy	308,328	1,2	2	22	1.49	P > .10
Athens, Greece	242	5	1	11	1.64	P > .10

* Data derived from Plate II supplemented by Fig. II of Luporini et al., 1971.

Table V. Body measurements in μm by length class and probable index of hybridization for specimens of the *Tetranchyroderma* spp. hybrid complex exceeding 350 μm total length, from Wood Neck Beach near Woods Hole, Massachusetts. See Table I for key to column headings.

Lt	LPh	LIn	LCFt	LHdTn	WBc	WNk	WTr	WFtBs	DTn	Tb A	Tb L	CFt	Probable scoring point range
393	143	180	19	abs	65	44	58	31	5	5	14	5	27–34
451	130	260	25	24	60	40	49	23	1	3	20	4	3–12
481	140	296	24	abs	67	51	55	31	7	5	31	7	27–34
483	150	303	22	abs	69	50	64	30	9	4	24	7	27–34
573	160	370	22	abs	75	58	68	34	7	7	33	7	27–34
661	171	400	20	abs	70	53	64	34	2	3	37	4	16–23

Tetranchyroderma papii has been found in each of 16 beaches I have surveyed in the northeastern U.S., ranging from the New Hampshire-Massachusetts border onto Long Island, New York (Hummon 1974c). In three of these beaches (Wood Neck; Mattakeset, inside; and Saquish, inside), all low energy beaches, *T. enallosa* was also found along with a broad spectrum cluster of hybrids indicating the presence of introgressive hybridization. Thus far *T. enallosa* has not been found either in the absence of *T. papii* or in significant numbers.

Returning to Table IV, four Mediterranean specimens of *T. papii*, three from the Tyrrhenian Sea and one from the Aegean Sea, were also compared morphometrically with the reference population from Crane's Beach. All four of the Mediterranean specimens compared favorably with the Crane's Beach specimens, though all of the former had more elongate lateral head tentacles than were found in the latter. The specimen from Loutsa Beach near Athens, Greece (Fig. 1c) was of some significance, however, in that it scored a probable 5 points, largely on the basis of its relatively few lateral adhesive tubes and the presence of two dorsal trunk tentacles. It would appear that this specimen too is a hybrid of basically *T. papii* stock. One other congeneric species, *T. antennatum* Luporini et al., 1970, was found on the same beach, but lacking dorsal trunk tentacles it could hardly have been the other parent species. Thus remains a mystery.

As preliminary as are these data, I believe they strongly support a case for introgressive hybridization between *T. papii* and *T. enallosa* on Wood Neck Beach, probable introgression between these same two species on two additional beaches in the northeastern U.S. and possible hybridization of some sort between *T. papii* and another congeneric species on Loutsa Beach in the Aegean. It may well be that *T. papii* is prone to such interaction with other congeners. One must remember, however, that this sort of hybridization, localized sympatric hybridization, according to Woodruff (1973), is a dynamic and possibly an ephemeral phenomenon, and that it was observed in mobile animals at low tide. What changes in distribution of parents and hybrids accompany the incoming tide remains to be seen. I believe with Mayr (1970) that this situation represents a local breakdown of reproductive isolation between two sympatric species and that it does not represent a significant evolutionary pathway in these animals. But one must allow with Lewontin & Birch (1966) that under certain circumstances the introgression of genetic material can lead to rapid adaptive evolution.

Acknowledgements

I wish to thank Mr. A. Eleftheriou, of the DAFS Marine Laboratory, Aberdeen, Scotland, for collecting material for me from several beaches near Athens, Greece.

Bibliography

Anderson, E.: Introgressive hybridization. Wiley, New York (1949).
–: The role of hybridization in evolution. In: Johnson & Steere (eds.): This is life, pp. 286–314. Holt, Rinehart and Winston, New York (1962).

Chandrasekhara Rao, G. & P. N. Ganapati: The interstitial fauna inhabiting the beach sands of Waltair Coast. Proc. Nat. Inst. Sci. India **34** (B), 82–125 (1968).

Delamare Deboutteville, C.: Monographie d'une espèce. L'écologie la al répartition du Mystacocaride *Derocheilocaris remanei* Delamare et Chappuis, en Mediterranée. Vie Milieu **4**, 321–380 (1953).

–: Eaux souterraines littorales de la Côte Catalane Française (Mise au point faunistique). Vie Milieu **5**, 408–451 (1954).

DeZio S. & P. Grimaldi: Analisi comparative del mesopsammon di due spiagge pugliesi rapporto ad alcuni fattori ecologici. Arch Bot. Biogeogr. Ital. 40 (4 ser.) **9**, 357–367 (1964).

Fize, A.: Premiers résultats des récoltes de microfaune des sables effectuées sur la côte languedocienne. Vie Milieu **8**, 377–381 (1957).

–: Contribution à l'étude de la microfaune des sables littoraux du Golfe d'Aigues-mortes. Vie Milieu **14**, 669–774 (1963).

Gerlach, S.: Gastrotrichen aus dem Küstengrundwasser des Mittelmeeres. Zool. Anz. **150**, 203–211 (1953).

–: Die Tierwelt des Küstengrundwassers von San Rossore (Tyrrhenisches Meer). Physiol. Compar. Oecol. **4**, 54–73. (1955).

Gorman, G. C., M. Soulé, S. Y. Yang & E. Nevo: Evolutionary genetics of insular Adriatic lizards. Evolution **29**, 52–71 (1975).

D'Hondt, J.-L.: Contribution à la microfaune interstitielle des plages de l'ouest algerien. Vie Milieu **23A**, 227–242 (1973).

Hummon, W.D.: Interstitial marine gastrotrichs from Woods Hole, Massachusetts. Biol. Bull. **133**, 452 (abstract) (1967).

–: Interstitial marine gastrotrichs from Woods Hole, Massachusetts. Part II. Biol. Bull. **135**, 423–424 (abstract) (1968).

–: Distributional ecology of marine interstitial Gastrotricha from Woods Hole Massachusetts, with taxonomic comments on previously described species. PhD Thesis, University of Massachusetts, Amherst, 117 pp (1969).

–: The marine and brackish-water Gastrotricha in perspective. In: N. C. Hulings, ed.: Proceedings of the First International Conference on Meiofauna. Smithson. Contr. Zool. **76**, 21–23 (1971).

–: Gastrotricha. In: A. C. Giese and J. S. Pearse (eds.): Reproduction of marine invertebrates, Vol. I. Acoelomate and pseudocoelomate metazoans, pp. 485–504. Academic Press, New York (1974a).

–: Some taxonomic revisions and nomenclatural notes concerning marine and brackish-water Gastrotricha. Trans. Amer. Microsc. Soc. **93**, 194–205 (1974b).

–: S_H·:A similarity index based on shared species diversity, used to assess temporal and spatial relations among intertidal marine Gastrotricha. Oecologia (Berl) **17**, 203–220 (1974c).

–: Habitat suitability and the ideal free distribution of Gastrotricha in a cyclic environment. In: H. Barnes (ed.): Proc. 9th Europ. Mar. Biol. Symp., pp. 495–525. Aberdeen University Press, Aberdeen (1975).

Levi, C.: Contribution à l'étude des Gastrotriches de la région de Roscoff. Arch. Zool. Exp. Gén. **87**, 31–42 (1950).

Lewontin, R. D. & L. C. Birch: Hybridization as a source of variation for adaptation to new environments. Evolution **20**, 315–336 (1966).

Luporini, P, G. Magagnini & P. Tongiorgi.: Gastrotrichi macrodasioidei delle coste della Toscana. Pubbl. Staz. Zool. Napoli **38**, 267–288 (1970).

–: Contribution à la connaissance des Gastrotriches des côtes de Toscane. Cah. Biol. Mar. **12**, 433–455 (1971).

Mayr, E.: Animal species and evolution. Harvard University Press, Cambridge (1963).

–: Populations, species and evolution. Harvard University Press, Cambridge (1970).

Nixon, D. E.: Dynamics of spatial patterns for a gastrotrich in the surface sand of high energy beaches. Intern. Rev. ges. Hydrobiol. (In press).

Pollock, L. W. & W. D. Hummon: Cyclic changes in interstitial water content, atmospheric exposure, and temperature in a marine beach. Limnol. Oceanogr. **16**, 522−535 (1971).

Remane, A.: Morphologie und Verwandtschaftsbeziehungen der aberranten Gastrotrichen. Z. Morph. Ökol. Tiere **5**, 625−754 (1926).

Remington, C. L.: Suture-zones of hybrid interaction between recently joined biotas. In: T. Dobzhansky, M. K. Hecht and W. C. Steere (eds.): Evolutionary biology, Vol. 2, pp. 321−428. Appleton-Century-Crofts, New York (1968).

Sibley, C. G.: Hybridization and isolating mechanisms. In: W. F. Blair (ed.): Vertebrate speciation, pp. 69−88. University of Texas Press, Austin (1961).

Swedmark, B.: Étude de la microfaune des sables marins de la région de Marseille. Arch. Zool. Exp. Gen. **93**, 70−95 (1956).

Thane-Fenchel, A.: Interstitial gastrotrichs in some south Florida beaches. Ophelia **7**, 113−138 (1970).

Westheide, W.: Räumliche und zeitliche Differenzierungen im Verteilungsmuster der marinen Interstitialfauna. Ver. dt. Zool. Ges. **65**, 23−32 (1972).

Wilke, U.: Mediterrane Gastrotrichen. Zool. Jahrb. (Abt. Syst.) **82**, 497−550 (1954).

Woodruff, D. S.: Natural hybridization and hybrid zones. Syst. Zool. **22**, 213−218 (1973).

| Mikrofauna Meeresboden | 61 | 137–151 | 1977 |

W. Sterrer & P. Ax (Eds.). The Meiofauna Species in Time and Space. Workshop Symposium, Bermuda Biological Station, 1975

Niche Fractionation Studies of Two Sympatric Species of Enhydrosoma (Copepoda, Harpacticoida)[1]

by

M. Susan Ivester[2] and Bruce C. Coull

Belle W. Baruch Institute for Marine Biology and Coastal Research and Department of Biology, University of South Carolina, Columbia, SC 29208, USA

Abstract

Niche breadth and niche overlap of two sympatric harpacticoid species (*Enhydrosoma propinquum* Brady and *E. baruchi* Coull) have been investigated combining statistical methods with a scanning microscope analysis of the mouth parts. While there is no doubt that the two species differ slightly but consistently in terms of temporal, spatial and morphological parameters, these alone are not sufficient to explain their coexistence. It is suggested that besides factors unknown, coexistence may be due to the fact that the competitively superior species (*E. baruchi*) occupies the narrower niche which is included in the wider niche of the inferior *E. propinquum*.

A. Introduction

One aspect of the continuing studies of the meiofauna of the North Inlet, South Carolina (33° 20′ N, 79° 10′ W) has been an attempt, through observational, statistical and morphological analyses, to determine niche dimensionality. We have been particularly concerned with niche breadth and niche overlap of harpacticoid copepods along a 5-step sediment gradient from shell gravel to fine mud as well as mechanisms responsible for harpacticoid niche fractionation within any one of these sedimentological regimes. It is not the purpose of this paper to elaborate on the mechanisms or explain the differences between stations, but rather to provide insight into the mechanisms allowing two sympatric species to co-exist. In the fine mud station of the gradient, we regularly encountered two species of *Enhydro-*

[1] Contribution No. 168 of the Belle W. Baruch Institute for Marine Biology and Coastal Research. This research was supported by the Oceanography Section, National Science Foundation, NSF Grant DES 72-01573 A01 and a Pre-Doctoral Fellowship to M. Susan Ivester from the Slocum-Lunz Foundation.

[2] Present address: Marine Science Program, University of Alabama, P. O. Box 386, Dauphin Island, AL 36528 USA.

soma, i. e. *E. propinquum* Brady and *E. baruchi* Coull. As early as 1934 Gause suggested that two species with identical ecologies cannot live together at the same place at the same time, a thesis accepted and expounded upon by most modern ecologists (e. g. Hutchinson 1957, 1965; Pianka 1974). From an observational point of view *E. propinquum* and *E. baruchi* appear to occupy the same niche, but from the theoretical point of view this cannot be so. What then are the mechanisms allowing this apparent contradiction of niche theory? In the following we will attempt to explain the apparent paradox involving *E. propinquum* and *E. baruchi* using temporal, spatial and morphological measurements.

B. Methods and materials

Monthly samples were collected by hand coring with 3.5 cm (inner diameter) core tubes within a m^2 area of the station marker in Bread and Butter Creek, North Inlet. Collections were made at the same time of the month (5th–12th) at a l m depth at low tide. Immediately upon withdrawal of the core tube from the sediment each sample was placed in a separate container with buffered formalin, seawater and Rose Bengal. Additional samples were collected and analyzed for sediment organic carbon and granulometric analysis following the methods of Buchanan and Kain (1971). Temperature, salinity, pH and Eh were always measured in the field at the time of sample collection. In the lab each sample was extracted by the shaking-supernatant technique of Wieser (1960). The copepods were counted, sorted to species and plotted as in Figure 1 (log scale since log transformation tends to normalize a non-normal distribution).

Niche breadth of the ith species was calculated by measuring the uniformity of the distribution of individuals of a species along the resource state using

$$\beta_1 = \frac{1}{\Sigma p_{ij}^2} \qquad\qquad j = 1, 2, 3 \ldots 32$$

and

$$\log \beta_2 = \sum_j p_{ij} \log p_{ij} \qquad\qquad j = 1, 2, 3 \ldots 32$$

where p_{ij} is the proportion of the individuals of species i which is associated with resource state j (Levins 1968). Proportional niche overlap is estimated by comparing distributions of two species among resource states using

$$C_{ih} = 1-0.5 \sum_j \left| p_{ij} - p_{hj} \right| \qquad\qquad j = 1, 2, 3 \ldots 32$$

where p_{ij} is the portion of the ith species and p_{hj} is the proportion of a second species on the resource state; (Levins 1968; Colwell & Futuyma 1971; Pielou 1972). Because the proportional method of calculating niche overlap has come under question (see Colwell & Futuyma 1971 for discussion) two other measures of niche overlap were also employed, i. e.

(1) Levins' (1968) measure of species interaction

$$\alpha_{ih} = \sum_j p_{ij}\, p_{hj} / \Sigma p_{ij}^2 \qquad\qquad j = 1, 2, 3 \ldots 32$$

where p_{ij} and p_{hj} are the same as above.

(2) Multiple discriminate analysis of the copepods of the entire sediment gradient was also used (Cooley & Lohnes 1971; Davis 1973). Multiple discriminate analysis begins with a data set consisting of n measurements on m parameters, each measurement associated with one of g

species. The analysis reduces the data set to n measurements on k new parameters which are linearly independent additive functions of the original parameters. The species are then separated in space by linear additive functions of the original ecological parameters. The BMD (1973) package Class M programs were employed in the multivariate analysis.

Additionally, we calculated Levins' (1968) stability criterion for coexistence, i. e. α_{ih} α_{hi}, where α_{ih} is the species interaction of h with i and α_{hi} is the interaction of i with h. For α_{ih} α_{hi} less than unity, competitive coexistence between two species is possible. The University of South Carolina's IBM 370/168 computer system was used for all computations.

Copepods used in the scanning electron microscope studies were dehydrated using the critical point dehydration method (Cohen et al. 1968). Organisms placed in small plankton net bags were dehydrated in an increasing series of EtOH and increasing series of EtOH – Freon TF solutions to 100% Freon TF. The Freon TF in the cells was then substituted with Freon 13 in a Bomar Critical Point Dehydration apparatus (SPC-900) and evaporated at high pressure (900 psi) and temperature (38 °C). The dehydrated specimens were attached to the carrying stub using double-sided tape and coated with a thin film of gold-palladium. A Joel JSM-U3 Scanning Electron Microscope was used to observe the specimens at 25 kv accelerating voltage.

B. Results

The station inhabited by *E. propinquum* and *E. baruchi* (Bread and Butter) is located just subtidally to an intertidal *Spartina alterniflora* mud flat. The sediment is a clay-mud (median grain size 2–32 μm) composed of fecal pellets, *Spartina* debris and clay particles. Sediment organic carbon, which ranges between 1.6 and 5.0% is highest in September-October. The overlying water column and upper 2 mm of sediment is oxidizing (Eh +10 to +70 mv), whereas below 2 mm the sediment is reducing (−40 to −270 mv). Overlying water column salinities range between 29–35 °/oo and water-sediment interface temperatures between 10.5–29.8 °C.

Figure 1 shows the temporal variation of the two sympatric *Enhydrosoma* over a 32 month period at the Bread and Butter station. Both species have the same temporal abundance pattern with maxima in late fall and late spring.

Table 1 summarizes the niche breadth and niche overlap data for *E. propinquum* and *E. baruchi* at the mud station based on the proportion of each species' abundance for each of the 32 months.

Ivester (1975), using multiple discriminant functions, has calculated coefficients of the spatial niche on the benthic copepods along our five step sediment gradient. On the basis of two discriminant functions (i. e. DF I, mean particle size-skewness vs. sorting; DF II, sediment organic content vs. sorting), functions which account for 77% of among-species variance, *E. propinquum* and *E. baruchi* are significantly correlated (Fig. 2). Figure 2 also indicates that the closest neighbors of the two *Enhydrosoma* are *Harpacticus* sp., *Microarthridion littorale* and *Paracyclopina* sp.

Morphological measurement of *E. propinquum* and *E. baruchi* based on scanning electron microscopy are summarized in Table 2.

Figures 3 and 4 are SEM montages of typical females of *E. propinquum* and *E. baruchi*, respectively. Both animals are the same size, 0.55 mm and 0.53 mm, differ-

M. Susan Ivester and Bruce C. Coull

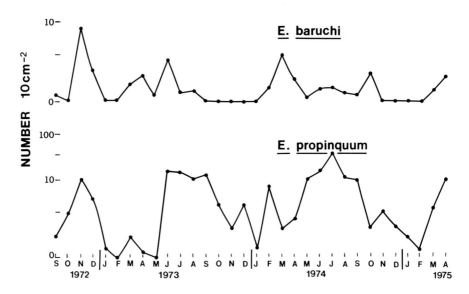

Fig. 1. Seasonal abundance of the two sympatric *Enhydrosoma* at the same location in the North Inlet, Georgetown, South Carolina.

Table 1. Summary of niche parameters for the two sympatric species of *Enhydrosoma*.

Niche Measurement	*E. propinquum*	*E. baruchi*
Niche breadth		
β_1	10.28	9.07
β_2	3.35	3.18
Niche overlap		
C	0.99	
α_{bp} *	0.41	
α_{pb} **	0.46	
Stability criterion for co-existence		
$\alpha_{bp}\,\alpha_{pb}$ ***	0.19	

* α_{bp} is the probability of *E. propinquum* overlapping *E. baruchi*.

** α_{pb} is the probability of *E. baruchi* overlapping *E. propinquum*.

*** if $\alpha_{bp}\,\alpha_{pb}$ is < 1 then competive co-existence occurs (Levins 1968).

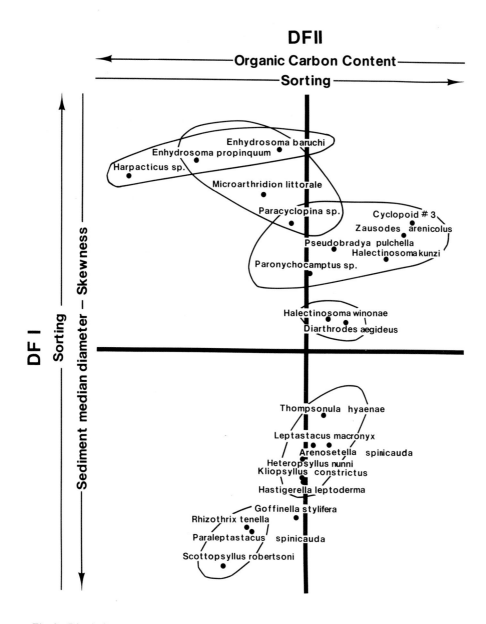

Fig. 2. Discriminant analysis of copepods over a five step sediment gradient. Discriminant function I is the interaction between sorting coefficient vs. sediment median diameter – skewness and DFII is organic carbon content vs. sorting. Note that of all 22 species the two sympatric *Enhydrosoma* are very closely associated.

M. Susan Ivester and Bruce C. Coull

Table 2. Summary of the salient mouthpart characters of the two sympatric *Enhydrosoma*.

Structure	*E. propinquum*	*E. baruchi*
Antennule (A$_2$)		
Exopod		
setae number	3	2
Endopod		
setae number	7	7
distal setae length	9.5–16.5 µm	10.5–28.5 µm
spine length	6 µm	8 µm
Labrum		
Shape	straight to tri-lobate	concave
Length	22.5 µm	19.0 µm
Width	21.3 µm	17.5 µm
Spine Length	5 µm	5–6 µm
Mandible		
Coxa-basis length	6 µm	6 µm
Setae length	12 µm	12 µm
Maxillula		
Precoxal arthrite		
structure	1 prominent spine	several slender spines
length	13.3 µm	12.5 µm
Maxilla		
Basis & spine length	17 µm	16.6 µm
Endopod length	7 µm	8.9 µm
Basis spine number	1 robust	2 slender
length	10 µm	9 µm
Endite spine length	4–7 µm	5.8–7.8 µm
Maxilliped		
1st segment length	14.3 µm	14.5 µm
Claw seta length	23 µm	23 µm
Minor seta length	7 µm	4.3 µm
Endopod spinule length	2 µm	5.1 µm
Basis spine	absent	present 9 µm
Coxa ornamentation	papillae and hairs	nude
Body		
distal segment	recurved flanges in urosome	segments rounded in urosome
swimming leg setae	longer	long
cephalothorax ornamentation	smooth	finely setose

ing in caudal rami length, urosome segment form (recurved flanges in *E. propinquum*; rounded in *E. baruchi*), length of the swimming leg setae and ornamenta-

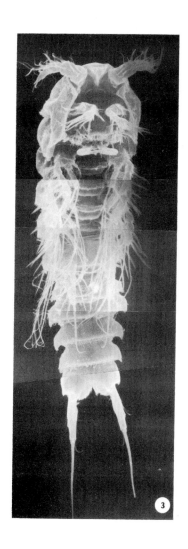

M. Susan Ivester and Bruce C. Coull

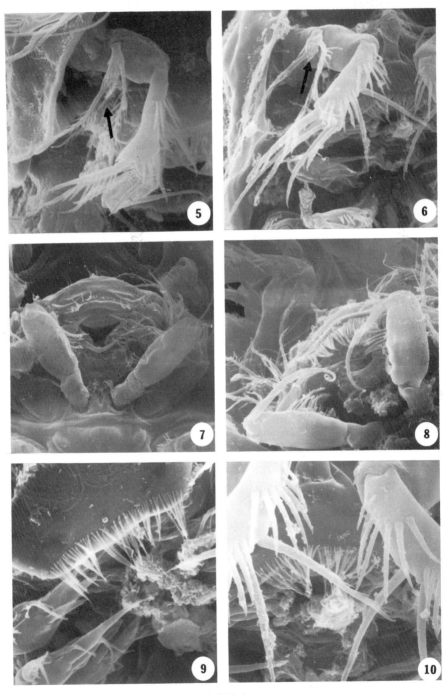

tion of the inner portion of the cephalothorax (smooth in *propinquum*; finely setose in *baruchi*).

The antennule (Figs. 5–6; Table 2), although not usually considered a feeding appendage, is different in exopod setation (3 setae in *propinquum*; 2 in *baruchi*) and shape, and all spinule setation in *baruchi* is coarser and longer than the equivalent setation in *propinquum*.

The labrum in the two species is quite different (Figs. 7–10; Table 2). The major differences are in setae number and size, shape (trilobed, with "hairs" proximally in *propinquum* and concave, and nude proximally in *baruchi*) and overall size (Table 2).

Mandible (Figs. 11–12; Table 2). The coxa-basis of both species is very similar in size and shape; endopod of each has three setae of similar length.

Maxillula (Figs. 13–16; Table 2). The maxillula of *baruchi* is larger than that of *propinquum*. The precoxal arthrite of *propinquum*, however, has one large broad spine with numerous accessory spines while *baruchi* has several similarily sized spines.

Maxilla (Figs. 17–18; Table 2). The maxillae are similarly shaped, both endopods have 2 setae. The basis in *propinquum* has one large spine, whereas in *baruchi* 2 slender spines are present. The endite spines of *propinquum* are longer and appear to be more robust than those in *baruchi*.

Fig. 3. SEM montage of *Enhydrosoma propinquum* habitus. X 850. – Fig. 4. SEM montage of *Enhydrosoma baruchi* habitus. X 850. – Fig. 5. Antennule (A₂) of *E. propinquum*. Exopod has three setae (arrow). X 1,800. – Fig. 6. Antennule (A₂) of *E. baruchi*. Exopod has two setae (arrow). X 1,800. – Fig. 7. Frontal view of labrum apparatus of *E. propinquum*. Note relationship of maxillae to the labrum. X 2,000. – Fig. 8. Oblique frontal view of labrum apparatus of *E. baruchi*. X 2,500. – Fig. 9. Dorsal view of labrum of *E. propinquum*. Note straight to trilobed shape of distal edge. X 4,000. – Fig. 10. Dorsal view of labrum of *E. baruchi*. Note concave shape of distal edge. X 3,200.

Fig. 11. Mandible of *E. propinquum*. Coxa-basis has three setae (one is missing). X 3,600. – Fig. 12. Mandible of *E. baruchi* (arrow). X 2,700. – Fig. 13. Maxillula of *E. propinquum*. Note intimate relationships of maxillula with oral opening. X 3,900. – Fig. 14. Maxillula of *E. baruchi*. Note relationship of maxillula and oral opening. Appears to be pushing materials into opening. X 2,100. – Fig. 15. Enlargement of maxillula of *E. propinquum* showing arrangement of one prominent spine and smaller accessory spines of precoxal arthrite. X 4,200. – Fig. 16. Enlargement of maxillula of *E. baruchi* showing arrangement of several similar spines of precoxal arthrite. X 4,000.

Fig. 17. Maxilla of *E. propinquum* with the one robust basis spine (arrow). Note ornamentation of endite spines. X 4,400. – Fig. 18. Maxilla of *E. baruchi* showing two slender basis spines (arrow). X 3,900. – Fig. 19. Maxilliped of *E. propinquum*. X 1,700. – Fig. 20. Maxilliped of *E. baruchi*. Note presence of long basis spine (arrow). X 2,300. – Fig. 21. Enlargement of maxilliped of *E. propinquum* showing absence of basis spine. X 3,900. – Fig. 22. Enlargement of maxilliped coxa of *E. propinquum* showing hairs (arrow) and papillae (arrow). X 5,300. – Fig. 23. Enlargement of maxilliped of *E. baruchi* showing structure of basis spine (arrow). X 3,200.

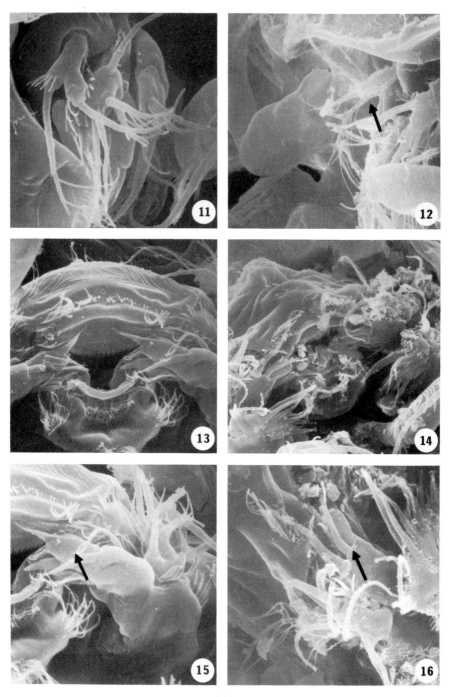

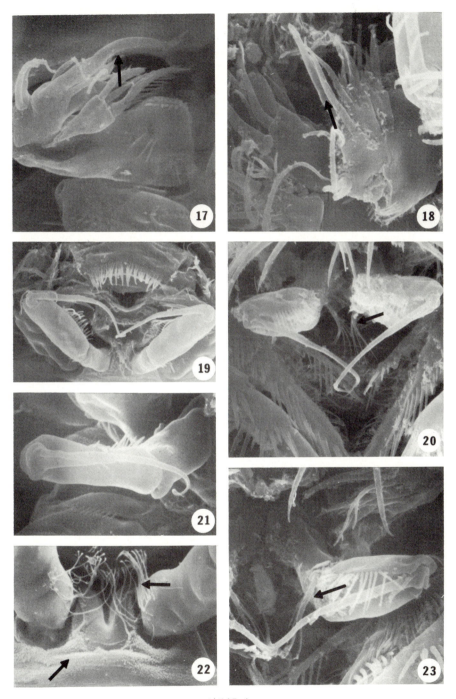

Maxilliped (Figs. 19–23; Table 2). The maxillipeds vary in size of first endopodite segment, length of endopod claw setae and secondary spine, length of spinules on endopod segment 1, and most drastically in the presence of an inner basis spine in *E. baruchi* but no inner basis spine in *E. propinquum*. However, in *propinquum* (the species without the basis spine) the coxa is papillary with associated "hairs" (Figure 22), whereas in *baruchi* (the species with the basis spine) the coxa is devoid of papillae and hairs.

C. Discussion

The coexistence of more than one species within the same habitat has posed problems to ecologists since the formulation of the Volterra-Gause theory of competitive exclusion in the early 1930's. Many different methods of niche separation or fractionation have been proposed, including facets of reproductive strategy; physiological and behavioral responses to adverse conditions; foods and feeding; and predation and adaptation (Milstead 1972).

Several authors have recently measured niche breadth and/or fractionation of various organisms (Price 1971 for parasitic insects; Krazanowski 1971 for bats; Young 1972 for butterflies; MacArthur & MacArthur 1961; Reller 1972; and Snow & Snow 1972 for birds), but to date we are unaware of attempts using niche theory with meiofauna. The meiofauna, although difficult to work with, provide an excellent assemblage with which one may easily collect the large numbers of animals requisite for population studies and, in fact, meiofaunologists often encounter closely related species in the same location.

We have no doubt that *Enhydrosoma propinquum* and *E. baruchi* are sympatric species in both the traditional sense (i. e. common ancestry) and sensu Croker (1967) who contends that taxonomically related species that coexist in the same habitat and have essentially the same reproduction periods are sympatric, i. e. "those species whose breeding individuals are within cruising range of each other regardless of the spatial overlap of habitat" (Croker 1967, p. 273). However, just because two closely related species live and reproduce in the same habitat does not mean they are necessarily sympatric. For example, the reproductive periods and therefore the abundance maxima could alternate and then temporally separate the species, as already pointed out by Coull & Vernberg (1975). This, however, is not the case with *E. propinquum* and *E. baruchi* since Figure 1 illustrates that both species are present at the same time and have abundance maxima simultaneously. Hutchinson (1967) contends that nonoverlapping trophic niches are possible if two copepod species differ in size by a factor of 1.35. *E. propinquum* is 0.55 mm whereas *E. baruchi* is 0.53 mm; a factor of 1.03, and therefore not sufficient to separate the trophic niches in the Hutchinsonian sense.

The data in Table 1 indicate that *E. propinquum's* niche breadth is wider than that of *E. baruchi* and since C (proportional niche overlap) is at 0.99, *E. baruchi's* niche (the smaller of the two) is contained almost entirely within that of *E. pro-*

pinquum. Furthermore the α values in Table 1 where α_{pb} is larger than α_{bp} support this premise since the probability of *baruchi* overlapping with *propinquum* is greater than the reverse.

Additionally, the multiple discriminate function analysis (Fig. 2) shows a close relationship between the two *Enhydrosoma* and *Harpacticus* sp., *Microarthridion littorale*, and *Paracyclopina* sp. However, since *Harpacticus* sp. and *Paracyclopina* are epifaunal algal associates, and since *Microarthridion* is so much larger (0.8 mm; differing in size by a factor 1.45–1.51 from the *Enhydrosoma*), this leaves only *E. propinquum* and *E. baruchi* as potential competitors. Due to these arguments and since Levins' (1968) stability criteria for co-existence (i. e. $\alpha_{bp} \, \alpha_{pb}$) is less than 1 (Table 1) we are assured that these species are not only sympatric but co-existing as well.

Since all our temporal and spatial approximations indicated niche similarity it seemed only appropriate to attempt an analysis of trophic preference and buccal morphology of these two species. Unfortunately, we have not been successful in observing the mode of feeding in the *Enhydrosoma* spp. and in fact, only Fahrenbach (1962) and Marcotte (1977) provide data on actual feeding in harpacticoid copepods. Fahrenbach's animal, *Diarthrodes cystoecus* and Marcotte's animal, *Tisbe furcata*, are algal dwellers and probably atypical in their feeding modes since they feed either on the algae (*D. cystoecus*) or on suspended particles (*Tisbe furcata*). Burrowing harpacticoids such as *Enhydrosoma* would obviously feed quite differently. Our gut content studies revealing only an amorphous mass containing some diatom frustules, detritus and sediment particles have shed no light upon types of food materials or the method of obtaining food.

In an earlier paper (Ivester & Coull 1975) we have shown that mouthpart structure can be an important factor in allowing sympatric haustoriid amphipods to co-exist, and thus we examined the mouthparts of the two sympatric *Enhydrosoma* to determine if character displacement is responsible for niche fractionation. We have included the antennule (A_2), not normally considered as a crustacean oral appendage, because it has been implicated in calanoid copepod feeding (Anraku & Omori 1963; Marshall & Orr 1972) and might be important in feeding of the *Enhydrosoma*.

As indicated in Figures 5–23 and Table 2, the mouthparts of the two species are basically similar with major differences in antennule setation, labrum structure, maxillula and maxilla spination and maxilliped structure: differences which may enable the species to feed on slightly different sized food. In *E. baruchi* the measured setation and structure is larger and/or coarser than the same setation or structure in *propinquum* except in the labrum, maxilla and maxillula. Further, the lack of a basis maxilliped seta along with the presence of minute coxa papillae in *E. propinquum* suggests to us that *baruchi* may capture and ingest different size food particles than does *propinquum*. Since comparisons of different mouthpart structure give different size factor ratios, it is futile to generalize on the Hutchinson 1.35

size factor for the trophic niche separation and in fact, this magic factor is probably an oversimplification. It is difficult therefore to determine if these structural modifications serve in niche fractionation between the two sympatric species or whether they are simply non-adaptive differences between the species. Maly and Maly (1974) state that size may perhaps be indicative of morphological differences in feeding appendages, but that the morphology of the feeding appendages and probably behavior are the major determinants allowing co-existence, just as beak morphology and behavior was to Darwin's finches.

On the basis of the temporal, spatial and morphological parameters we are not able to sufficiently explain the co-existing nature of *E. propinquum* and *E. baruchi* (our original premise). So then what are the alternatives?

We have shown that the niche of *E. baruchi* lies within that of *E. propinquum*, and since Pianka (1974) states that two competitive species may coexist provided the competitively superior one occupies the narrower included niche, it is not unreasonable to suspect that *E. baruchi* is competively superior to *E. propinquum*. To infer competitive superiority would necessitate resource utilization specialization and this may well be the case with our sympatric species. For example, the mouthparts of *E. baruchi* are geared primarily to handling a narrow range of food particles (i. e. resource utilization specialization), whereas *E. propinquum* with a varied setation (e. g. the absence of a Mxp. basis spine, but with Mxp. papillae and "hairs", and with coarse setae on the Mx. and Mxl. but relatively fine setae elsewhere) suggests a broader range of food materials and thus a more generalistic approach to feeding.

Of course, other factors such as feeding behavior and/or biochemical differences may possibly be responsible for allowing these two species to coexist. However, they are beyond the scope of this paper and can only be supposed. Even though niche studies on meiofauna may appear difficult and time consuming they are essential if we are to understand the complex interactions that occur in the sediment biocenose.

Bibliography

Anraku, M. & M. Omori: Preliminary survey of the relationships between the feeding habit and the structure of the mouthparts of marine copepods. Limnol. Oceanogr. 8(1), 116–126 (1963).

BMD (Biomedical Computer Programs): W. J. Dixon, Ed. University of California Press. 773 pp (1973).

Buchanan, J. B. & J. M. Kain: Measurement of the physical and chemical environment. In: Methods of the Study of Marine Benthos. IBP Handbook No. 16 (N. A. Holme and A. D. McIntyre eds.) Blackwell Scientific Publications. Oxford. pp. 30–51 (1971).

Cohen, Q. L., D. P. Marolow & G. E. Garner: A rapid critical point method using fluorocarbons ("Freons") as intermediate and transitional fluids. J. Microscopy 7(3), 331–342 (1968).

Colwell, R. K. & D. J. Futuyma: On the measurement of niche breadth and overlap. Ecology 52(4), 567–576 (1971).

Cooley, W. W. & P. R. Lohnes: Multivariate data analysis. Wiley, New York. 364 pp (1971).

Coull, B. C. & W. B. Vernberg: Reproductive periodicity of meiobenthic copepods: Seasonal or continuous? Mar. Biol. **32**, 289–293 (1975).

Croker, R. A.: Niche diversity in five sympatric species of intertidal amphipods (Crustacea: Haustoriidae). Ecol. Monogr. **37**(3), 173–200 (1967).

Davis, J. C.: Statistic and Data Analysis in Geology. J. Wiley and Sons, Inc. 550 pp (1973).

Fahrenbach, W. A.: The biology of a harpacticoid copepod. La Cellule **62**(3),303–374 (1962).

Hutchinson, G. E.: Concluding remarks. Cold Spring Harbor Symp. Quant. Biol. **22**, 415–427 (1957).

–: The Ecological Theater and the Evolutionary Play. Yale Univ. Press, New Haven, Conn. 139 pp (1965).

–: A Treatise on Limnology. Vol. II. Introduction to Lake Biology and the Limnoplankton. Wiley, New York. 1115 pp (1967).

Ivester, M. S.: Ecological diversification within benthic harpacticoid copepods. Ph. D. dissertation, University of South Carolina (1975).

– & B. C. Coull: Comparative study of ultrastructural morphology of some mouthparts of four haustoriid amphipods. Canadian J. Zool. **53**(4), 408–417 (1975).

Krazanowski, A.: Niche and species diversity in temperate zone bats (Chiroptera). Acta Zool. Cracov. **16**(15), 346–354 (1971).

Levins, R.: Evolution in Changing Environments. Princeton University Press, Princeton, N. J. 120 pp (1968).

MacArthur, R. & J. MacArthur: On bird species diversity. Ecology **42**, 594–598 (1961).

Maly, E. J. & M. P. Maly: Dietary differences between two co-occurring calanoid copepods species. Oecologia **17**, 325–333 (1974).

Marcotte, B. M.: An introduction to the architecture and kinematics of harpacticoid (Copepoda) feeding: *Tisbe furcata* (Baird, 1837). Mikrofauna Meeresboden **61**, 183–196 (1977).

Marshall, S. M. & A. P. Orr: The Biology of a Marine Copepod. Oliver and Boyd, London. 195 pp. (reprinted by Springer-Verlag) (1972).

Milstead, W. W.: Toward a quantification of the ecological niche. Amer. Midl. Nat. **87**(2), 346–354 (1972).

Pianka, E. R.: Evolutionary Ecology. Harper and Row, New York. 356 pp. (1974).

Pielou, E. C.: Niche width and niche overlap: a method for measuring them. Ecology **53**, 687–692 (1972).

Price, P. W.: Niche breadth and dominance of parasitic insects sharing the same host species. Ecology **52**(4), 587–596 (1971).

Reller, A. W.: Aspects of behavioral ecology of red-headed and red-bellied woodpeckers. Amer. Midl. Nat. **88**(2), 270–290 (1972).

Snow, B. K. & D. W. Snow: Feeding niches of hummingbirds in a Trinidad valley. J. Anim. Ecol. **41**(2), 471–485 (1972).

Wieser, W.: Benthic studies in Buzzards Bay. II. The meiofauna. Limnol. Oceanogr. **5**(2), 121–137 (1960).

Young, A. M.: Community ecology of some tropical rain forest butterflies. Amer. Midl. Nat. **87**(1), 146–157 (1972).

| Mikrofauna Meeresboden | 61 | 153—165 | 1977 |

W. Sterrer & P. Ax (Eds.). The Meiofauna Species in Time and Space. Workshop Symposium, Bermuda Biological Station, 1975.

Taxonomy, Phylogeny and Biogeography of the Genus Austrorhynchus Karling (Turbellaria, Polycystididae)

by

Tor G. Karling

Section of Invertebrate Zoology, State Museum of Natural History, Stockholm, Sweden

Abstract

The taxonomy and phylogeny of the genus *Austrorhynchus* Karling, 1952 are discussed principally on the basis of the complex male cuticular apparatus. In the world-wide distribution of the genus the Atlantic gap and the rather large number of species in the Antarctic and the Mediterranean areas are especially considered. The following *Austrorhynchus* species are diagnosed: *A. pectatus* Karling, *A. magnificus* Karling, *A. calcareus* sp.n., *A. parapectatus* sp.n. and *A. spinosus* sp.n. from the Scotia Arc; *A. pacificus* sp.n., *A. californicus* sp.n. and *A. hawaiiensis* sp.n. from the Pacific; *A. maldivarum* sp.n. from the Indian Ocean and *A. bruneti* sp.n. from the Mediterranean.

A. Introduction

The genus *Austrorhynchus* was first described from material picked out from samples taken in the coastal areas of the Scotia Arc (Tierra del Fuego, the Falklands, South Georgia) by the Swedish Antarctic Expedition 1901–1903 (Karling 1952, pp. 8–18). The type species *A. pectatus* was divided into the two subspecies *A. p. pectatus* and *A. p. magnificus* and the nominate subspecies into four "forms" based on the shape of the prostatic stylet. In 1965 Brunet described two new *Austrorhynchus* species from the Mediterranean and discussed the taxonomy of the Antarctic species. (For practical reasons I use the name "Antarctic" though T. del Fuego and the Falklands belong to the Magellanic region.) *A. elixus* Marcus, 1954, must be excluded from the genus *Austrorhynchus*. The new facts and several new *Austrorhynchus* phena from the Pacific and the Indian oceans call for a re-evaluation of the taxonomy of this genus, and the rather large material from different parts of the world motivates some reflections concerning the phylogeny and biogeography of the genus. In this paper, too, the Antarctic species are the main subjects for my study, and I agree with Dell (1972, p. 172) that "studies" on the systematics and distribution of Antarctic organisms are of basic importance, not only to a narrow range of specialists, but to all biologists concerned with biogeography".

B. The cuticular copulatory apparatus

The cuticular copulatory apparatus consists of two organs, the stylet and the accessory organ (here abbreviated A-organ). The stylet consists of a funnel enclosing the distal part of the prostatic vesicle, a narrow tube for discharge of the prostatic secretion and in some species a pointed hook beside the tube. A longitudinal strand of granular substance evidently indicates a previous groove-shape of the stylet (cf. Karling 1956, p. 204). The A-organ is a complex cuticular structure without adjacent glands or vesicles but joined to the prostatic vesicle by means of strong muscles. It consists mostly of several parts, the differentiation of which is taxonomically important. In the type species *A. pectatus* it consists of a plate distally provided with a comb running out on the finger- or threadlike flagellum (Fig. 1). The plate is provided with two proximal processes for attachment of the muscles, the foot and the style. Is in some species the A-organ simplified (reduced?), in extreme cases consisting of a longish hook only — with or without comb — emerging from the foot. The generic identification (as *Austrorhynchus*) is mostly facilitated by remainders of the lacking parts — and of course by the stylet and other structural features. The function of the A-organ is unknown. Similar organs are known in many turbellarian taxa and regarded as sexual stimulators.

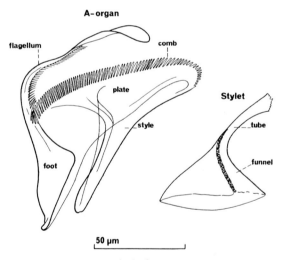

Fig. 1. *Austrorhynchus pectatus*, male cuticular apparatus.

C. The taxonomic level of the Austrorhynchus phena

My reconsideration of the taxonomy of the Antarctic *Austrorhynchus* phena is based on the fact that in a large material a given fixed structure of the A-organ is combined with a fixed structure of the stylet (see further below). The differences between these phena in regard to the whole complex male cuticular apparatus are

(154)

quite on the same level as between well established species in e.g. the genera *Prome-sostoma, Proxenetes, Provortex*. In a sample from the Falklands (Port Louis) *A. pectatus* was found together with *A. parapectatus* and *A. calcareus*; in another sample (Falklands, Berkeley Sound) it was found together with *A. spinosus*. Thus, structural as well as geographical evidence speaks in favour of these four Antarctic phena as species, not forms (variants) or subspecies (cf. Mayr 1969, p. 145).

A. magnificus is doubtlessly a distinct species (cf. Brunet 1965, p. 151), considering its prominent structural characteristics (see below, phylogeny) and its sublittoral occurrence, the other Antarctic species being found in the littoral zone.

A. pacificus and *A. californicus* can be identified by their stylets as well as by their A-organs. At least in one locality in California (Pacific Grove, Point Pinos, "Great tidepool") and in one locality in Oregon (Yaquina head, *Zostera*) they have been found together.

A. californicus, A. hawaiiensis, A. maldivarum and *A. bruneti*, all of them with A-organ of similar construction (more or less transversal comb, more or less pronounced style) and with hook-bearing stylet, could perhaps be regarded as points on a clinal line (subspecies), but today a specific separation appears practical, considering not only their occurrence but also the differences in shape and size of the different parts of the cuticular organs.

The Mediterranean phena *A. scoparius* and *A. karlingi* will not be discussed here as regards their taxonomic state, but Brunet's opinion that they must be regarded as species appears justified (1965, p. 151).

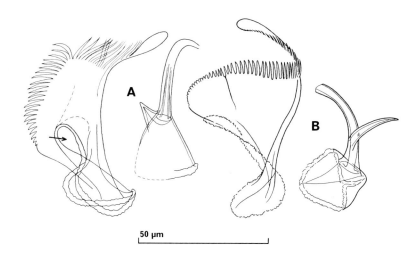

50 µm

Fig. 2. Male cuticular apparatus. A. *Austrorhynchus calcareus*. B. *A. parapectatus*. The arrow shows the "window" in the plate.

D. Phylogenetic viewpoints

Here mainly the evolution of the male cuticular apparatus and its bearing on the phylogeny of the *Austrorhynchus* species will be discussed. Other anatomical features will be touched upon only with reference to *A. magnificus*.

I have found it impossible to coordinate the evolution of the cuticular stylet with that of the A-organ. Considering that complicated structures are most suitable for phylogenetic analyses I will focus the attention on the A-organ. In my opinion the existence of a stylet of fixed type in combination with different types of accessory organs (e.g. Figs. 2B and 5D) reveals a trend towards differentiation of a hook in the stylet. Considering both structures the *Austrorhynchus* genus reveals a case of mosaic evolution.

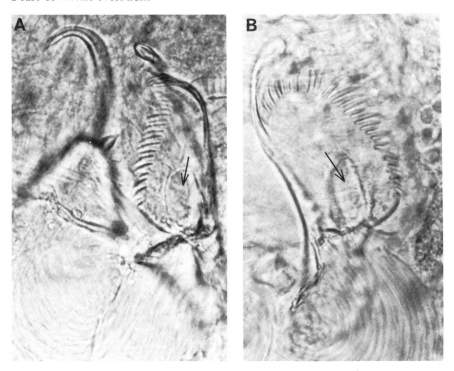

Fig. 3. *Austrorhynchus calcareus*. A. Male cuticular apparatus. B. A-organ from another specimen. The arrows show the "window" in the plate.

The species *A. pectatus, A. hawaiiensis, A. maldivarum, A. bruneti, A. pacificus* and *A. californicus* constitute a species group (the *pectatus*-group, e.g. Fig. 1), where the A-organ has a well differentiated plate with a more or less transversal comb and with two separate processes for muscular attachment, the foot and the style, the latter weakly differentiated in the last two species mentioned. In *A.*

calcareus and *A. parapectatus* (the *calcareus*-group, Figs. 2A, 2B) the plate resembles that of the *pectatus*-group, but there is no separate style, the foot being broad, more or less folded. The rest of the species (group 3) resembles the two species of the *calcareus*-group in the lack of a style, showing otherwise no unambiguous common features.

Very simple A-organs have been described in the Mediterranean species *A. scoparius* Brunet and *A. karlingi* Brunet. In the latter species the A-organ somewhat resembles a simple cuticular hook, a structure from which many complex structures can be derived (Karling 1956, p. 204). However, I find it difficult to place it at the root of the evolutionary lines of the A-organs in *Austrorhynchus*. In my opinion it reveals a high degree of specialization, and this is also the case in regard to the A-organ in *A. scoparius* (see further below).

Apparently the A-organ of *A. calcareus* gives the clue to the understanding of one of the main steps in the evolution of this organ in general (Figs. 2A, 3A,B). The "window" in its plate is evidently the remains of the deep incision between foot and style in *A. pectatus*, demonstrating that its complex foot is the result of a fusion of foot and style. The A-organ of *A. parapectatus* evidently shows a further step, no "window" being found (Fig. 2B). (However, it also somewhat resembles the A-organ of *A. pacificus* and *A. californicus*.) I thus regard the *pectatus*-group as ancestral in regard to the A-organ. In this group the A-organs of *A. pacificus* and *A. californicus* appear rather primitive judging by their weakly differentiated style (Figs. 5A+B, 6A+B).

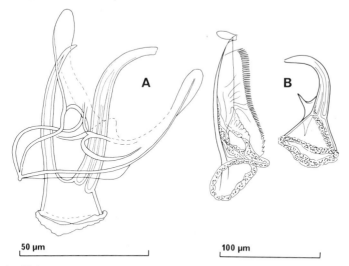

50 μm

100 μm

Fig. 4.　Male cuticular apparatus. A. *Austrorhynchus spinosus*. B. *A. magnificus*.

The A-organ of *A. spinosus* can be derived from that of the *calcareus*-group by reduction of plate+comb (Fig. 4A). In fact, in one preparation a small group of comb teeth was indistinctly seen in the gap between the two fingers. Further reduction along similar lines could result in the simple organ of *A. karlingi* — a flagellum without teeth, emerging from a foot provided with an indistinctly delimited process. Another possibility is to derive the A-organ in *A. karlingi* from that of *A. maldivarum* by further reduction of plate and style. For biogeographical reasons this alternative appears more probable (see below). The line leading to the simple styled brush of *A. scoparius* is as yet obscure.

   A. magnificus is in several ways of phylogenetic interest. Its body length and the size of its cuticular apparatus distinctly exceed those of the other species, in regard to the cuticular apparatus by about 100%. In the A-organ (Fig. 4B) the style-part of the plate is proximally withdrawn, evidently in connection with a fusion of style and foot, like in *A. calcareus* leaving an open window in the proximal part of

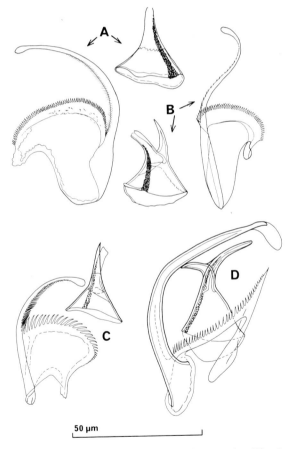

50 µm

Fig. 5.   Male cuticular apparatus. A. *Austrorhynchus pacificus*. B. *A. californicus*. C. *A. hawaiiensis*. D. *A. maldivarum*, hook of stylet evidently bent through pressure against the A-organ.

the plate. The number of the internal longitudinal muscles in the pharynx is about twice that in the other species from which information in this respect is available (approx. 44/22—24). I think the possibility that *A. magnificus* is a result of hybridization deserves attention.

The number of *Austrorhynchus* species in the Antarctic area (and of course also in other areas) will certainly increase with intensified studies. In the available material two phena with aberrant cuticular organs were found (Figs. 8A,B). Due to the insufficient material — one specimen of each — I hesitate to regard them as species. Together with the described species they reveal the high potential of speciation in the Antarctic *Austrorhynchus*. The stylets of these two phena demonstrate two patterns of hook formation not described heretofore. Together with other aberrant hooks found in my Pacific material they strengthen my theory concerning a trend in the formation of the hook.

Fig. 7 shows the hypothetic evolution of some types of A-organ. The indicated evolutionary lines parallel possible phylogenetic lines within the genus *Austrorhynchus*.

E. Biogeographic viewpoints

The large *Austrorhynchus* material picked out from the Antarctic samples indicates a great abundance of most species of this genus in the collecting areas. The same is true for the occurrence of the other species. The single specimen from the occasional sampling on the Maldives does not contradict this statement. Thus, total absence of the genus appears very likely in the European N. Atlantic, the most intensively studied area in regard to marine turbellarians. There are no reported finds of *Austrorhynchus* species from the other parts of the Atlantic (with the exception of T. del Fuego and the Falklands). Thus, in the world-wide distribution of *Austrorhynchus* there seems to be an (N.) Atlantic gap (Fig. 9).

In that context the rich occurrence of *Austrorhynchus* in the Mediterranean deserves attention against the background of the otherwise rather homogeneous Atlantic-Mediterranean turbellarian fauna. It is possible, that — due to climatic changes — the *Austrorhynchus* genus has disappeared from the Atlantic, together with other previously well represented taxa, but persisted in the Mediterranean (Ekman 1953, pp. 66, 84). Another possibility is that the Mediterranean element of the genus is a relict from the Tethys Sea. Then, *A. maldivarum* can be regarded as a link between the Mediterranean and the Pacific populations of the genus.

All the Pacific species belong to the *pectatus*-group. *A. maldivarum* from the Indian Ocean, *A. bruneti* from the Mediterranean and the Antarctic *A. pectatus* belong to the same group. The *calcareus*-group is represented in the Antarctic area only, and the most specialized group (group 3) is known from this area and the Mediterranean. This pattern of distribution supports my opinion that the *pectatus*-group is the "ancestral" one from which the other groups have evolved. It furthermore indicates that this evolution has been locally restricted, i.e. that the *calcareus*-

group has evolved in the Antarctic area and that the representatives of group 3 in the Mediterranean and the Antarctic area belong to separate evolutionary lines (cf. above *A. spinosus* and *A. karlingi*). The presumed primitiveness of the A-organ in *A. pacificus* and *A. californicus* can indicate that the Pacific is the original evolutionary center of the genus *Austrorhynchus*.

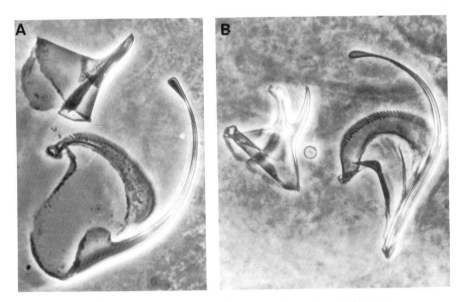

Fig. 6.   Male cuticular apparatus. A. *Austrorhynchus pacificus*. B. *A. californicus*.

Five *Austrorhynchus* species are known from the Scotia Arc, but the number of species will certainly increase. The geographic disintegration of this area without doubt favours a local speciation, but the marked difference between the species (at least in regard to the cuticular apparatus) indicates an ancient origin of the species. In contrast to the Galapagos Islands (cf. Ax 1974) the islands of the Scotia Arc have a continental origin and their fauna consists in part of "refugees" from ancient continental bridges (cf. Dell 1972, p. 166).

A. *magnificus* appears today as the only endemic species of its genus in S. Georgia. *A. pectatus* and *A. parapectatus* are common to S. Georgia and the Falklands. *A. calcareus* is endemic on the Falklands and *A. spinosus* is common to these islands and T. del Fuego. These conclusions corroborate Dell's assumption (op. cit., p. 104), that the number of species common to the Scotia Arc will increase "in the light of modern systematic practice."

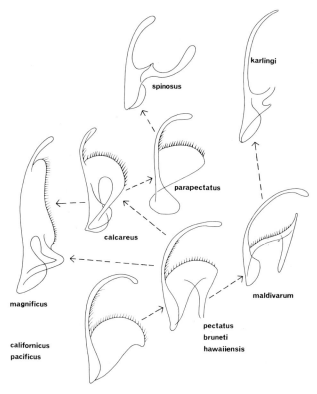

Fig. 7. Hypothetic evolution of the A-organs in the genus *Austrorhynchus*, together with hypothetic phylogeny of the species (-groups).

F. Diagnoses, types and distribution

All material, including the types, is deposited in the Swedish Museum of Natural History, Section of Invertebrate Zoology, Stockholm. The exact body length of the live species has been impossible to determine for the material available in preserved condition only. It is evidently approx. 1—1.5 mm in all species except in *A. magnificus,* where it may be about 2.5 mm. All species except *A. magnificus* have paired eyes, the pigment of which can be diffuse.

A. pectatus Karling, 1952. Fig. 1
A. pectatus pectatus Karling, 1952, "forma Einfach"
A. pectatus: Brunet, 1965
Stylet 45—80 μm (11 specimens) without hook, with tube and basal funnel of equal length. A.-organ 115—150 μm with longish separate style, deep proximal incision between style and foot. Transversal comb and flagellum with fine teeth. – Lectotype No. 2688 a whole mount, leg. Sw. Ant. Exp.
Distribution. S. Georgia and Falkland Islands, seaweed, sand and shell, 1—16 m; type locality,

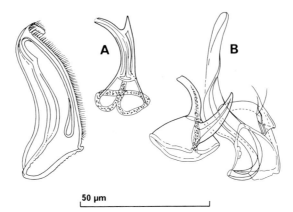

Fig. 8.   Male cuticular apparatus of aberrant shape in two undescribed Antarctic phena.

S. Georgia, Cumberland Bay, May Creek, tidal zone, seaweed, 1902-05-02.

A. magnificus Karling, 1952. Fig. 4B
A. pectatus magnificus Karling, 1952
A. magnificus: Brunet, 1965
Stylet 130–170 μm (5 specimens) with a short (12–15 μm) hook, with basal funnel half the length of the tube. A-organ 190–220 μm without separate style, with folded foot, proximal "window" in the plate and longitudinal comb with fine teeth. The short flagellum without teeth. No pigment eyes. – Lectotype No. 2690 a whole mount, leg. Sw. Ant. Exp.
Distribution. S. Georgia, Pot Bay, clay and seaweed, 22 m, 1902-05-30 (type locality) and Cumberland Bay, clay and stones, 250–310 m, 1902-06-05.

A. calcareus sp.n. Fig. 2A, 3A, B
A. pectatus pectatus Karling, 1952, "forma Sporn"
Stylet 70–80 μm (8 specimens) with long tube and short (6–9 μm) hook; basal funnel slightly shorter than the tube. A-organ 87–104 μm without separate style, with folded foot and a proximal "window" in the plate. Curved comb and flagellum with coarse teeth. – Holotype No. 2689 a whole mount, leg. Sw. Ant. Exp.
Distribution. Falkland Islands,Port Louis, sandy beach, 1 m, 1902-08-09 (type locality).

A. parapectatus sp.n. Fig. 2B
A. pectatus pectatus Karling, 1952, "forma Doppelt"
Stylet 60–70 μm (4 specimens) with tube and hook; hook and basal funnel slightly shorter than the tube. A-organ 87–96 μm without separate style. Transversal comb and flagellum with coarse teeth. – Holotype No. 2679 a whole mount, leg. Sw. Ant. Exp.
Distribution. Falkland Islands, Port Louis, sandy beach, 1 m, 1902-08-09 (type locality); S. Georgia, Cumberland Bay, May Creek, tidal zone, seaweed, 1902–05–02.

A. spinosus sp.n. Fig. 4A
A. pectatus pectatus Karling, 1952, "forma Stachel"
Stylet (basal funnel + tube) 116–130 μm (3 specimens), with long hook (slightly shorter than the tube) and comparatively short basal funnel (1/4 of the whole size). A-organ of the same size as the stylet, two-branched without plate and with folded foot. – Holotype No. 2691 a whole mount, leg. Sw. Ant. Exp.
Distribution. Falkland Islands, Berkely Sound, gravel, shell and seaweed, 16 m, 1902-07-19 (type locality); Tierra del Fuego, Ushuaia, mud, 6 m, 1902-10-16.

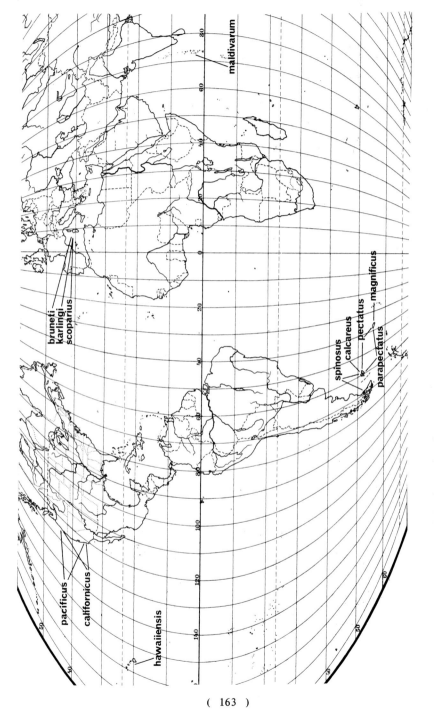

Fig. 9.   Geographic distribution of the *Austrorhynchus* species.

A. pacificus sp.n. Fig. 5A, 6A

Stylet 22–44 μm (6 specimens) without hook, with basal funnel and tube of equal length. A-organ 87–116 μm with S-shaped basal line and weakly differentiated style. Curving comb and flagellum with fine teeth. - Holotype No. 2680 a whole mount, leg. Karling.

Distribution. Pacific coast of N. America (Monterey area, Calif. and Newport area, Oregon), tidal zone, sand, gravel and seaweed. Type locality, Pacific Grove, Calif., Point Pinos, "Great tidepool," 1969-03-28.

A. californicus sp.n. Fig. 5B, 6B

Stylet 35–50 μm (4 specimens), with basal funnel, hook and tube of equal length. A-organ 87–130 μm with triangular, proximally pointed plate and weakly differentiated style. Transversal curving comb and flagellum with fine teeth. – Holotype No. 2682 a whole mount, leg. Karling.

Distribution. Pacific coast of N. America (Monterey area, Calif. and Newport area, Oregon), 0–10 m, sand and seaweed. Type locality, Monterey Bay, fine sand in the kelp zone, 10 m, 1969-04-03.

A. hawaiiensis sp.n. Fig. 5C

A. pectatus Karling et al. 1972

Stylet 35–38 μm (4 specimens), with basal funnel, hook and tube of equal length. A-organ 74–78 μm with square plate and short style. Transversal curving comb with coarse teeth. Flagellum with fine teeth. – Holotype No. 2683 a whole mount, leg. Karling.

Distribution. Hawaii, Oahu, Coconut Island, sand and coral reefs in the tidal zone, 1969-04-16-19 (type locality).

A. maldivarum sp.n. Fig. 5D

Stylet 72 μm (one specimen) with tube and basal funnel of almost equal length and with a hook slightly shorter than the tube. A-organ 130 μm with longish separate style, ribbon-like plate and transversal comb with fine teeth. Flagellum without teeth. – Holotype No. 2685 a whole mount, leg. Anne-Marie Nilsson.

Distribution. Maldive Islands, Vihamana Fushi, coral and gravel, 1–1.5 m, 194-02-25 (type locality).

A. bruneti sp.n.

A. pectatus: Brunet 1965

Stylet 56–66 μm with tube and hook of almost the same length and the basal funnel slightly shorter than these. A-organ about 100 μm with short separate style and transversal comb with fine teeth. Differs from *A. hawaiiensis* in the following respects: tube and hook of the stylet finer, plate of accessory organ transversally extended with fine teeth. – Holotype No. 2778 a whole mount, leg. M. Brunet.

Distribution. Mediterranean, Marseille area, Amphioxus sand.

A. scoparius Brunet and *A. karlingi* Brunet, see Brunet (1965).

G. Conclusions

The *Austrorhynchus* phena from the Antarctic area, previously described as subspecies and forms have now been regarded as separate species due to their stable characteristics in regard to the male cuticular apparatus, the rather large number of specimens of each phenon and the occurrence of several of them in the same localities (*A. pectatus, A. magnificus, A. calcareus, A. parapectatus, A. spinosus*). Two species previously identified with *A. pectatus* have now been considered separate species (*A. bruneti, A. hawaiiensis*). Two new species from the Pacific (*A. pacificus, A. californicus*) and one species from the Indian Ocean (*A. maldivarum*) have been described.

The accessory organ (A-organ) in the male cuticular apparatus is the basis for the species taxonomy. The evolution of this organ and the phylogeny of the species have been discussed and presented in the scheme Fig. 7.

There is evidently a trend towards evolution of a hook on the cuticular stylet. The joint evolution of this organ and the A-organ can be interpreted as a mosaic evolution.

The *pectatus* group (A-organ with well differentiated plate, comb and style) has the widest geographic distribution and has been regarded as phylogenetically ancestral. Adaptive radiation resulting in speciation has possibly occurred separately in the Mediterranean and the Antarctic areas.

In the geographic distribution of the genus *Austrorhynchus* there seems to be an Atlantic gap. The Mediterranean *Austrorhynchus* fauna originates either from a fauna with world-wide distribution, the Atlantic part of which has disappeared, or from the Tethys fauna. The original evolutionary center of the genus *Austrorhynchus* was possibly located in the Pacific.

Acknowledgements

On behalf of the Swedish Museum of Natural History, Section of Invertebrate Zoology, I thank Dr. M. Brunet, Marseille, for presenting his valuable collection of slides of Turbellaria Kalyptorhynchia to this institution. I thank Miss Anne-Marie Nilsson, Stockholm, for the material of one species, and Dr. Ernest Schockaert, Diepenbeek, Belgium, for letting me use an unpublished species name proposed by him. – My participation in the Bermuda symposium was supported by grant R 2328-009 from the Swedish Natural Science Research Council.

Bibliography

Ax, P.: Zur Evolution der marinen Mikrofauna von Galapagos. Akad. Wiss. Lit. Mainz 1949–1974, 90–105 (1974).

Brunet, M.: Turbellariés Calyptorhynques de substrats meubles de la région de Marseille. Recl. Trav. Stn mar. Endoume **39**, 127–219 (1965).

Dell, R. K.: Antarctic benthos. In: Advances in Marine Biology (ed. F. S. Russell & M. Yonge) **10**, 1–216 (1972).

Ekman, S.: Zoogeography of the sea. Sidgwick and Jackson, London (1953).

Karling, T. G.: Kalyptorhynchia (Turbellaria). Further zool. Results Swed. Antarct. Exped. 1901–1903 **4**, **9**, 1–50 (1952).

–: Morphologisch-histologische Untersuchungen an den männlichen Atrialorganen der Kalyptorhynchia (Turbellaria). Ark. Zool. (2), **9**, 187–279 (1956).

– V. Mack-Fira & J. Dörjes: First report on marine microturbellarians from Hawaii. Zool. Scr. **1**, 251–269 (1972).

Marcus, E.: Turbellaria Brasileiros XI. Papéis Dep. Zool. S. Paulo **11** (24), 419–489 (1954).

Mayr, E.: Principles of systematic zoology. MacGraw-Hill, New York (1969).

Mikrofauna Meeresboden	61	167–181	1977

W. Sterrer & P. Ax (Eds.). The Meiofauna Species in Time and Space. Workshop Symposium, Bermuda Biological Station, 1975

Remarks on Taxonomy and Geographic Distribution of the Genus Ototyphlonemertes Diesing (Nemertina, Monostilifera)

by

Ernst Kirsteuer

American Museum of Natural History, New York, USA

Abstract

Based on observations made on a large collection of *Ototyphlonemertes* from the western Atlantic and Caribbean Sea the intraspecific variability or stability of various morphological details (sensory cirri, adhesive plate, statoliths, armature and other proboscis related features) is briefly discussed and the results are compared with previous descriptions of species. It is concluded that neither cosmopolitan species nor amphiatlantic or amphiamerican species of *Ototyphlonemertes* are presently known. The distribution of the species in the western hemisphere is illustrated.

A. Introduction

The genus *Ototyphlonemertes* was erected by Diesing in 1863 for a monostiliferous hoplonemertean with statocysts and lacking eyes, which Keferstein (1862) had originally described as *Oerstedia pallida*, based on a single specimen collected at St. Vaast la Hougue, European North Atlantic. Since then more than a dozen species became known from European waters, the western Atlantic and the west coast of North America, and quite recently several species of *Ototyphlonemertes* were also reported from the Galapagos Islands (Mock & Schmidt 1975). In addition, unidentified congeneres have been found in the Red Sea (Remane & Schulz 1964) and on the coast of India (Rao & Ganapati 1968); a worldwide distribution of the genus is thus documented.

Ototyphlonemertes is comprised of small and slender species with a range in length of mature animals from 3 mm to 50 mm and a body width not exceeding 0.5 mm. They are all marine, mesopsammic forms inhabiting intertidal and shallow subtidal sands.

The majority of taxa is incompletely described and inadequately illustrated, and some of the diagnostic features employed for distinction of species are subject to confusion. The taxonomic problems are further compounded by technical difficulties in observing morphological details and by the fact that type specimens are not available to verify many of the older species accounts. There has been also little attention paid to within-population variability and geographic variation and hence it is occasionally difficult to correctly identify specimens, particularly when material from new and remote places is at hand.

A revision of the genus *Ototyphlonemertes* is presently under way (Kirsteuer, in preparation) based on a critical review of the literature as well as on collections obtained during the past few years from a wide range of western Atlantic and Caribbean localities. Live observation of several hundred animals and examination of a considerable number of sectioned specimens disclosed that there is a much greater morphological diversity within the genus than previously recognized, and also provided evidence in regard to variability or stability of conventional taxonomic criteria. Some of the findings are here briefly outlined and discussed in connection with zoogeographic aspects of presumable immediate interest.

The West Atlantic-Caribbean study collection contained specimens of *O. erneba* Corrêa, 1950, and *O. lactea* Corrêa, 1954, and of four new species not yet formally described and therefore indicated by roman numerals whenever referred to in the following text.

B. Observations and comparisons

Sensory cirri (Fig. 1). — Cirri, sometimes also termed hairs or bristles, have been described for several species and usually occur around the anterior and posterior margin of the body. In living animals they are conspicuous by being four to eight times longer than the epidermal ciliation (Fig. 1A, B). Each cirrus consists of a tuft of cilia which separate under prolonged pressure in squeeze preparation (Fig. 1C, D) and disintegrate completely during fixation.

Corrêa (1954, p. 35) stated that in all species of *Ototyphlonemertes* encountered in Brasil, i.e., *evelinae, brevis, erneba, parmula, fila*, and *lactea*, both ends of the body are provided with cirri. The same situation was observed in all the animals examined during the present survey, and had also been regularly found by Müller (1968a) in about 200 specimens of *O. antipai*. Cirri are scattered over the whole body in *O. cirrula* Mock & Schmidt (1975) and are altogether lacking in *O. aurantiaca* (sensu Gerner, 1969). Intraspecific variation is so far only indicated for. *O. americana* which Gerner (1969) described from Washington as being devoid of cirri whereas the latter were found on the posterior end of presumably conspecific animals from the Galapagos Islands (Mock & Schmidt 1975). It should be mentioned, however, that it is not clear whether Gerner's (1969) observations were made on living or preserved specimens.

Fig. 1: Cephalic cirri in *Ototyphlonemertes*. A) *Ototyphlonemertes* spec. I; B) *O. erneba* (phot. Gibson); C) *O.* spec. I; D, E) *O.* spec. II; F) *O. erneba*; G) *O.* spec. III; (E–G traced from photomicrographs). Further explanations in text.

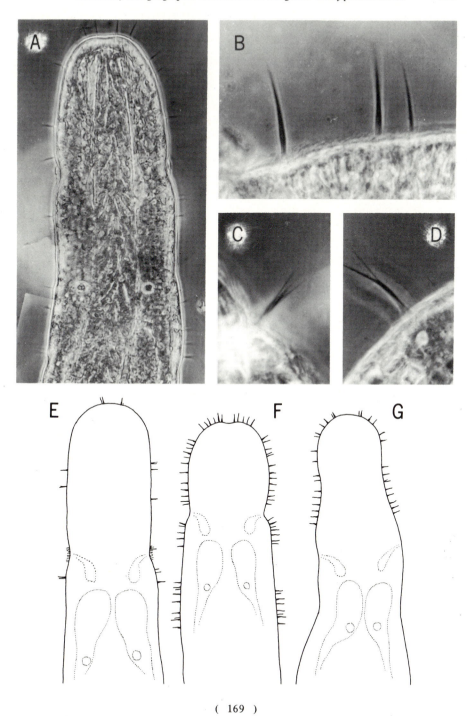

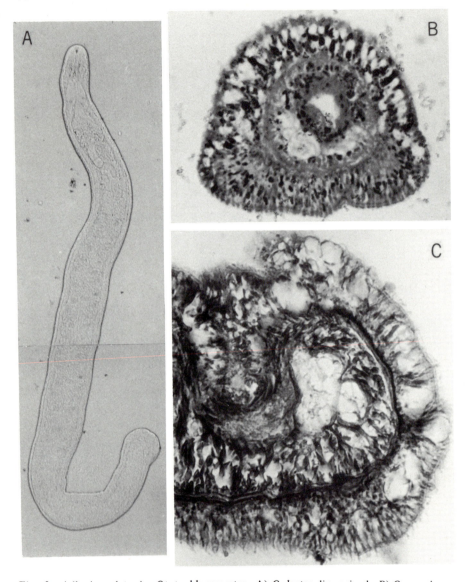

Fig. 2: Adhesive plate in *Ototyphlonemertes*. A) *O. lactea*, live animal; B) *O. americana;*
C) *O. erneba*. Further explanations in text.

The arrangement of the anterior cirri appears to be of taxonomic significance.
Müller (1968a, p. 344, fig. 1A) was the first to notice a specific pattern in *O. antipai*
and subsequently interspecific differences in the placement of anterior cirri were
shown by Mock & Schmidt (1975, figs. 2A, 8A) for two of the Galapagos taxa.
Similarly, the present study demonstrated that in the various species (e.g., Fig. 1A,

E–G) the cirri are either confined to the area in front of the cerebral organs or occur along the entire precerebral region, and that in some species also postcerebral cephalic cirri are present. No intraspecific variation regarding these patterns was found. The distribution of cirri in *O. erneba* as depicted in Fig. 1F has invariably been observed in the specimens from Brasil (Gibson, pers. comm.) and Colombia but is quite different from the one described by Mock & Schmidt (1975, fig. 8A) for the specimens from the Galapagos Islands. This, however, does not exemplify a case of geographic variation because comparisons of other features evidence that the Galapagos animals are not conspecific.

Adhesive plate (Fig. 2). – It was experienced during this study that the presence of a posteroventral adhesive plate is sometimes already detectable in living animals owing to a temporary widening of the hind end of the body when attached to the substratum (Fig. 2A) and that in sectioned specimens it is always distinct and easily discerned by an accumulation of epidermal gland cells in the subanal region (Fig. 2B, C), yet the epidermis of the plate is not necessarily always thicker than that in the adjacent areas of the body. Of the species examined all but *Ototyphlonemertes* spec. I and *O.* spec. IV had such a histologically differentiated plate and there was no indication that the presence or absence of this structure is intraspecifically variable.

For several species the possession or lack of an adhesive plate has still to be verified. According to Corrêa (1948; 1950), it occurs as a high epidermal cushion in *O. brevis* and *O. parmula* but not in *O. evelinae* and *O. erneba* although the latter two species are said to be also capable of attaching with the posterior end of the body. Müller (1968b) therefore assumed that in these instances adhesion is accomplished by aggregations of gland cells, and at least for *O. erneba* this can now be confirmed. Another species for which the available information is ambiguous is *O. fila*, originally described from Brasil by Corrêa (1953, p. 549) who noted that the animals react haptic with the posterior end but stated concomitantly that it is "desprovida de estrutura fixadora especial". Gerner (1969, p. 99) then employed the lack of an adhesive plate in *O. fila* as a differential diagnostic criterion for delineating *O. americana*. The fact that in animals from the Galapagos Islands assigned to *O. fila* by Mock & Schmidt (1975) a well developed, prominent adhesive plate is present does not help to clarify the situation because due to other differences between the Pacific and Atlantic animals, the postulated identity cannot be accepted.

Statocysts and statoliths (Fig. 3). – Statocysts are characteristic for all species of *Ototyphlonemertes* and as a rule two of these organs are present. Duplications of statocysts have been observed (Keferstein 1862; Corrêa 1948; Gibson, pers. comm.) but are extremely rare individual aberrations. The shape of the statocysts varies from spherical to slightly ovoid depending on whether animals are gliding or contracting, and stronger changes, i.e., pronounced elongations, are induced under increasing pressure in squeeze preparation (compare Figs. 1A, 3A).

172   Ernst Kirsteuer

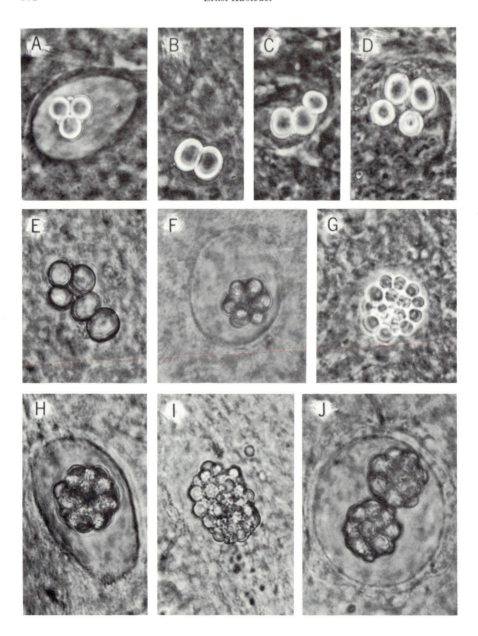

Fig. 3: Statocysts and statoliths in *Ototyphlonemertes*. A) *Ototyphlonemertes* spec. I; B–E)
O. spec. II; F) *O.* spec. IV; G) same as F but more strongly squeezed; H) *O. lactea*; I) same as H
but more strongly squeezed; J) statolith duplication in *O. lactea*; (all photomicrographs from
live animals). Further explanations in text.

There is generally one composite statolith contained in each statocyst. The number of granules forming the statoliths is intraspecifically variable but two basic types of statoliths are distinguishable and proved to be species specific. In the oligogranular type the number of granules can vary from 2 to 5 as was observed in Belize in a population of *Ototyphlonemertes* spec. II (Fig. 3B–E) whereby any combination also occurred asymmetrically in individuals. Other examples are *O. aurantiaca* (cf. Gerner 1969) and *O. antipai* (Müller 1968a) with 2 to 4 granules, and *O. evelinae* (Corrêa 1948), *O. pallida* (cf. Friedrich 1950), and *O. erneba* (cf. Gibson, pers. comm.) with 2 or 3 granules per statolith. In contrast, the polygranular statoliths always consist of a much larger number of comparatively small granules and are of morula-like appearance (Fig. 3F–J). Due to overlapping of granules it is difficult to make exact counts and in insufficiently squeezed animals it may seem as if only 7 or 8 granules are forming the statolith (Fig. 3F, H); however, exertion of more pressure always revealed that about 20 to 30 granules are present (Fig. 3G, I). This undoubtedly explains why in older descriptions these statoliths are frequently depicted as a rosette of 7 granules (e.g., Corrêa 1948, fig. 12; 1950, fig. 33; 1953, fig. 3; 1954, fig. 32) and also why Coe (1940; 1943) noted a variation of only 8, 12, or 16 statolith granules in *O. spiralis* and *O. pellucida*.

Oligogranular statoliths are always confined to species with smooth proboscis stylets whereas polygranular statoliths are characteristic for all species with spirally structured stylets, including *O. macintoshi* (see Bürger 1895, pl. 29, fig. 14), and only *O. brevis* Corrêa, 1948, and *O. cirrula* Mock & Schmidt 1975, are known to have polygranular statoliths and smooth stylets. Implementation of actual numbers of statolith granules as key characters (e.g., Corrêa 1960, p. 22) is not warranted. It is also not justified to use the number of statoliths for the discrimination of species (e.g., Gerner 1969, p. 103) because in some older descriptions each granule has been treated as a statolith and the implied differences are merely reflections of semantic discrepancies. Genuine duplication of statoliths (Fig. 3J) is a very rare intraspecific phenomenon and has no taxonomic value.

Proboscis (Figs. 4, 5, 6). – The condition of the musculature, i.e., the arrangement of fibers in the proboscis itself and in the proboscis insertion (precerebral septum) as well as in the wall of the proboscis sheath, is constant in a given species. This has also been sufficiently documented in the past for many other taxa of Monostilifera and does not require any further elaboration. Attention is here only drawn to the fact that in *O. erneba* the precerebral septum is incomplete (Corrêa 1950, p. 214; present material) whereas in the animals from the Galapagos Islands assigned to this species (Mock & Schmidt 1975, p. 17) a complete, closed septum is developed.

The proboscis is divisible into anterior chamber, diaphragm with stylet apparatus, vesicle, and posterior chamber. The epithelium of the anterior chamber forms papillae the shape of which (e.g., Figs., 4A–C, 6C) is species specific and useful

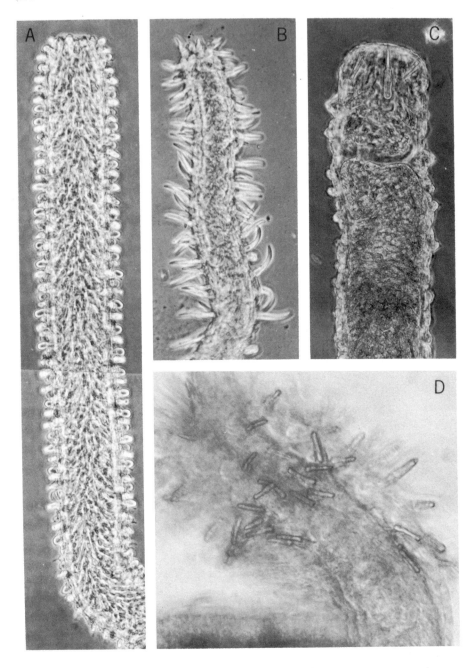

Fig. 4: Proboscis papillae in *Ototyphlonemertes*. A) *Ototyphlonemertes* spec. IV; B) *O.* spec.
I; C) *O.* spec. II; D) rhabdites in papillae of *O. erneba*; (all photomicrographs from live animals).

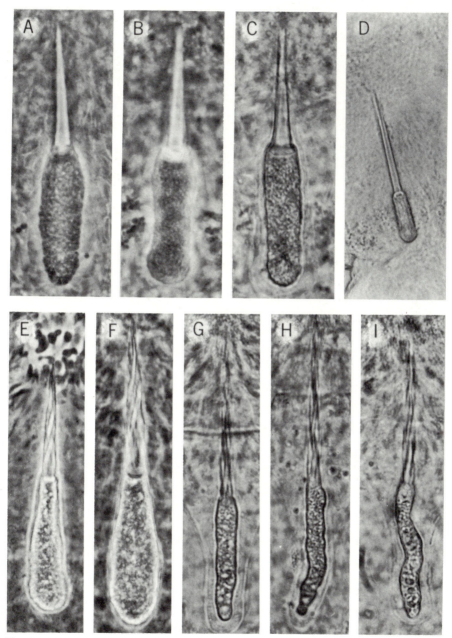

Fig. 5: Proboscis stylet and stylet base in *Ototyphlonemertes*. A–C) *Ototyphlonemertes* spec.
II, same population; D) *O. erneba*, stylet base without constricution; E, F) *O.* spec. IV; G–I)
O. lactea, same population; (all photomicrographs from live animals). Further explanation
in text.

for diagnostic purposes. Conspicuously large rhabdites are known to occur in the papillae of *O. parmula* and *O. erneba* (Corrêa 1950, p. 211, 212, figs. 23, 24) and have been also regularly encountered in the specimens of the latter species examined during this survey (Fig. 4D). In the animals from the Galapagos Islands assigned to *O. erneba* these rhabdites are lacking (Mock & Schmidt 1975, p. 21).

The proboscis armature consists of the main stylet resting on the stylet base, and usually two accessory stylet pouches (a third pouch was found only once in a specimen of *Ototyphlonemertes* spec. IV). The stylets are either smooth (e.g., Fig. 5A–D) or spirally structured (e.g., Fig. 5E–I) and each type is species specific. Examinations with phase contrast optics (Gerner 1969; Mock & Schmidt 1975; present study) have shown that the spirally structured stylets always consist of intertwined rods, and it is safe to assume that illustrations in older descriptions depicting just wavy or diagonal lines on the stylets express the same condition.

The shape of the stylet base (e.g., cylindrical with or without a median constriction; pyriform) has been customarily used as a diagnostic feature but a considerable within-population variability was found during this survey (Fig. 5).

The size of stylet and base of mature animals varies within a population (e.g., Fig. 5E, F), and there is also evidence for geographic variation as indicated by two different, not overlapping sets of measurements obtained for *O. erneba* from Brasil (Gibson, pers. comm.) and Colombia. However, in all instances (including *O. erneba*) the proportions of the armature, i.e., ratios of stylet length to base length, stylet length to stylet width, and base length to base width, showed only a slight intraspecific variation and differences between species turn out to be significant (at the 95 percent level) when the relevant data are compared with the t-test.

Posterior to the armature-bearing diaphragm follows the highly muscular vesicle which, depending on the species, is present in the form of a globose bulb (e.g., Fig. 6A, D) or a comparatively long tube (e.g., Fig. 6E). Only for *O. fila* the vesicle was described by Corrêa (1953, p. 549) as being an elongate bulb but unfortunately no further details were mentioned. In vivo observations of *O. lactea* have shown that a globose vesicle may stretch to some degree during eversion of the proboscis (Fig. 6B, C). However, this is a temporary condition and on the invaginated, inactive proboscis (Fig. 6A) as well as after fixation (Fig. 6D) the vesicle is always unmistakably globose. Since Corrêa's (1953) description was based on several specimens of which at least one had the proboscis invaginated (loc. cit., p. 552, fig. 7), and since, furthermore, Corrêa (loc. cit., p. 551, 554) used the elongate shape of the vesicle in *O. fila* as a main characteristic to distinguish this species from the earlier described *O. parmula* Corrêa, 1950, for which a globose vesicle had been reported, it must be presumed in accordance with Müller (1968b, p. 446) that tangible differences exist. The animals from the Galapagos Islands referred to *O. fila* by Mock & Schmidt (1975) have a vesicle which, though described as elongate (loc. cit., p. 13), is slightly wider than long in fixed specimens as well

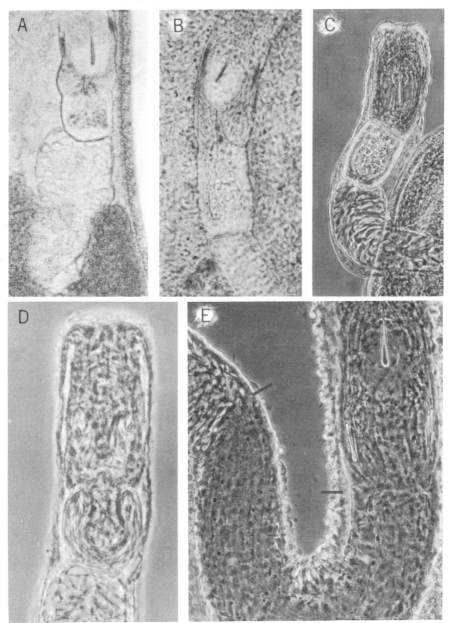

Fig. 6: Configuration of proboscis vesicle in *Ototyphlonemertes*. A) Globose bulbous vesicle on invaginated proboscis of *O. lactea*; B) Temporary elongation of vesicle during eversion of proboscis in *O. lactea*; C) same as B but with proboscis almost completely everted; D) fixed and cleared proboscis of *O. lactea*; E) tubular vesicle (marked by indication lines) in *O.* spec. IV; (A–C, E, photomicrographs from live animals).

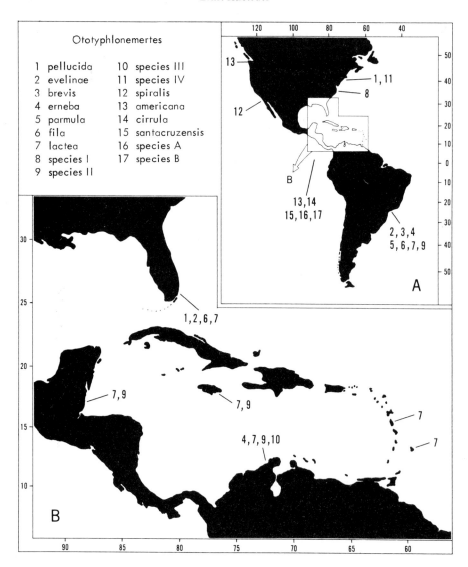

Fig. 7: Distribution of *Ototyphlonemertes* in the western hemisphere.

as in the everted proboscis of live animals (loc. cit., fig. 2A, B, 3B), and hence it is distinctly globose and the condition thus not comparable to the one characteristic for *O. fila* Corrêa. Comparison of the descriptions provided by Corrêa (1953) and Mock & Schmidt (1975) also brings to light additional differences (manifested in the orientation of the cephalic grooves, extent of the cephalic gland, mode of transition of foregut to intestine, and development of intestinal diverticula), which lend further support for excluding the possibility that the Galapagos specimens are conspecific with *O. fila* Corrêa.

C. Discussion and conclusion

Scrutiny of previous morphological accounts and results from present observations (which inter alia also evidenced tetraneury, interwoven proboscis sheath musculature, and excretory organs) allow to distinguish 24 species of *Ototyphlonemertes*. Seven of these species belong to the European fauna and have been listed together with distribution data in an earlier paper (Kirsteuer 1971). Of the remaining species, eleven (including *Ototyphlonemertes* spec. I–IV) occur in the western Atlantic-Caribbean region (Fig. 7) and six in the eastern Pacific (Fig. 7). The animals from the Galapagos Islands referred by Mock & Schmidt (1975) to *O. fila* Corrêa, and *O. erneba* Corrêa, have been obviously misidentified and represent two new species which will be formally described elsewhere as soon as the relevant materials are received for re-examination. However, the information contained in the descriptions and illustrations provided by Mock & Schmidt (1975) permits to show already here (see Appendix) the respective combinations of characteristics by which these two entities can be discriminated from all other recognized taxa in the genus.

The problems attending zoogeographic analysis of interstitial faunae were recently enumerated by Sterrer (1973) and clearly apply also unrestricted to *Ototyphlonemertes*. Far-reaching conclusions regarding the distribution patterns of species must be considered premature as long as large gaps in the records exist (e.g., west coast of Africa, west coast of the Americas, Indo-Pacific region). However, it can be stated with a considerable degree of confidence that at this point in time neither cosmopolitan species nor amphiatlantic or amphiamerican species of *Ototyphlonemertes* are known.

Acknowledgements

Field work conducted in Barbados, Colombia, and Jamaica was supported by U.S. National Science Foundation Grant GB-7952, whereas investigations in Dominica and Belize were sponsored by the U.S. National Museum of Natural History, Smithsonian Institution (Belize-IMSWE Contribution No. 27). Collecting in North Carolina was made possible through funds from NSF Grant XA 120–09 to Dr. R. Riedl, University of North Carolina. Appreciation is expressed in particular to Dr. P. Ax, University of Göttingen, for lending the types of *O. americana* Gerner, and for agreeing to initiate a loan of the Galapagos material in the near future, to Dr. R. Gibson, Liverpool Polytechnic, for providing information and specimens from Brasil, and to Dr. N. Riser, Northeastern University, for sending live specimens from Massachusetts.

Appendix

With the provision that the genus *Ototyphlonemertes* Diesing, 1863, (syn. *Typhlonemertes* Plessis, 1891), is defined as Hoplonemertini Monostilifera with statocysts and without eyes, a species A (for animals referred by Mock & Schmidt 1975, p. 7, to *O. fila* Corrêa, 1953), and a species B (for animals referred by Mock & Schmidt 1975, p. 15, to *O. erneba* Corrêa, 1950) can be delineated as follows:

Species A: *Ototyphlonemertes* with polygranular statoliths, spirally structured stylet, globose bulbous **proboscis** vesicle, cerebral organs, and frontal organ; lateral sense organs lacking; cirri on both ends of body, cephalic cirri in precerebral and cerebral region; dorsal branches of cephalic grooves pointing posteriad; adhesive plate present; cephalic gland extending into gonad region; precerebral septum incomplete; proboscis sheath 1/3 of body length; length ratio of stylet to base approx. 1:1; proboscis papillae without rhabdites; wall of posterior proboscis chamber without musculature; transition of foregut to midgut without pyloric differentiation; intestinal caecum lacking; deep midgut diverticula present; excretory organs lacking.

Species B: *Ototyphlonemertes* with oligogranular statoliths, smooth stylet, length ratio of stylet to base at least 2:1, and closed precerebral septum; cirri on both ends of body, cephalic cirri only anterior to cerebral organs; dorsal branches of cephalic grooves pointing posteriad; adhesive plate present; cephalic gland lacking; proboscis sheath 1/2 of body length; proboscis papillae without rhabdites; proboscis vesicle globose bulbous; transition of foregut to midgut without pyloric differentiation; intestine without caecum and distinct diverticula; excretory organs lacking.

Bibliography

Bürger, O.: Nemertinen. Fauna and Flora des Golfs von Neapel **22**, 1–743 (1895).

Coe, W. R.: Revision of the nemertean fauna of the Pacific coasts of North, Central, and northern South America. Allan Hancock Pacif. Exped. **2**, 247–323 (1940).

–: Biology of the nemerteans of the Atlantic coast of North America. Trans. Conn. Acad. Arts Sci. **35**, 129–328 (1943).

Corrêa, D. D.: *Ototyphlonemertes* from the Brazilian coast. Comun. zool. Mus. Hist. Nat. Montevideo 2(49), 1–12 (**1948**).

–: Sôbre *Ototyphlonemertes* do Brasil. Bol. Fac. Filos. Ciênc. Letr. Univ. S. Paulo (Zool.) **15**, 203–234 (1950).

–: Sôbre neurofisiologia locomotora de hoplonemertinos e a taxonomia de *Ototyphlonemertes*. An. Acad. Brasil. Ciênc. **25**, 545–555 (1953).

–: Nemertinos do litoral brasileiro. Bol. Fac. Filos. Ciênc. Letr. Univ. S. Paulo (Zool.) **19**, 1–122 (1954).

–: Nemerteans from Florida and Virgin Islands. Bull. mar. Sci. Gulf Caribb. **11**, 1–44 (1960).

Diesing, K. M.: Nachträge zur Revision der Turbellarien. Sitzber. math. nat. Kl., Akad. Wiss. Wien **46**, 173–188 (1863).

Friedrich, H.: Zwei neue Bestandteile in der Fauna der Nordsee. Neue Ergebn. Probl. Zool. (Klatt-Festschrift), 171–177 (1950).

Gerner, L.: Nemertinen der Gattungen *Cephalothrix* und *Ototyphlonemertes* aus dem marinen Mesopsammal. Helgoländer wiss. Meeresunters. **19**, 68–110 (1969).

Keferstein, W.: Untersuchungen über niedere Seethiere. Z. wiss. Zool. **12**, 1–147 (1862).

Kirsteuer, E.: The interstitial nemertean fauna of marine sand. Smithson. Contr. Zool. **76**, 17–19 (1971).

Mock, H. & P. Schmidt: Interstitielle Fauna von Galapagos. XIII. *Ototyphlonemertes* Diesing (Nemertini, Hoplonemertini). Mikrofauna Meeresboden **51**, 1–40 (1975).

Müller, G. J.: *Ototyphlonemertes antipai* n. sp., ein neues Mitglied des mediolitoralen Mesopsammals des Schwarzen Meeres.Trav. Mus. Hist. Nat. „Grigore Antipa" **8**, 343–348 (1968a).

–: Betrachtungen über die Gattung *Ototyphlonemertes* Diesing, 1863, nebst Bestimmungsschlüssel der validen Arten (Hoplonemertini). Senckenbergiana biol. **49**, 461–468 (1968b).

Plessis, G. du,: Sur une nouvelle *Oerstedia* aveugle mais portant une paire de vésicules auditives (otocystes). Zool. Anz. **14**, 413–416 (1891).

Rao, G. C. & P. N. Ganapati: The interstitial fauna inhabiting the beach sands of Waltair coast. Proc. Natl. Inst. Sci. India **34**, 82–125 (1968).

Remane, A. & E. Schulz: Die Strandzonen des Roten Meeres und ihre Tierwelt. Kieler Meeresforsch. **20**, 5–17 (1964).

Sterrer, W.: Plate tectonics as a mechanism for dispersal and speciation in interstitial sand fauna. Neth. J. Sea Res. **7**, 200–222 (1973).

Mikrofauna Meeresboden	61	183—196	1977

W. Sterrer & P. Ax (Eds.). The Meiofauna Species in Time and Space. Workshop Symposium, Bermuda Biological Station, 1975

An Introduction to the Architecture and Kinematics of Harpacticoid (Copepoda) Feeding: Tisbe furcata (Baird, 1837)

by
Brian M. Marcotte
Department of Biology, Dalhousie University Halifax, Nova Scotia, Canada

Abstract

The functional morphology of the mouthparts, and the behaviour of feeding in the harpacticoid copepod *Tisbe furcata* (Baird, 1837) was studied by means of slow-motion videotape. The species is a raptorial feeder which uses its second antenna and second maxilla for initial food manipulation, the first maxilla for food movement within the oral frustum, and the mandible for crushing. A critical review of known feeding behaviour in the three orders of freeliving copepods defines filter, raptorial and mixed feeders. It is suggested that the size, shape, physical behaviour and acceptance behaviour of harpacticoid feeding may all be important for the specific diversification of the order.

Introduction

There are three orders of mostly freeliving Copepoda: Calanoida, Cyclopoida and Harpacticoida. The functional morphology of feeding in the first two of these orders has been studied at some length. However, the architecture and kinematics of harpacticoid feeding is largely unknown. The present essay will: 1) present data concerning feeding by the harpacticoid *Tisbe furcata* (Baird, 1837), 2) integrate these data into the corpus of information known concerning feeding in the other orders of copepods and 3) provide some tentative comments on the ecological and evolutionary significance of feeding in this harpacticoid.

B. Methods and materials

Live copepods were collected from marine paludal sediments and algal scrapings in West Lawrencetown, Nova Scotia. Populations were separated to species and kept overnight in food-free sea water at 10 °C. The following morning, active individuals were placed in depression slides containing one of the following substrates.
1. Pure sea water
2. Sea water and methyl-cellulose
3. Sea water and a culture of the diatom *Thallassiosira fluviatilis*
4. Sea water, methyl-cellulose and the diatom *T. fluviatilis*

184					Brian M. Marcotte

The copepods were observed by means of a Zeiss Universal Microscope fitted with a mercury vapour light source and a darkfield epi-illuminator. The feeding and locomotor behaviour of the copepods was video-taped by means of a Sony television camera mounted atop the microscope. The image was recorded at a rate of fifty frames per second on a Sony video-tape recorder. These tapes were later viewed in slow motion with repeated stop-motion examination of individual appendages. Velocities of appendicular movement were calculated only for individuals in the pure seawater preparations.

Anatomical results presented in this essay were obtained from the same individuals as were the video-tape studies. After the video-taping session, individuals were fixed in 4 % formalin and examined with phase contrast microscopy. The ventral and lateral aspects of the oral region were drawn from a whole mount preparation using a drawing tube. Subsequently, the oral region of these same specimens was divided between the first and   second maxillae and mounted on a glass slide with the posterior surface upward. Individual appendages were then drawn. If details were obscured in these second mounts, further dissection followed, but only after the proper position and general shape of the individual mouth parts had been recorded. Examination of at least twenty additional individuals confirmed that the oral anatomy of the single specimens studied in this manner was usual for the West Lawrencetown populations.

In situ observations were also made of individuals living in laboratory mud cultures and feeding on large detritus particles and nematode prey. These observations further clarified the function of oral movements observed in the video-tape studies.

The names of the oral appendages are abbreviated as follows: First Antenna (A_1 = antennule), Second Antenna (A_2 = antenna), Mandible (Md), First Maxilla (Mx_1 = maxillula), Second Maxilla (Mx_2 = maxilla), and Maxillipede (Mxp).

The anatomy and movement of the mouth parts were recorded in three series of figures. Figures 1 to 7 are posterior views of the oral appendages, figures 8 to 13 are ventral views and figures 14 to 19 are lateral views. The arrows on the figures indicate the direction and extent of appendicular movement from their figured positions.

C. Results

I. Architectural Aspects

The harpacticoid copepod, *Tisbe furcata* (Baird, 1837) is a raptorial feeder --- it grasps for food rather than filter it from the water. Its morphology is adapted to grab, crush and poke food into the animal's mouth.

The appendages of the oral region circumscribe a frustal space ventral to the mouth (Fig. 13). This space is bordered: 1) anteriorly by the proximal segments of the A_1 (Figs. 1 and 8), A_2 basis and endopod (Figs. 2 and 9) and the labrum (Figs. 3 and 8), 2) laterally by the lateral and distal setae of the A_2 endopod (Fig. 15), mandibular palps (Fig. 16) and the distal portions of the Mx_1 protopod (Fig. 17) the body of the Mx_2 (Fig. 18) and the lateral margins of the carapace (Figs. 8 and 14), 3) posteriorly by the proximal region of the Mx_2 (Figs. 6 and 12) and Mxp (Figs. 7 and 13) protopods and the labium (Fig. 8) and 4) ventrally by the distal and medial setae of the A_2 endopod (Fig. 9) and the claws of the Mx_2 and Mxp (Figs. 12 and 13). The volume of this frustum of a pyramid is approximately 4.7×10^5 μm^3, a volume similar to that of a sphere 100 μm in diameter or the size of a fine grain of sand. This volume varies with the size of the individual copepod and with the position of the mouth parts.

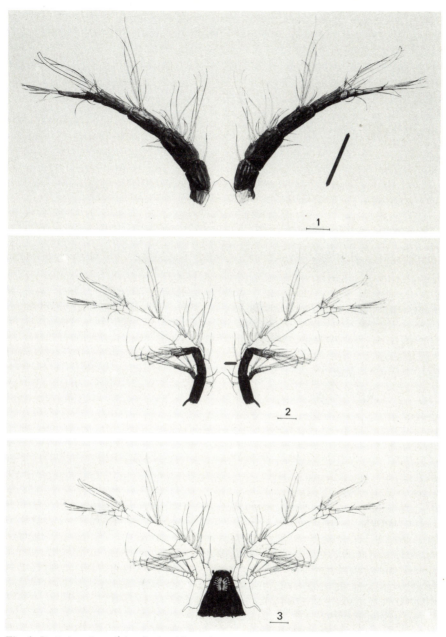

Fig. 1. Posterior view of A_1. Scale: 40 μm. Arrows indicate the direction and extent of appendicular movement from the illustrated position. – Fig. 2. Posterior view of A_2. Scale: 40 μm. Arrows indicate the direction and extent of appendicular movement from the illustrated position. – Fig. 3. Posterior view of Labrum. Scale: 40 μm.

The A_1 seems to be primarily a sensory and swimming organ. All setae of the A_1 are tapered and smooth (Fig. 1). There are no setae on the posterior surface of the A_1 (Figs. 1, 8 and 14), i.e. no setae project toward the mouth.

The A_2 is used directly in the feeding process. Its endopod arcs up to 75 μm below the ventral surface of the head (Fig. 15) and is equipped with long, strong, naked setae which recurve medio-ventrally (Figs. 2, 9 and 15). A spur lies on the median, distal corner of the A_2 endopod (Fig. 2). The A_2 exopod closes the gap beneath the arch of the A_2 endopod (Figs. 2, 9 and 15).

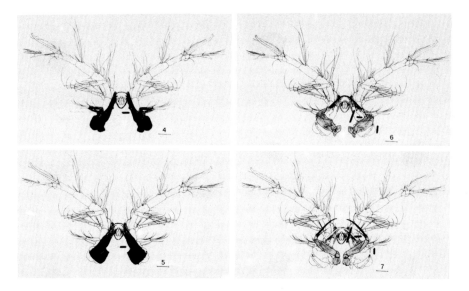

Fig. 4. Posterior view of Md. Scale: 40 μm. Arrows indicate the direction and extent of appendicular movement from the illustrated positions. – Fig. 5. Posterior view of Mx_1. Scale: 40 μm. – Fig. 6. Posterior view of Mx_2. Scale: 40 μm. Arrows indicate the direction and extent of appendicular movement from the illustrated positions.– Fig. 7. Posterior view of Mxp. Scale: 40 μm.

The labrum   (Figs. 3, 8 and 14) projects 33 μm below the ventral surface of the head. In the space immediately below and posterior to the labrum, the Md and Mx_1 are compactly arranged. The cutting or biting edges of the Md appose directly posterior to an area of reinforced exoskeleton on the inner surface of the labrum. This area has two crescent-shaped buttresses and a system of ridges and grooves similar to those on a file or washboard.

The posterior surface of the mandibular protopod (Fig. 4) is grooved to fit the surface setae of the Mx_1 (Fig. 5). The flat surface of the distal blade of the Mx_1 protopod is turned laterally and is couched in the hollow of the Md palps (Figs. 4, 5, 10, 11, 16 and 17). The proximal portion of the Mx_1 protopod is medially concave, creating a cradle for objects entering the mandibular area.

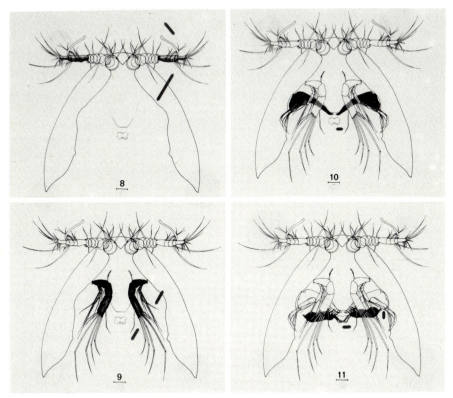

Fig. 8. Ventral view of A_1 and Labrum. Scale: 20 μm. Arrows indicate the direction and extent of appendicular movement. – Fig. 9. Ventral view of A_2. Scale: 20 μm. – Fig. 10. Ventral view of Md. Scale: 20 μm. Arrows indicate the direction and extent of appendicular movement from the illustrated positions. – Fig. 11. Ventral view of Mx_1. Scale: 10 μm.

The Mx_2 is equipped with a large prehensile claw which curves in all three dimensions (Figs. 6, 12 and 18). This claw is fitted with a stout, spinulose seta at its base. This seta seems to act as a pressure sensitive stop or hilt (see kinematics, below). There is a fine, lash-like seta projecting from an endite on the Mx_2 protopod. This seta projects anterio-laterally from the resting Mx_2 and may serve a sensory function.

The Mxp, with its prehensile claw, sits on a basal segment which projects from the midline of the head. The volume swept clear by this claw lies behind and ventral to that circumscribed by the claw of the Mx_2.

II. Kinematic Aspects

When a food particle was first encountered, it was either: 1) impaled or grabbed and pushed to the mouth directly by the Mx_2 or 2) was embraced by the body and

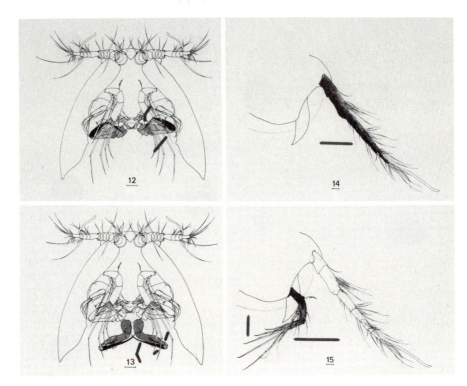

Fig. 12. Ventral view of Mx_2. Scale: 20 μm. Arrows indicate the direction and extent of appendicular movement from the illustrated positions. – Fig. 13. Ventral view of Mxp. Scale: 20 μm. – Fig. 14. Lateral view of A_1 and Labrum. Scale: 10 μm. Arrows indicate the direction and extent of appendicular movement. – Fig. 15. Lateral view of A_2. Scale: 10 μm.

setae of the A_2 and secondarily manipulated to the mouth by the Mx_2. In either case the A_2 was important in holding particles within the oral frustum. When a food item was first encountered, there was no activity in the Md or Mx_1. The Mxp seemed oddly unimportant in the initial operations of food capture. This may have been an artifact of the small size of the diatoms and detritus used in this study. The Mxp seemed most useful in guarding the posterior and ventral-most margins of the oral frustum against escape of "reluctant" food. The Mxp also helped the Mx_2 by pressing food to the dorsal surface of the mouth. At no time was the claw of the Mxp fully apposed to the body of the appendage. Throughout the feeding process the Mx_2 seemed most dextrous and useful, its body could pivot on its basal segment circumscribing an arc which extended from the base of the Mx_1 to the lateral edge of the basal stalk of the Mxp. At the same time the claw of the Mx_2 could be swung anteriorly and ventrally to the medial edge of the Md palps. When used to grasp or impale an object, the body of the Mx_2 was swung to its posterior-most position. The claw was then closed in a strong dorso-posterior thrust. The extent

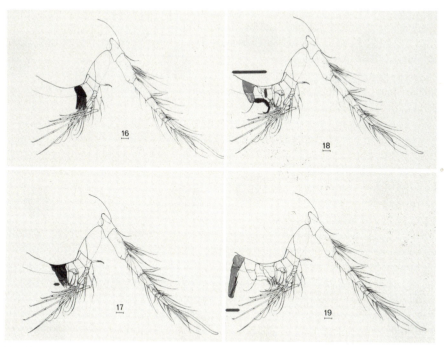

Fig. 16. Lateral view of Md. Scale: 10 μm. – Fig. 17. Lateral view of Mx$_1$. Scale: 10μm. Arrow indicate the direction and extent of appendicular movement from the illustrated position. – Fig. 18. Lateral view of Mx$_2$. Scale: 10 μm. Arrows indicate the direction and extent of appendicular movement from the illustrated positions. – Fig. 19. Lateral view of Mxp. Scale: 10 μm.

of this lunge seemed limited by the presence of the seta at the base of the Mx$_2$ claw. After the claw was closed, the body of the Mx$_2$ was pivoted forward, ramming the food item into the heretofore quiescent Mx$_1$ and Md.

Once a food particle was oriented with its long axis parallel to the main axis of the copepod's body, the Md and Mx$_1$ were triggered into action. The sensory physiology of this event is unknown. However an analysis of the structure of the oral region suggests that the two spinulose setae on the medio-posterior margin of the Mx$_1$ arthrite (Figs. 5 and 11) and/or the spinulose seta on the medio-dorsal face of the Md cutting edge (Fig. 4) are the most likely candidates for sensory organs in this area.

The Md and Mx$_1$ always act in concert. The Mx$_1$ moved rapidly and repeatedly through an elliptical path, the major axis of which was transverse to the major axis of the copepod's body. The effect of this movement was to direct and, perhaps, propel food directly into the biting edges of the Md.

The movement of the Md was very difficult to define. Most of its movement was in the transverse direction. In addition to transverse biting, the cutting edges appeared to roll across each other just as they were fully apposed. From a study of the anatomy of the Md and labrum, four masticatory events seem likely.

1) The slight roll may crush and squeeze food items between the cutting edges and pars molaris of the Md.

2) The transverse apposition of the biting edges may shear large food items into smaller ones. This shearing action may be either facilitated by the rolling motion or may, in fact cause the apparent roll.

3) The transverse biting may abrade anteriorly directed food particles against the ridges on the posterior face of the labrum, thus crushing them further.

4) If the partes molaris of the Md's are fully apposed during transverse movements of the Md, food items dorsal to the cutting edge of the Md could be squeezed dorsally to the oesophagus. Food particles once crushed were swallowed by powerful foregut peristalsis. These waves of contraction moved posteriorly throughout the ingestion process.

Tisbe furcata seemed able to reject a prospective food item at three times in the ingestion process: 1) when approaching stationary detritus or plant particles, the copepod could reject the item after it had apprehended the particle with its A_1, 2) after the particle had made contact with the Mx_1 and Md and 3) after mastication. The last two rejection events may have an identical sensory basis. Rejection after mastication was facilitated by reverse peristaltic contractions of the foregut, rapid transverse movement of the Md and Mx_1 and very delicate manipulation of the tip of the Mx_2 claw at the ventral border of the labrum to remove the unwanted particle from between the blades of the Md. The ecological significance of this last type of rejection is unknown. The peristaltic contractions which attended this event seemed violent and nearly arhythmic. This activity seemed very strenuous and may be an exceptional occurrence.

D. Discussion

In order to appreciate the phylogenetic and ecological significance of the present study, it is necessary to review what is known concerning feeding in all the orders of free-living copepods.

Copepod feeding may be classified either with regard to the process of food capture and mastication, or with regard to the objects of the feeding process. In the first scheme one can recognize three types of feeding: filter, raptorial and mixed. In the second, one can define herbivores, carnivores and omnivores. Each of these categories has a correlative morphology, the elements of which overlap between the two schemata. The following review and discussion will concentrate on the first of the two.

I. Filter feeding

Easterley (1916), Cannon (1928), Lowndes (1935), Marshall & Orr (1955), Anraku & Omori (1963), Gauld (1966) and Conover (1966) have described the filter feeding habits of *Calanus* spp., planktonic calanoid copepods.

The head appendages of *Calanus* spp. form an oral chamber bordered dorsally by the ventral body wall, anteriorly by the labrum, laterally by the Mx_2 and the margins of the carapace and posteriorly by the Mxp and swimming legs. Using the exopods of the Mx_1, *Calanus* spp. generates a region of low water pressure lateral to the oral chamber. Water is drawn into the oral chamber from under the swimming legs and between the Mxp's. The water is then sucked out of the chamber into the region of low pressure. Food particles are filtered from the water by Mx_2 setae as the water leaves the oral chamber. The food collected on the Mx_2 setae is scraped off and moved to the mouth by setae of the Mx_1 endites or by the long setae on the basal joints of the Mxp's.

Arashkevich (1968) has reviewed the morphological characters associated with filter feeding in *Calanus*. Anraku & Omori (1963) have studied filter feeding in a wide variety of other copepod genera. These studies indicate that the exopod and endopod of the mandible in filter feeders are long, powerful and armed with densely spinulose setae. The biting edge of the Md is equipped with short, sometimes multicuspid teeth. Beklemishev (1954) has shown that the crowns of these teeth are reinforced with silica. Sullivan et al. (1975) have confirmed the existence of these silicious crowns. This type of biting edge is adapted to crushing the valves of diatoms. The maxillae and Mxp's are all richly ornamented with spinulose setae.

The size of particles which can be filtered by the Mx_2 "net" depends upon the distance between adjacent setae and spinules on the Mx_2. Marshall & Orr (1956) have measured this distance in *Calanus* spp.. The lower limit of particle sizes is 5 μm. Particles larger than 50 μm cannot be filtered. This upper limit is set by the separation of some setae on the Mx_2 which guard the posterior entrance to the oral chamber. However, *Calanus* is able to spread these setae and feed on larger particles (Conover 1966).

Conover (1966) has described the process by which *Calanus hyperboreus* ingests large particles. This type of feeding may be termed "pseudo-raptorial" (Porter 1973). In this process, the Mx_2 catches large particles by spreading its guard setae, allowing the particle to slip between the setae and then, closing the setae, it flicks the particle toward the mouth. This process differs from true filter feeding insofar as large particles must be manipulated so that the greatest dimension of the particle is parallel to the long axis of the copepod's body. The Mxp, Mx_2 and Mx_1 act in concert to orient the particle. The Mx_1 is, finally, used to press the particle against the biting edges of the Md until the particle is consumed. Conover (1966) has also reported that *C. hyperboreus* can suck out the contents of large diatoms and discard the empty test (though in most cases the entire diatom was consumed).

Conover (1966) reported that, in consuming *Artemia* nauplii, *C. hyperboreus* oriented its prey using a procedure similar to that for manipulating large diatoms. However, once a nauplius was oriented, all movement of the oral appendages ceased except for the blades of the Md and the endites (arthrites?) of the Mx_1, which alternately crushed and pushed the prey into the mouth.

II. Raptorial feeding

The second mode of food-getting among Copepoda is raptorial feeding. In this type food is seized, not filtered. Raptorial feeding is characteristic of some calanoids (Lowndes 1935; Gauld 1966; Anraku & Omori 1963) and all cyclopoids (Fryer 1957a, 1957b) and the harpacticoid *Tisbe furcata* (present report).

In an excellent account, Fryer (1957b) has described the feeding behaviour and mechanisms of raptorial (carnivorous) cyclopoids. Cyclopoids seized food as it was encountered; no searching pattern was observed by Fryer. Animal prey was initially clasped between the Mx_1. The Mx_2 and Mxp joined in holding the prey if opportunity permitted. Throughout the early capture process the Mx_1 were the most important clasping organs. The distal denticles of the Mx_1 were clamped into the prey. The palps of the Mx_1 embraced the prey, and the setae of the palps were splayed over the surface of the prey. Some of these setae were sensitive to prey movement, initiating a biting response if the prey struggled to escape. The embrace of the Mx_2 and Mxp restricted the movement of active prey.

Fryer studied the feeding habits and oral morphology of a large number of cyclopoids. He indicated that it is not possible to find a one-to-one correspondence between trophic group (carnivore, herbivore, etc.) and mouth part morphology among raptorial cyclopoids.

III. Mixed feeders

Arashkevich (1968), Arashkevich & Timonin (1970), Gauld (1966) and Anraku & Omori (1963) have recognized a third feeding type among the Copepoda: the mixed feeders. This type shares characters common to both filter feeders and raptorial feeders. Gauld (1966), for example, listed calanoids in the genera *Anomalocera, Labidocera* and *Centropages* as mixed feeders. These animals may be recognized by the structure of the Mx_2 which is proximally like that of a filter feeder and distally fitted with claw-like structures appropriate for seizing large food items. Some of the studies just cited suffer from a lack of functional verification of raptorial feeding in these groups. Lowndes (1935) has shown that the calanoids in the family Centropagidae are indeed raptorial feeders.

IV. Tentative notes on the evolutionary ecology of the Copepoda

Gurney (1931) concluded from a morphological study of calanoid copepods that the family Centropagidae is of a primitive stock. As Lowndes (1935) showed, this family exhibits raptorial feeding. Prompted by these studies and his own extensive observations of cyclopoid feeding, Fryer (1957b) has speculated that present day calanoid filter feeders and cyclopoid (raptorial) feeders diverged from a common group of raptorial feeders similar to those in the family Centropagidae.

This conclusion is contrary to the opinions of Lang (1946, 1948a, 1948b) who suggested that filter feeding is a primitive character. Manton's (1964) monumental contribution to our knowledge of mandibular mechanisms in Arthropoda supports the view that filter feeding is primitive. The present essay cannot resolve this issue.

No matter where it came from, it was probably a primitive raptorial feeder that gave rise to both the cyclopoid and harpacticoid copepods. Noodt (1971) has indicated that species of the harpacticoid genus $Tisbe$ probably resemble this primitive condition best. However, the similarity between $Tisbe$ and the (primitive) cyclopoids may testify as well to convergence. The present study indicates that there are many differences in the functional oral morphology separating the cyclopoids (described by Fryer) and the harpacticoid $Tisbe$. For example, cyclopoids use their Mx_1 as the principle appendage of food capture. This seems to be the most dexterous organ on the cyclopoid head. (Calanoids also use their Mx_1 extensively: to generate currents of water to filter and in scraping food off the Mx_2 net.) In $Tisbe$ $furcata$ the Mx_2 is by far the most important and dexterous oral appendage. The Mx_1 plays a relatively minor role in $Tisbe$ feeding. Further, in cyclopoids, the palps of the Mx_1 are splayed over a food item to restrain it. In $Tisbe$ this function has passed to the A_2 and Mxp.

The Mx_1 of cyclopoids and $Tisbe$ $furcata$ are similar in one important respect: the pressure sensitivity of some of their setae. In both copepods this sensitivity is used to trigger a biting response.

It is of phylogenetic interest to note that in $Tisbe$ $furcata$ the biting response of the Mx_1 is linked to that of the Md. Further, as I have already noted, the Md and Mx_1 seem to form a compact unit both structurally and functionally. They exhibit a kind of lock and key arrangement wherein a groove on one cradles a protrusion on the other. It could be speculated that such an arrangement could lead to conservative or perhaps linked evolution of these two organs in $Tisbe$ since a morphological change in one appendage might necessitate a corresponding change in the other.

Besides morphological and functional homologies between the three orders of Copepoda, there also seem to be behavioural homologues. These fall into two categories. 1) Physical behaviour refers to the types and order of appendicular movements used when an animal is manipulating a food item. 2) Acceptance behaviour refers to the sensory apprehension of a food item and the judgement to accept or reject the item as a desirable source of nutrition. Two examples of these behaviours can be adduced from the present work. Of the first type, the manipulations used by $Tisbe$ $furcata$ in feeding are ample example. Of the second, recall the three types of food rejection exhibited by $Tisbe$. The first of these rejection events seems surely to be an example of acceptance behaviour. (The other two rejection events could simply represent an inability to masticate a particle, though I suspect this reason does not exhaust the possible significance of these two rejection events vis à vis acceptance behaviours.)

Mayr (1970) has reviewed some evolutionary issues surrounding behavioural responses. He pointed out that behaviour can be both a subject of evolution and an initiator of new selection pressures. The present study suggests that copepods in general and harpacticoids in particular have a rich potential for behavioural diversification both in food preferences and in oral manipulations. For example, Conover (1966) has shown that *C. hyperboreus* is able to crush large diatoms, suck out their contents and discard the empty frustule. As I have pointed out, *Tisbe furcata* is able to physically eject food items from its mouth. These physical behaviours may be homologous. Further, one can imagine that some such primitive, physical behaviour may have conditioned the origin and diversification of acceptance behaviours among derived species of harpacticoids by providing both a new order of selection pressures and a new character substrate for evolution.

In short, the present study suggests that the size, shape, physical behaviour and acceptance behaviour of harpacticoid feeding may all be important for the specific diversification of the order. This may be of particular importance with regard to the systematics of species in the genus *Tisbe* - - - a genus encumbered with several sets of sibling and incipient species (e.g. Volkmann 1972, 1973, 1977). It could be possible for co-occurring sibling species, for example, to exhibit different acceptance behaviours and thus avoid competitive displacement. Further, the present study suggests that one must know the three-dimensional structure of an appendage before 1) the ecological significance of the organ can be known and 2) before morphological similitude can be affirmed. For example, when describing species in the genus *Tisbe,* workers (e.g. Marcotte 1974) have followed Lang (1965) in figuring all oral appendages from their anterior surface only. Three-dimensional descriptions have not been provided. That an appendage can differ widely in shape through its three dimensions will be manifest with a glance at the structure of the important Mx_2 in *Tisbe furcata* (Figs. 6, 12 and 18). It may be important, therefore, to re-examine the morphological similarity of sibling species in light of their three-dimensional morphologies.

The ecological significance of an appendage is hopelessly obscured without reference to its three-dimensional shape and function. For example, the purpose of the robust Mx_2 of *Tisbe furcata* remained unsuspected by this author until it was seen in three dimensions and in action. *Tisbe furcata* can be a voracious carnivore capable of devouring larval nematodes in laboratory cultures (McClelland, personal communication). Garstang (1900) has reported that groups of individual *Tisbe furcata* are able to attack and eat larval fish. They do so by first eating the fins, thus immobilizing the fish and then devouring the body. This carnivorous habit may help explain the inverse relationships that have been reported between the densities of *Tisbe furcata* and nematodes (e.g. Marcotte & Coull 1974; Marcotte, study in progress).

Acknowledgements

I acknowledge with much gratitude the people of Canada for their financial support. I thank Alex Wilson and Paul Sampson for their indispensable technical assistance, and Dr. B. Volkmann for identifying the species of *Tisbe*. This essay profited from the comments of Drs. E. L. Mills, F. Wells and B. C. Coull.

Bibliography

Anraku, M. & M. Omori: Preliminary survey of the relationship between the feeding habit and the structure of the mouth parts of marine copepods. Limnol. Oceanogr. **8**, 116–126 (1963).

Arashkevich, Ye. G.: The food and feeding of copepods in the Northwestern Pacific. Oceanology (Moscow) **9**, 695–709 (1968).

–: & A. G. Timonin: Copepod feeding in the tropical Pacific. Dokl. Akad. Nauk SSSR **191** (4), 935–938 (1970).

Beklemishev, K. V.: The discovery of silicious formations in the epidermis of lower Crustacea. Dokl. Akad. Nauk SSSR **97**, 543–155 (1954).

Cannon, H. G.: On the feeding mechanism of the copepods *Calanus finmarchicus* and *Diaptomus gracilis* J. exp. Biol. **6** 131–144 (1928).

Conover, R. J.: Feeding on large particles by *Calanus hyperboreus* (Kroyer). In Some Contemporary Studies in Marine Science, Ed. by H. Barnes. Allen and Unwin, London, 187–194 (1966).

Esterley, C. O.: The feeding habits and food of pelagic copepods and the question of nutrition by organic substances in solution in the water. Un. California Publ. in Zool. **16**, 171–184 (1916).

Fryer, G.: The feeding mechanism of some freshwater cyclopoid copepods. Proc. Zool. Soc. London **129** 1–27 (1957a).

–: The food of some freshwater cyclopoid copepods and its ecological significance. J. Anim. Ecol. **26**, 161–286 (1957b).

Garstang, W.: Preliminary experiments on the rearing of sea fish larvae. J. Mar. Biol. Ass. U. K. **6**, 70–93 (1900).

Gauld, D. T.: The swimming and feeding of planktonic copepods. In Some Contemporary Studies in Marine Science, Ed. by H. Barnes. Allen and Unwin, London, 313–334 (1966).

Gurney, R.: British Freshwater Copepoda 1. Ray Society, London. pp. 211–238 (1931).

Lang, K.: A contribution to the question of the mouth parts of the Copepoda. Ark. for Zool. **38** (5), 1–24 (1946).

–: Copepoda Notodelphyoida from the Swedish West-Coast with an outline on the systematics of the copepods. Ark. för Zool. **40** (14), 1–36 (1948a).

–: Monographie der Harpacticiden. Håkan Ohlssons Bokt. Vol 1 & 2, 1682 pp. (1948b).

–: Copepoda Harpacticoidea from the California Pacific Coast. Almquist & Wiksell. Stockholm 560 pp. (1965).

Lowndes, A. G.: The swimming and feeding of certain calanoid copepods. Proc. Zool. Soc. London. 1935, 687–715 (1935).

Manton, S. M.: Mandibular mechanisms and the evolution of arthropods. Phil. Trans. R. Soc. London **B247** (737), 1–187 (1964).

Marcotte, B. M.: Two new harpacticoid copepods from the North Adriatic and a revision of the genus *Paramphiascella*. Zool. J. Linn. Soc. **55** (1), 65–82 (1974).

– and B. C. Coull: Pollution, diversity and meiobenthic communities in the North Adriatic (Bay of Piran, Yugoslavia). Vie Milieu. **24**, 281–300 (1974).

Marshall, S. M. & A. P. Orr: The Biology of a Marine Copepod. Oliver & Boyd. London. 195 pp. (1955)

–: On the biology of *Calanus finmarchicus* IX. Feeding and digestion in the young stages. J. Mar. Biol. Ass. U. K. **35**, 587–603 (1956).

Mayr, E.: Evolution und Verhalten. Verh. Dtsch. Zool. Ges. **64**, 322–336 (1970).

Noodt, W.: Ecology of the Copepoda. Smithsonian Contr. Zool. **76**, 97–102 (1971).

Porter, K. G.: The selective effects of grazing by zooplankton on the phytoplankton of Fuller Pond, Kent, Connecticut. Ph. D. Thesis Yale University. 160 pp. (1973).

Sullivan, B. K., C. B. Miller, W. T. Peterson & A. H. Soeldner: A scanning electron microscope study of the mandibular morphology of boreal copepods. Mar. Biol. **30**, 175–182 (1975).

Volkmann, B.: A new case of sibling species in the genus *Tisbe*. Fifth European Marine Biological Symposium, 67–80 (1972).

–: Étude de quatre éspèces jumelles du groupe *Tisbe reticulata* Bocquet. Arch. Zool. Exp. & Gen. **114** (3), 317–348 (1973).

–: Geographic and reproductive isolation in the genus *Tisbe* (Copepoda, Harpacticoida). Mikrofauna Meeresboden **61**, 313–314 (1977)

| Mikrofauna Meeresboden | 61 | 197—216 | 1977 |

W. Sterrer & P. Ax (Eds.). The Meiofauna Species in Time and Space. Workshop Symposium, Bermuda Biological Station, 1975.

The Relationship of Character Variability and Morphological Complexity in Copulatory Structures of Turbellaria-Macrostomida and -Haplopharyngida[1]

by

Reinhard M. Rieger

Department of Zoology, University of North Carolina, Chapel Hill, North Carolina 27514, USA

Abstract

Within- and between-population variability of shape and dimensions of copulatory stylets has been studied in 7 species of macrostomid and one species of haplopharyngid turbellarian, found in the area of the northern Atlantic Ocean. Both overall shape and size have been found to vary extremely little within and between spatially isolated populations. The coefficient of variability of stylet length ranges usually between 5 %—10 % even if geographically distant populations of one species are compared. Members of the family Dolichomacrostomidae show generally less variability of stylet length than members of the family Macrostomidae. This difference is suggested to be due to differences in stylet formation. In comparing geographically isolated populations of species with simply shaped stylets versus species with extremely complicated ones, no clear evidence could be found that more complex structures would, as a rule, vary significantly more over space and time than simple ones. This is also emphasized in the comparison of the structurally similar stylets of the different species of the macrostomid genus *Paromalostomum* and of the *Messoplana falcata* subspecies group.

A. Introduction

In many groups of Turbellaria the shape and dimensions of cuticularized copulatory structures are the most important characters for species recognition. Special studies assessing the variability or stability of such characters within single populations and between spatially separated populations are, however, essentially absent from the literature. In trying to fill this gap, the present paper focuses on variability and stability of copulatory structures within selected species of Turbellaria-Macrostomida and -Haplopharyngida from the North Atlantic Ocean. At the same time the study attempts to provide insights into the relationship between complexity of structures and the morphologic rates of change in evolution, which has received special attention in recent studies (see Schopf et al. 1975).

---

[1] Contribution No. 699 from the Bermuda Biological Station for Research, Inc.

B. Material and Methods

The material used was collected over several years on both sides of the northern Atlantic Ocean. To characterize variations of total shape of stylets, several stylets from a single population were drawn with the Wild drawing tube and are illustrated at the same scale. Measurements of various dimensions were taken from these drawings. The distances measured are either indicated with small arrows on the respective drawings or, in the case of extremely convoluted tubes (see Rieger 1971a), were measured with a distance measurer as it is used in cartography. For some of the species used in this paper, brief diagnoses are given in an Appendix (see p. 212). Only fully mature specimens were used in this study. The following species were considered (see also Fig. 1):

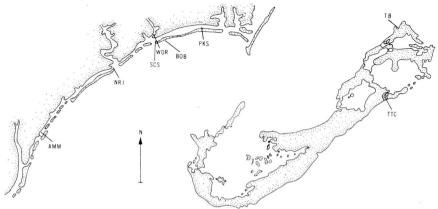

Fig. 1. Maps illustrating sample sites in North Carolina and Bermuda. In North Carolina: PKS = Pine Knoll Shores; BOB = Bogue Banks; SCS = Swansboro Coast Guard Station; WOR = White Oak River sand flat; NRI = New River Inlet; AMMF = Anne McCrary's mud flat. In Bermuda: TOB = Tobacco Bay; TTC = Tuckerstown Cove.

Macrostomum hystricinum Beklemischev, 1951, subspecies *marinum*, (n. ssp.; see Appendix) collected from:
a) North Carolina: Onslow Bay area, White Oak River (WOR) sand flat (see Crezée 1975, Oct. 1975; Pine Knoll Shore (PKS) sand flat on the inside of Bogue Banks at Pine Knoll Shore golf course, Oct. 1975; Anne McCrary's mud flat (AMM) at Wrightsville Beach (see Riedl 1970), Jan. 1970;
b) Florida Keys: in sand patches at LTL at Pigeon Key, May 1971;
c) Fiascherino, Italy: in sand from shallow subtidal on the landward side of the island Palmaria, Aug. 1967;

Paromalostomum dubium (Beauchamp, 1927) from:
Arcachon, France (see Rieger 1971b for further information);

Paromalostomum atratum Rieger, 1971 from:
Fiascherino, Italy (see Rieger 1971b for further information);

Paromalostomum sp. (see Tyler 1976) from:
Bogue Banks (BOB), North Carolina, in the open ocean beach between LTL and shallow subtidal, Aug. 1971: New River Inlet (NRI), North Carolina, at and slightly below LTL, July 1971;

Cylindromacrostomum mediterraneum (Ax, 1955) (Venice form) from:

Lido of Venice at the public beach of Alberoni at LTL and in the shallow subtidal, May 1966;

Austromacrostomum sp. from:

Rovinj, Istria (same location as type locality of *M. bistylifera* Rieger, 1968);

Paramyozonaria riegeri Sopott-Ehlers & Schmidt, 1974 (Caribbean form) from:

Carrie-Bow Cay (British Honduras), coarse *Halimeda* sand between coral heads, 2 m water depth, Dec. 1974.

Paramyozonaria bermudensis (nov. spec.) from:

a) Bermuda: Tobacco Bay (TOB) in various places in the shallow subtidal (~ 1–2 m water depth), July 1973; Tuckerstown Cove (TTC), in the shallow subtidal,

b) Florida Keys: Bahia Honda Key (same location as described for *Myozonaria jenneri* in Rieger & Tyler (1974), May 1971;

Haplopharynx quadristimulus Ax, 1971 (Carolina form) from:

Bogue Banks (BOB), North Carolina, in the open ocean beach between LTL and MTL in 10–30 cm sediment depth and on a low energy beach on the inside of Bogue Banks at the Swansboro Coast Guard Station (SCS) at HTL in 5–20 cm sediment depth, July 1971.

C. Results

Two possible sources of variability in copulatory structures were considered: 1) variation in total shape (e.g. variability in relative position and dimension of substructures), and 2) variation in size. These two aspects have been studied within single populations and between populations over various geographic distances.

1) Variability within a single population.

Within-population variability of shape of copulatory structures is extremely low, as illustrated for *M. hystricinum* (Fig. 2a-1), *C. mediterraneum* (Fig. 4a), *P. bermudensis* (Fig. 2b) and *P. riegeri* (Fig. 4c). This appears to be true for simple (e.g. *M. hystricinum*) as well as complex shapes (*C. mediterraneum* and the two *Paramyozonaria* species).

For the simple hook-shaped stylet of *M. hystricinum*, the only noticeable source of variation is in the shape and thickness of the distal hook. Other differences in shape, such as the shape of the proximal end of the stylet, are due to differences in squeezing conditions or contractions of muscles surrounding the stylet. The shortened stylet in Fig. 2f is drawn from a senile specimen.

The extraordinary consistency of shape in the three species with complex stylets is more difficult to demonstrate. Observed variations are often due to differences in squeezing conditions and to different views (e.g. ventral or dorsal) of the stylets. Eight stylets from specimens of a single population of *P. bermudensis* in Tobacco Bay, Bermuda, were carefully drawn and compared (4 of these are shown in Fig. 4b). In Fig. 5a,b two stylets of this species are drawn as viewed from the dorsal and ventral sides. The only significant structural variations independent of squeezing conditions seen in all 8 specimens were the number of tips on the distal end of the proximal spine (x in Fig. 5a), the number of circular wrinkles (y in Fig. 5b), the

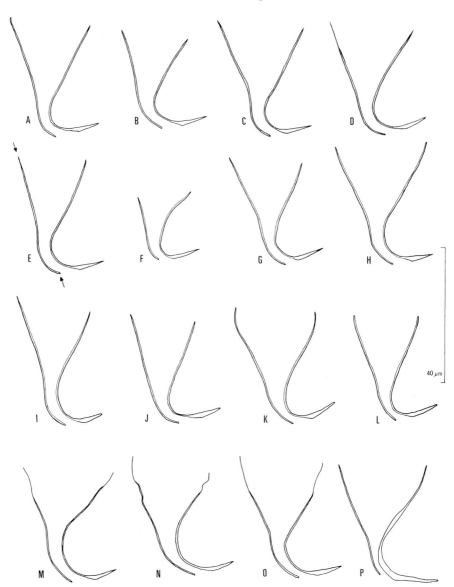

Fig. 2. Stylets of different populations of *Macrostomum hystricinum*. a-1 from a population Anne McCrary's mud flat near Wrightville Beach, N.C. m-o from a population on Pigeon Key, Florida. p from Tobacco Bay, Bermuda.

extent of vacuolization in some areas of the stylet (Fig. 5a, b), and the shape of the tip of the distal hook (w in Fig. 5a,b) and the proximal hook (z. in Fig. 5a,b). All the other features indicated in Fig. 5a,b have been found essentially unvaried. In

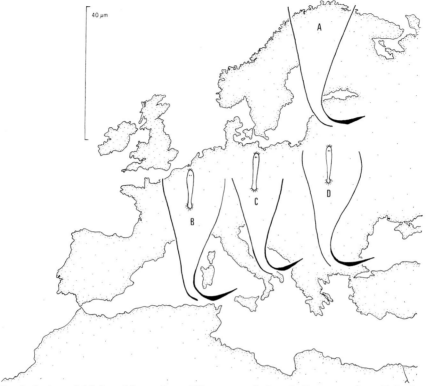

Fig. 3. Stylets of *M. hystricinum* from different populations in Europe. a from Tvärminne, Finland (After Luther 1960). b from Fiascherino, Italy. c from Venice, Italy. d from Dubrovnik, Yugoslavia.

the new form of *P. riegeri* from a single population in British Honduras, the most significant variation in shape is seen in the number (4–9) and shape of the proximal hooks (see Fig. 5c). From looking at this figure, it might appear that the amount and position of the various folds in the distal stylet area is much more variable than in *P. bermudensis*. With the exception of the variations seen in number of proximal hooks (4–9), however, these differences are correlated with different squeezing conditions. Furthermore, whereas each stylet of *P. bermudensis* in the Tobacco Bay population in Bermuda was studied for about half a day, all four stylets available from the *P. riegeri* populations in British Honduras were examined for a total of 4 hours. This should indicate that in cases of extremely complex stylets, longer and more careful observations show that the extent of variation is actually much smaller than seen at a first brief examination.

Similarly, significant variations in shape are also essentially absent in other macrostomid species with complex stylets (see Fig. 4a for *C. mediterraneum*, and Rieger 1971b).

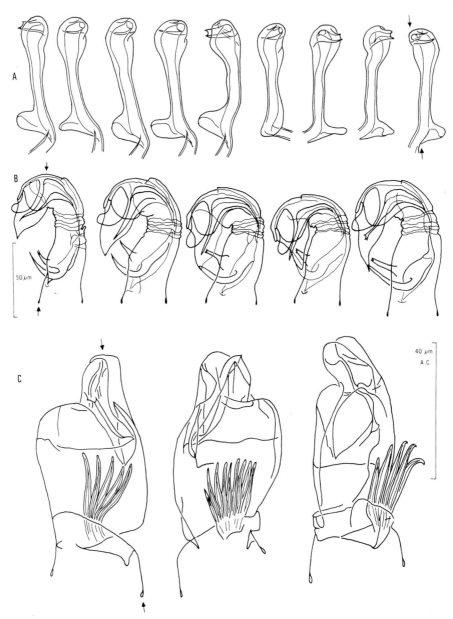

Fig. 4. a: Straight distal end piece of penis stylet of *Cylindromacrostomum mediterraneum* from a population in Venice, Italy. b: Penis stylets from *Paramyozonaria bermudensis*. The first four were drawn after specimens from Tobacco Bay population, Bermuda; the last one on the right was drawn from a population at Tuckerstown Cove. c: Penis stylets of *Paramyozonaria riegeri* from British Honduras.

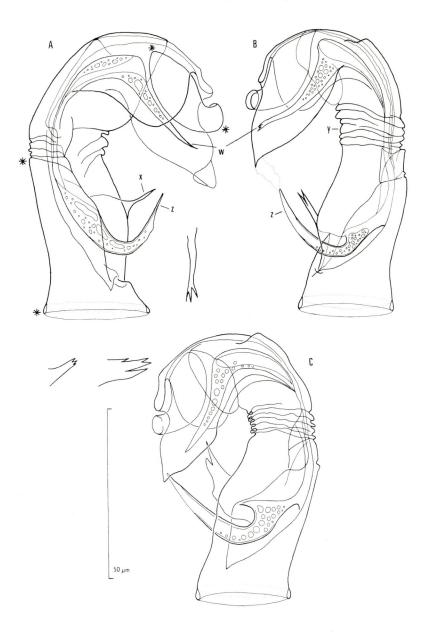

Fig. 5. Penis stylets of *Paramyozonaria bermudensis*. a,b after two specimens from Tobacco Bay, Bermuda, seen from different sides. c after a specimen from Bahia Honda Key, Florida. The insets next to a indicate differences in the number of tips on the proximal spine (x) and the distal hook (w). For further explanation, see text.

Variation in shape appears generally more noticeable in cuticularized parts of the female system (Fig. 6). But again, most of the illustrated differences can be explained by squeezing conditions and differences due to different views on these structures. However, the middle piece of the bursal copulatory apparatus in *P.*

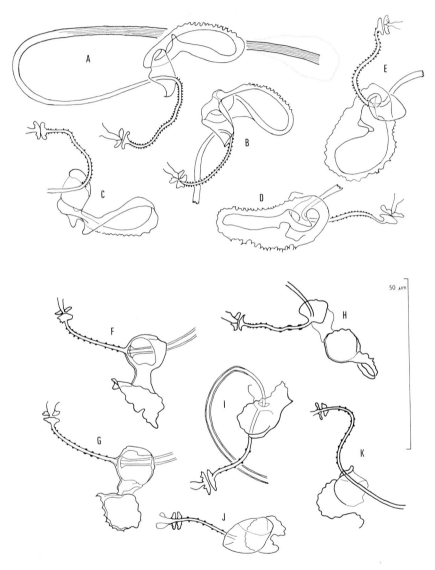

Fig. 6. a–e: Copulatory apparatus of bursal organ of *Paramyozonaria riegeri* from British Honduras. f–k: Copulatory apparatus of bursal organ of *Paramyozonaria bermudensis* from the Tobacco Bay population.

bermudensis shows rather distinct structural variations if viewed under similar squeezing conditions (Fig. 6f,g and 6h). In *P. riegeri*, much less variation was found in the same structure (see Fig. 6a,b,c).

Thus overall, the shape of copulatory structures in marine macrostomids is extremely stable within single populations, whether one considers a simple or a complex form.

In looking at within-population variability in dimensions (total stylet length, or measurements of parts), it was found in all cases studied here that the coefficient of variability (Mayr 1969) is always less than 10 % (see Table 1), again independent of the absolute length and the complexity of the stylets. Complex three-dimensionally wound tubes such as the penis stylet in Dolichomacrostominae show the same relative amount of variation in total stylet length as does the simple hook-shaped stylet of *M. hystricinum* Table 1). A complex stylet should be expected to show a greater amount of length variation than a simple one, due to the observational error induced by studying squeeze preparations of three-dimensional structures. Since this is not the case in the data shown in Table 1, it can be concluded that the actual amount of variation in dimensions of the complex stylets in dolichomacrostomids must be less than that in the simple stylet of *M. hystricinum*. In order to prove this, measurements of certain parts of the complex forms were made which are not affected by the amount of squeezing. In *P. bermudensis* and *C. mediterraneum*, such measurements have a coefficient of variability of around or less than 5 %, which is almost half of the variation seen in total stylet length. As I have shown earlier, dolichomacrostomids have a relatively shorter period of stylet growth during development than do representatives of the family Macrostomidae (see Rieger 1971b). In addition, dolichomacrostomids show a clear "end differentiation" (proximal circular swelling) when stylet formation is completed, and stylet length does not change afterwards with age or starvation. In *M. hystricinum*, on the other hand, one can clearly observe during starvation and aging a shortening of the stylet from the proximal end (see the one short stylet in Fig. 2f). Thus variations in dimensions of stylets in macrostomids are not related to complexity of structure but to the mode of stylet formation.

2) Variability within spatially separated populations.

In *M. hystricinum*, specimens were compared from spatially separated populations in the Onslow Bay area (see Fig. 1 PKS, WOR, AMM, NRJ, SCS), from Pigeon Key, Florida, from Bermuda, from Tvärminne, Finland, from Fiascherino and Venice in Italy and from Dubrovnik, Yugoslavia. No significant variations in total shape were detected in the copulatory stylets of the five populations in the Onslow Bay area or from the three populations in Europe (see Figs. 2, 3). This is best expressed by the statement that it is clearly impossible to separate these populations on the basis of stylet shape. The four specimens collected of a population on the Florida Keys all appeared consistently different in the amount of cuticularization at the proximal end of the stylet (see Fig. 2m–o). The single specimen

Table 1

species and locality	date collected	dimension	number of individuals	M̄	S.D.	coefficient of variability S.D. × 100 / M̄
Macrostomum hystricinum						
—U. S. Atlantic coast:						
Pine Knoll Shores, NC	9/75	total stylet length	21	33.48	2.79	8.34
White Oak River, NC	9/75	total stylet length	8	33.63	2.73	8.12
Anne McCrary's mud flat, NC	10/70	total stylet length	9	36.78	2.86	7.78
Pigeon Key, FL	5/11	total stylet length	4	41.00	1.63	3.98
—Europe:						
Fiascherino, Italy	8/67	total stylet length	4	35.75	2.50	6.99
Paramyozonaria bermudensis						
Tobacco Bay, Bermuda	7/73	total stylet length	8	78.13	6.73	8.61
Tobacco Bay, Bermuda	7/73	distal horn*	8	29.63	1.60	5.40
Tobacco Bay, Bermuda	7/73	proximal part*	8	47.00	2.62	5.57
Paromalostomum sp.						
Bogue Banks, NC	8/70	penis stylet, total	22	167.64	12.19	7.27
Bogue Banks, NC	8/70	acc. stylet, total	22	78.18	6.89	8.81
New River Inlet, NC	7/71	penis stylet, total	5	161.80	9.76	6.03
New River Inlet, NC	7/71	acc. stylet, total	5	78.00	3.08	3.95
Paromalostomum atratum						
Fiascherino, Italy	8/67	penis stylet, total	10	132.90	5.46	4.11
Fiascherino, Italy	8/67	acc. stylet, total	10	55.2	2.20	3.99
Paromalostomum dubium						
Arcachon, France	9/68	penis stylet, total	13	183.62	9.72	5.29
Arcachon, France	9/68	acc. stylet, total	13	103.92	5.28	5.08
Cylindromacrostomum mediterraneum						
Venice, Italy	8/66	penis stylet, total	10	153.7	9.49	6.17
Venice, Italy	8/66	acc. stylet, total	10	135.8	9.84	7.25
Venice, Italy	8/66	penis stylet, distal end part*	9	34.56	0.88	2.55
Austromacrostomum sp.						
Rovinj, Istria	3/69	penis stylet, total	8	198.75	11.25	5.67
Rovinj, Istria	3/69	acc. stylet, total	8	168.87	5.79	3.13
Haplopharynx quadristimulus						
Bogue Inlet mud flat, N. C.	6/71	main stylet, total	5	50.0	3.08	6.16
		acc. stylet, total	5	41.8	1.79	4.28

* distances marked in drawings

obtained from Bermuda also appeared different but here in the amount of cuticularization of the distal part of the stylet (see Fig. 2n).

For looking at between-population variability of complex stylets, the following information is available. Aside from the Tobacco Bay population of *P. bermudensis* on Bermuda, two specimens of geographically separated populations are available for comparison. One specimen was found in the shallow subtidal of Tuckerstown Cove (TTC), an area within Castle Harbor Basin which is far away from Tobacco Bay (TOB) and on the other side of the island (see Fig. 1). The second specimen was obtained from the shallow subtidal at Bahia Honda Key, Florida. In both cases there is no significant difference in the total shape of the stylet if compared with the Tobacco Bay population (compare Fig. 5b and 5c); that is, within-population variability is greater than between-population variability. This lack of significant interpopulation variability in stylet shape over large geographic ranges in dolichomacrostomids has been emphasized before for several species (*P. dubium, M. bistylifera, P. fusculum*: see Rieger 1971a,b). In *P. riegeri* the stylets of specimens from the population in British Honduras (Fig. 4c) appear to be more different from those of specimens from the type locality on Galapagos than in the species mentioned above (see Sopott-Ehlers & Schmidt 1974). With the exception that the number of proximal hooks range between 8 and 12 in the Galapagos specimens, and from 4 to 9 in the British Honduras specimens, the differences seen in comparing the Figures 4c and 6a–d of this paper with Figures 3b and c of Sopott-Ehlers & Schmidt's paper are definitely due more to the different drawing techniques than to any actual structural differences. Closer re-examination of these two populations will be necessary to clearly establish differences between within- and between-population variability.

In summary, one can say that the overall shape of stylets seems remarkably unvariable between spatially separated populations, independent of complexity and even if the likelihood of any genetic exchange is extremely low (e.g. *P. bermudensis* from Florida and Bermuda).

The present material suggests, however, some interesting changes of total stylet length over spatially distant populations (see Table 1). In the case of *M. hystricinum*, a t-test comparing the mean stylet length of populations from Pine Knoll Shores and White Oak River (both in the Bogue Sound area) indicated that there was no significant difference in the mean. However, a similar comparison between the populations from White Oak River and Pine Knoll Shores with the population from Wrightsville Beach indicated a significant difference at an a level of 0.005; the same was found in comparing the Bogue Sound populations with the few specimens of the Florida population. Although more measurements will be needed to substantiate these indications, the present material suggests a gradual increase of stylet size with decrease in geographic latitude. A similar phenomenon was noted by Rieger (1971c) for *Bradynectes sterreri*.

For complex stylets of dolichomacrostomids, Rieger (1971a,b) has pointed out some stylet length variation in geographically distant populations in *M. bistylifera*, *P. fusculum* and *P. dubium* (see also Appendix, p. 212). However, their statistical significance cannot be evaluated due to the scarcity of data. Two populations of a new species of the genus *Paromalostomum* from North Carolina showed no indication of a significant difference in the $\overline{M}$ at a small geographic scale (see Table 1). Also, in *P. riegeri* the stylets of the four specimens from British Honduras were 70, 73, 75 and 75 μm long, which closely approaches the average value given for the Galapagos population of "about 80 μm". In *Paramyozonaria bermudensis*, the one specimen from Tuckerstown had a stylet of 91 μm length, which was only slightly larger than any of the measurements in the Tobacco Bay populations, whereas the stylet of the specimen from Florida measured 79 μm, almost exactly the $\overline{M}$ value for the Tobacco Bay population.

If all stylets of different populations of what can be considered a species on the basis of equality of shape of stylet are lumped together, the coefficient of variability for one species may not change at all from the within-single-population value (e.g. *P. bermudensis*) or may increase (e.g. *M. hystricinum*). Even in the latter case the coefficient of variability for one species remains close to 10 %, that is, only slightly above the value characteristic for within-population variability.

Thus, the data suggest that between-population variability in total shape is insignificant, and also that variability is not proportional to the complexity of structures. This is particularly obvious in the stylets of *M. hystricinum* and *P. bermudensis* in the same geographic range from Florida to Bermuda.

In terms of using copulatory stylets as a diagnostic feature for species recognition, the data presented here suggest that in overall stylet shape, spatially separated populations show no significant difference, but that total length of stylet may vary to some degree with geographical distance. Dependent upon the mode of stylet formation, the coefficient of variability seems generally higher in single populations of the family Macrostomidae (and most likely also the Microstomidae) than in the Dolichomacrostomidae.

3) Geographic ranges of North Atlantic macrostomid and haplopharyngid species.

In using equality of shape of copulatory stylets but allowing some small differences in total length of stylet (around 12 %) for morphological recognition of macrostomid species (or groups of sibling species) in the North Atlantic, the following picture of geographic species ranges emerges. Within the Macrostomidae (see *M. hystricinum*) and most likely in Microstomidae and Haplopharyngida, amphiatlantic distribution patterns are not uncommon. The Haplopharyngida (see Karling 1974) are clearly the most closely related group to the Macrostomida (see Tyler 1976).

I have several observations on the species *H. rostratus* from Europe (Rovinj, Istria; Fiascherino, Italy; Portaferry, Ireland; Robin Hood's Bay, England), and from the American Atlantic coast (off Beaufort, N. C., and Bermuda). Though more careful study is necessary, the material so far suggests only extremely small

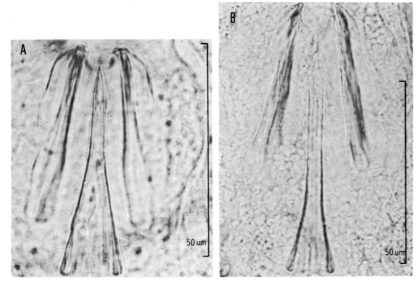

Fig. 7. Male copulatory apparatus of *Haplopharynx quadristimulus*. a after specimen from a sandflat at the Swansboro Coast Guard Station, N.C. b after specimen from the high energy beach on Bogue Banks.

differences between populations on the two sides of the Atlantic, at least in the structure of the copulatory stylet. Ax (1971a) described a new species, *H. quadristimulus*, from France. This species is distinctly different from *H. rostratus* in the construction of the copulatory organ. With this report one could speculate (see Ax 1971a) that from the amphiatlantic species of *H. rostratus*, the new species *H. quadristimulus* may have been derived from speciation processes in the European Mediterranean. However, similar to *H. rostratus*, I have also found *H. quadristimulus* in intertidal beaches of North Carolina (Fig. 7). There appear to be slight but significant differences between the North Carolina and the French form, as well as between populations in N. C.; however, there is no question that the N. C. specimens are extremely similar morphologically to the French form of this species. Thus, both species (or groups of sibling species) of this genus appear to have an amphiatlantic distribution pattern and with this an extremely wide range of distribution of one very similar phenotype, in spite of the fact that the complexity of the compared structures is considerably greater in this genus than in the amphiatlantic Macrostomidae (e.g. *M. hystricinum*) and Microstomidae.

On the other hand, morphologically distinct species in the Dolichomacrostomidae are restricted in their distribution to either side of the Atlantic. The number of morphologically distinct species in the amphiatlantic genus *Paromalostomum* is very high (14 on the American coast and 10 on the European coast; see Figs. 8, 9). This high number of distinct species seems to fit well into the concept that complex

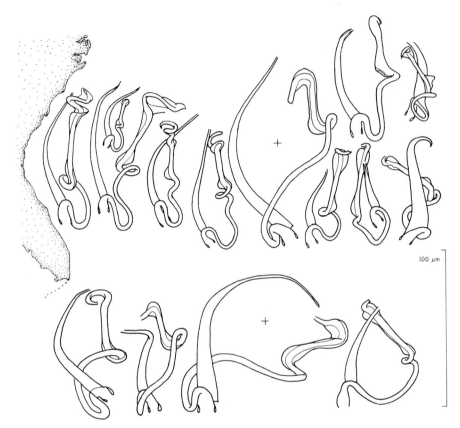

Fig. 8. Male copulatory apparatus of different species of the genus *Paromalostomum* in North Carolina (A) and Florida (B). Each stylet represents a different species, except the two stylets marked * (= one species); all species are new and will be desbribed at a later date.

structures have a higher probability of change than simple structures (see Schopf et al. (1975). However, I would like to mention that *Messoplana falcata*, a marine typhloplanoid with world-wide distribution (North Sea, Mediterranean, North Carolina and Galapagos) shows much less morphological proliferation in the northern Atlantic, although in terms of complexity of its stylet *M. falcata* can be regarded as equal to *Paromalostomum* (both with two tubes, each of which has distinct regions; see Figs. 8, 9, 10). So far only two subspecies are distinguished in this species, one from Europe and one from Galapagos (Ehlers & Ax 1974). The North Carolina specimens are extremely similar to the Galapagos form, but they may represent another subspecies based on the structure of the bursa mouth piece. It could be argued that the genus *Messoplana* contains three more Atlantic species (see Ax 1971b, Den Hartog 1966) which also should be considered in such a comparison of copulatory stylets with the genus *Paromalostomum*. This, however, must be re-

Fig. 9. Male copulatory apparatus of different species of the genus *Paromalostomum* in Europe. a and c: *P. dubium* from Sylt and Arcachon. b: *P. fusculum* from Sylt and Robin Hood's Bay. d: *P. massiliensis* from Marseille. e and f: New undescribed species from Marseille area (material collected by Dr. M. Brunet). g: *P. atratum* (Fiascherino). h: *P. minutum* (Venice), *P. parvum* (Venice). j: New species from Tunisia (collected by Dr. M. Crezée). This latter species is very similar to (*P. dubium* but differs in the accessory stylet, clearly suggesting it to be a different species.

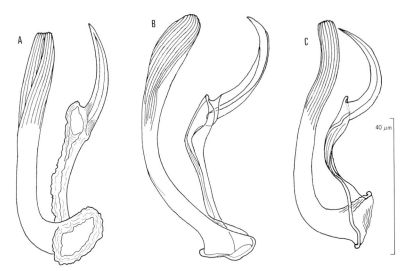

Fig. 10. Male stylet of the typhloplanoid turbellarian, *Messoplana falcata*. a after specimen from Sylt, Europe (after Ehlers & Ax 1974). b from Anne McCrary's mud flat, N.C. c from Galapagos (after Ehlers & Ax 1974).

jected since the amount of difference in stylets between the *M. falcata* subspecies group and the other species of this genus is only comparable to the amount of difference in stylets of different genera within the subfamily Dolichomacrostominae. Thus, it appears that the observed rates of morphological change seen as a function of zoogeographic distribution pattern (Sterrer, 1973) are not necessarily proportional to structural complexity. The stylets of *Messoplana falcata* and *Paromalostomum* appear to be an ideal system to further investigate the question of the relationship of morphological complexity and the speed of morphological differentiation over geologic time.

Acknowledgements

I would like to thank Dr. Sterrer for providing material of *Paramyozonaria riegeri* (British Honduras) and *P. bermudensis* (Tuckerstown Cove). Financial support was provided by NSF grants # DES 74-23781 and DES 75-20227 as well as a grant from the Exxon Corporation (Smithsonian IMSWE Contribution # 31). The drawings were executed after my designs by Ms. Mary White.

Appendix

Macrostomum hystricinum Beklemischev, 1951 subspecies *marinum* n.s.sp. Based on a number of characters but particularly on the structure of the stylet, one can distinguish a closely related species group in the genus *Macrostomum*. The following species belong to this group: *M. hystricinum* Beklemischev, 1951,

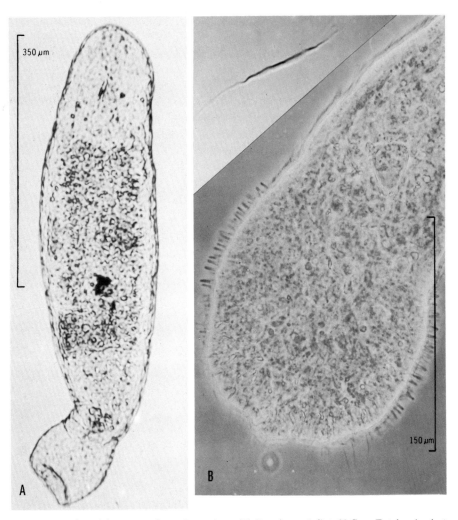

Fig. 11. *M. hystricinum*, specimen from Anne McCrary's mud flat, N.C. a Total animal. b: Enlargement of tail plate. Inset = sperm.

M. beaufortensis Ferguson, 1937, *M. peteraxi* Mack-Fira, 1971, *M. rubrocinctum* Ax, 1951, and *M. pusillum* Ax, 1951. *M. rubrocinctum* and *M. beaufortensis* are characterized by a ring of red pigmentation in the epidermis just anterior to the brain. *M. pusillum* is particularly characterized by the presence of long sensory hairs all the way surrounding the lateral margins of the animal. Although sensory hairs do occur all over the body in the other species as well, there is a clear concentration of the sensory hairs at the anterior tip and the post-adhesive plate in the other species. *M. peteraxi* appears in many features similar to *M. hystricinum*, but the drawings of the stylet of this species are too imprecise to allow any real comparison. From my sampling in Europe and in North America, I know that on sheltered beaches and in the shallow subtidal off fine sand flats there occur two extremely similar *Macrostomum* species. The larger one resembles *M. hystricinum*; the smaller one resembles *M. pusillum*. In this paper I have used stylet data of the larger species. Based on the complete identity of the copulatory stylet and based on the agreement in measurements of the various rhabdites and the construction of the reproductive system, I have identified this form as a new subspecies of *M. hystricinum*. There are differences in size and in distribution (I have never found this larger species at average salinities lower than 25⁰/oo which necessitate a distinction from the brackwater form of *M. hystricinum*). I include here a brief diagnosis of this subspecies, which I call *M. hystricinum marinum* form. A more detailed comparison with *hystricinum* will be given at a later date.

Diagnosis. Mean size of mature specimens: 0.5—1 mm. With two eyes, sometimes with brownish pigment granules in the epidermis. Body length to body width ~4(5) : 1. With a distinct adhesive plate. Sensory hairs concentrated on the anterior end, here 5—10 μm long, and along the margin of the adhesive plate (here often in groups and about 20 μm long). Rhammites 5—10 μm long, slightly more tipped on the posterior end. Rhabdites ± rounded on both ends, usually 1 μm thick and 4—10 μm long, particularly concentrated on the dorsal epidermis of the adhesive plate. Glands of the pharyngeal gland ring of two kinds: one with tiny (< 1μm) round secretory granules and one with 2—5 μm long slender sticks. Adhesive papillae average 8 μm in length. Arrangement of adhesive papillae: the papillae form a U-shaped field on the adhesive plate. On the two lateral sides most of the papillae point laterally; these two lateral fields are connected by a broad band of ventrally pointed papillae along the posterior margin of the adhesive plate. The relative size of the adhesive plate can vary somewhat within and between populations. Although the data are not conclusive, it appears that the plate is generally somewhat smaller in the European population than in the North American populations. Similarly, the total number of adhesive tubules seems to be higher in the American populations (~300 in adults) than in Europe (~100 in one count of a species from Fiascherino, Italy). Sperms without lateral processes. In the North American populations sperm 30 μm long. The subspecies appears to feed exclusively on diatoms. For the brackish water form of this species, I suggest the name of *M. hystricinum hystricinum.*

Cylindromacrostomum mediterraneum (Ax, 1955) (Venice form).

In the original description of this species from Banyuls-sur-Mer, France, Ax (1955) reports the length of the accessory stylet to be 42 μm; later Ax (1959) identifies the same species in the Black Sea, but here the accessory stylet was reported to be 172 μm long. I have studied specimens from two populations, one from Venice and one from Marseille (courtesy Dr. M. Brunet). As shown in Table 1, the mean stylet length in the Venice specimen was 153.7 (penis stylet) and 135.8 (accessory stylet). The stylets of the one specimen from Marseille were 175 (penis stylet) and 158 (accessory stylet). Based on the details of the construction of female and male copulatory apparatus, the Marseille specimen definitely belongs to the same species as the Venice population. Since the construction of all parts of the bursal copulatory apparatus is not completely known in the population from Banyuls-sur-Mer and from the Black Sea, the species identity with the other two populations is only based on basic shape equality of the male copulatory apparatus. If further studies prove the identity of shape also in the bursal apparatus, then *C. mediterraneum* would be an interesting species to investigate further with respect to size variations between local populations.

Paramyozonaria bermudensis nov. spec.

Diagnosis. Without eyes, stretched and unsqueezed between 1–1.5 mm long and 0.2 mm wide (in region of ovary). The shape of the cuticular parts is diagnostic for this new species (see Figs. 5a, c and 6f–k). A more detailed description is included in Rieger (in preparation).

Type locality. Bermuda, Tobacco Bay, shallow subtidal in 1–2 m water depth; in sand free of sea grass at several locations within the Bay.

Type material. Deposited under the number AMNH 905–909 in the American Museum of Natural History, New York.

Paramyozonaria riegeri Sopott-Ehlers & Schmidt, 1974 (Caribbean form)

Only squeezed specimens are available from the British Honduras population. Therefore, a comparison of the total length of the animals with the values of the original description is not very meaningful. The squeezed specimens from British Honduras ranged in length from 1.5–2.5 mm with a maximum width of 0.25–0.4 mm. The species identification is based on the agreement in shape and measurements of the penis stylet and the bursal copulatory structure. The number of proximal hooks on the stylet appears lower (4–9) in the British Honduras material.

Haplopharynx quadristimulus Ax, 1971 (Carolina form)

Mature unsqueezed specimens from the high energy beach at Bogue Banks and from the Neuse River Inlet measured 2–3 mm in length, whereas specimens collected at a sand flat near the Swansboro Coast Guard Station on the inside of Bogue Banks measured generally only 1 mm in length. As in the specimens from France, all North Carolina specimens have paired seminal vesicles and only 4 accessory spines in the male copulatory organ. The length of the stylet (about 50 μm in the Swansboro population and 60 μm in the Bogue Banks and Neuse River populations) and the accessory spines (about 40–50 μm) is distinctly less than in the Tunisian form (80–82 μm stylet, 65–67 accessory spines). Also, the proximal endings of the accessory spines are less spatulated in the North Carolina specimens. In addition to the difference in total body size between the Swansboro population and the Bogue Banks and Neuse River populations, these two populations appear also slightly different in the shape of the main stylet (compare Fig. 7a, b).

Bibliography

Ax, P.: Die Turbellarien des Eulitorals der Kieler Bucht. Zool. Jhb. Syst. **80**, 277–378 (1951).

–: Studien über psammobionte Turbellaria Macrostomida. III. *Paromalostomum mediterraneum* nov. spec. Vie Milieu **6**, 67–73 (1955).

–: Zur Systematik, Ökologie und Tiergeographie der Turbellarienfauna in den Ponto-Kaspischen Brackwassermeeren. Zool. Jhb. Syst. **87**, 43–184 (1959).

–: Neue interstitielle Macrostomida (Turbellaria) der Gattungen *Acanthomacrostomum* und *Haplopharynx*. Mikrofauna Meeresboden **9**, 1–14 (1971a).

–: Zur Systematik und Phylogenie der Trigonostominae (Turbellaria-Neorhabdocoela). Mikrofauna Meeresboden **4**, 1–84 (1971b).

Beklemischev, W.: Die Arten der Gattung *Macrostomum* (Turbellaria, Rhabdocoela) der Sowjetunion. Zschr. Moskauer Naturwiss. Ges. **56**, 31–40 (1951).

Crezée, M.: Monograph of the Solenofilomorphidae (Turbellaria-Acoela). Int. Revue ges. Hydrobiol. **60** (6), 769–845 (1975).

Den Hartog, C.: A preliminary revision of the *Proxenetes* group (Trigonostomidae. Turbellaria). X. Proc. Ned. Akad. Wetensch. **69**, 155–163 (1966).

Ehlers, U. & Ax, P.: Interstitielle Fauna von Galapagos. VIII. Trigonostominae (Turbellaria, Typhloplanoida). Mikrofauna Meeresboden **30**, 1–33 (1974).

Ferguson, F. F.: The morphology and taxonomy of *Macrostomum beaufortensis* n. sp. Zool. Anz. **120**, 230–235 (1937).

Karling, T.: On the anatomy and affinities of the turbellarian orders. In: Biology of the Turbellaria: N. W. Riser & M. P. Morse, eds. McGraw-Hill, N.Y. 1–16. (1974).

Luther, A.: Die Turbellarien Ostfennoskandiens, Part I. Fauna Fennica 7, 1–155 (1960).

Mack-Fira, V.: Deux Turbellariés de la mer noire. Rev. Roumaine Biol. Série de Zoologie **16**, 233–240 (1971).

Mayr, E.: Principles of Systematic Zoology. McGraw-Hill, N. Y. 428 pp. (1969).

Riedl, R.: On *Labidognathia longicollis* nov. gen. nov. spec. from the West Atlantic Coast. Int. Rev. ges. Hydrobiol. **55**, 227–242 (1970).

Rieger, R. M.: *Myozonaria bistylifera* nov. gen. nov. spec.: Ein Vertreter der Turbellarienordnung Macrostomida aus dem Verwandtschaftskreis von *Dolichomacrostomum* Luther mit einem Muskeldarm. Zool. Anz. **180**, 1–22 (1968).

–: Die Turbellarienfamilie Dolichomacrostomidae nov. fam. (Macrostomida). I. Teil. Zool. Jhb. Syst. **98**, 236–314 (1971a).

–: Die Turbellarienfamilie Dolichomacrostomidae Rieger. II. Teil. Zool. Jhb. Syst. **98**, 569–703 (1971b).

–: *Bradynectes sterreri* nov. gen. nov. spec., eine neue psammobionte Macrostomide (Turbellaria). Zool. Jhb. Syst., 205–235 (1971c).

–: Die Turbellarienfamilie Dolichomacrostomidae Rieger, 1971. III. Teil. Zool. Jhb. Syst. (in prep.).

– & S. Tyler: A new glandular sensory organ in interstitial Macrostomida. I. Ultrastructure. Mikrofauna Meeresboden **42**, 1–41 (1974).

Schopf, T. J. M., D. M. Raup, S. J. Gould & D. S. Simberloff: Genomic versus morphologic rates of evolution: influence of morphologic complexity. Paleobiolgy 1, 63–70 (1975).

Sopott-Ehlers, B. & P. Schmidt: Interstitielle Fauna von Galapagos. IX. Dolichomacrostomidae. Mikrofauna Meeresboden **34**, 1–19 (1974).

Sterrer, W.: Plate tectonics as a mechanism for dispersal and speciation in interstitial sand fauna. Neth. J. Sea Res. 7: 200–222 (1973).

Tyler, S.: Comparative ultrastructure of adhesive systems in the Turbellaria. Zoomorphologie **84**, 1–76 (1976).

Mikrofauna Meeresboden	61	217–230	1977

W. Sterrer & P. Ax (Eds.). The Meiofauna Species in Time and Space. Workshop Symposium, Bermuda Biological Station, 1975

Causal Aspects of Nematode Evolution:
Relations between Structure, Function, Habitat and Evolution

by

Franz Riemann

Institut für Meeresforschung, D-2850 Bremerhaven, Fed. Rep. Germany

Abstract

In reconstructing a phylogeny the distinction of primitive and derived characters is essential. Since the more traditional methods in ascertaining primitive structures of nematodes (e.g. by using the criteria of systematic and ontogenetic character precedence) have yielded controversial results, an attempt is made here to introduce a causal analytical approach. The enormous ecological success of nematodes is referred to their serpent-like locomotion, made possible by the evolution of a pressure-resistant flexible cuticle, a condition which results in a generation of an internal pressure as found by Harris & Crofton (1957). Nematodes without supporting structures in their cuticle are considered as primitive. These nematodes (*Trichodorus* spp. and similar nematodes) have a unique spicular apparatus; the supposed mechanism of their spicule protrusion cannot be associated with a high turgor pressure in the body cavity. It is concluded, therefore, that if the turgor pressure resistance of the nematode cuticle is a measure for evolutionary progress, the terrestrial and limnetic taxa Trichodoroidea, Diphtherophoroidea, Onchulinae and *Tripyla* are the most primitive nematodes, as has been independently postulated by other authors and with other arguments. – In addition, the oldest bifurcation step in the phylogeny of nematodes is examined, and the Adenophorea/Secernentea and the Enoplia/Chromadoria concept discussed.

A. Phylogenetic criteria

If we accept the so-called method of reciprocal illumination as a sound basis for phylogenetical considerations, and if we agree that this definition by successive approximation, in spite of logical difficulties, does not involve vicious circularity (see Hull 1967) then we can try to reconstruct the phylogeny of an organism group in the following way:

1. We recognize transformation series of homologous characters (see Osche 1973, but compare critical discussions by Peters 1973, Peters & Gutmann 1973).

2. We try to determine within these transformation series the direction of the evolutionary course and distinguish primitive, or conservative or plesiomorphic (Hennig) characters on one side and derived, progressive or apomorphic characters on the other.

3. Then, in a puzzle-like manner, we determine the relative chronological sequence of the appearance of the apomorphic characters, so we can construct a dendrogram of the hierarchy of taxa reflecting the "recency of common ancestry" (Bigelow 1956, in Hennig 1966).

The second task — judgment of the direction of the evolutionary progression (in German: Lesrichtungsbestimmung) — is the most difficult and problematic task in phylogenetic work (see Illies 1967, Peters & Gutmann 1971), and Gutmann, the pugnacious German phylogeneticist, in a recent paper (1973, see also Franzen et al. 1973) denies at all the possibility of getting a solution by means of using certain socalled phylogenetic rules. His reasoning introduces intuition leading to biomechanical models the values of which, in terms of adaptation and selection, are then discussed.

I think that causal aspects in reconstructing the nematode evolution are useful to supplement comparative morphological studies, but before giving examples I want to briefly list some more traditional phylogenetic methods and their respective success in nematology.

1. "Criterion of systematic character precedence". We search for the relatives of nematodes, for the next related sister-group in Hennigean terms and for the higher category which embraces the nematodes. Homologous characters common to Aschelminthes are to be regarded as primitive in nematodes — this corresponds to the criterion of systematic character precedence of typologists (see Remane 1952, p. 154 and the discussion in Brundin 1972, p. 117). Gastrotrichs are the next related sister-group of nematodes (Teuchert 1968, p. 412, see also the extensive comparison by Paramonov 1968); accordingly, gastrotrich characters in nematodes have been considered as indicating primitiveness (Inglis 1962, p. 217). In using such a criterion we rely on the credibility of an earlier accepted system. However, the taxonomic status of aschelminths is even today the object of controversies, and I want to remind of the ingenious theory of Rauther (1909) that nematodes could be highly degenerated neotenic descendants of terrestrial arthropods. Rauther first called the attention to the free combination of many single nematode characters and concluded (p. 515) that this is an expression of the fact that the nematode ancestors had a more complete assemblage of these characters. Continuing this kind of reasoning including kinorhynchs, gastrotrichs, tardigrades, pentastomids and advanced arthropods the author assumed a higher organized arthropod as common ancestor (p. 587–588). Of course, the relationship of nematodes and gastrotrichs would appear in quite another perspective if we adopt Rauther's theory in contrast to the presently accepted system.

2. "Criterion of geological character precedence". We know only few fossil nematodes. The oldest, to my knowledge, is *Nemavermes* from the Illinois Pennsylvanian (Oberes Karbon) described by Schram (1973). These fossil nematodes are nearly useless for comparative morphological considerations, but we should not be

too sad about this fact. The value of fossils for phylogenetical reconstructions is often overestimated (see Schlee 1971, p. 38, 45; Franzen et al. 1973).

3. "Chorological criteria". Global distribution patterns, to my knowledge, have not been considered in context with nematode evolution. It would, however, be promising to look for vicariance patterns in geographical respect (see Hennig 1966), but also in ecological respect concerning the colonization of different habitats. Why, for instance, are terrestrial invertebrates the hosts to numbers of adult parasitic nematodes, in contrast to marine invertebrates in which adult nematodes have practically never been found (Osche 1966)?

The habitat of nematodes has sometimes been discussed in context with the origin of the class. Marine nematodes have been considered as primitive (last Maggenti 1971, Gadea 1972) sometimes in analogy to other animal groups. Fenchel & Riedl (1970) and Boaden (1975, 1977) regard sulfidic sediment layers as an archaic habitat, refugium for a primitive fauna. The first papers on nematodes (Ott 1972, 1972a; Wieser et al. 1974) in this environment revealed members of different taxa, first of all Chromadorida, a group which is considered as primitive by Inglis (1962); however, I think that a conclusive statement cannot be made as yet, except that modern groups (according to our presently accepted system) are missing, e.g. the Enoploidea. I want to remind of another anaerobic habitat, the masses of decaying seaweed. The monhysterids (*Monhystera disjuncta* Bastian, 1865) living here are probably not the most primitive nematodes, but they are standing near the root of a highly expanding taxon, the Rhabditida (compare Riemann 1968). Micoletzky (1922, p. 111) assumed that the primitive nematodes were saprozoa.

4. "Criterion of ontogenetic character precedence" (see Remane 1952, Hennig 1966, Osche 1965/1966). This has been by far our most valuable criterion in recognizing primitive characters in nematodes. There is, for instance, the widespread occurrence of a rhabditoid stoma in nematode larvae (Adenophorea example: Riemann 1970, p. 421). Juvenile Adenophorea sometimes have distantly arranged circles of head setae which later on fuse together (see e.g. Wieser 1954, p. 187). The observation of a molting stage by Lorenzen (1971) promoted our knowledge of Demoscolecidae. This list of larva observations with phylogenetic significance today is very long. However, the notion that neoteny may play a role in nematode phylogeny can make our clues doubtful (see De Coninck 1965, p. 600). The answer of Chitwood (1937, p. 70; see also 1950, p. 192 within a magnificent compilation of nematode relationships) to advocates of neoteny may be taken as a document of a certain helplessness ("ontogeny supports the view that few-celled forms are more primitive than many-celled forms. In the writer's opinion, the converse assumption removes the study of nemic phylogeny from the realm of logical thought").

5. "Teratological criterion". This criterion for primitiveness I borrowed from the botanists (Zimmermann 1967, p. 106). Singular anomalies demonstrate, as a

rule, an ancestral condition. In this respect Osche (1955) discussed certain anomalies (three-pointed tail tips) in Nematoda Secernentea as reflection of a primitive condition. Rauther, when promoting his insect-ancestor hypothesis, mentioned (1909, p. 563) an anomal occurrence of three female gonads in *Ascaris*.

6. There are some other old phylogenetic rules enumerated by Remane (1952) as supplementary methods of phylogeny (Zahlenreduktionsgesetz, Differenzierungs-gesetz, Internationsgesetz, Gesetz der Konzentration, Spezialisationsgesetz) which need not particularly be mentioned here, their weaknesses having been discussed by Remane himself. Hennig (1966, p. 96) indicated a criterion of the correlation of series of transformation based on the assumption that if the direction of evolution of one character is known, other correlated ones could be analyzed. At last, it should be mentioned that in the case of parasitic nematodes, correlations of the phylogeny of the hosts and their parasites sometimes can be noted (see Osche 1966).

B. Functional aspects of nematode evolution

The results of applying these methods to the recognition of the primitive nematode have proven controversial. Members of very different families have been discussed as primitive (e.g. Cephalobidae, Plectidae, Monhysteridae, Tripylidae, Leptosomatidae: see Andrassy 1962, Gerlach 1966, Maggenti 1970, Chitwood in Gerlach & Riemann 1971). For better decision-making I think the extension of our methods to a causal analysis is useful. The resulting credibility is higher if we can present not only the probable history of the course of phylogeny but additionally can give an explanation for the historical events. Peters & Gutmann (1971 p. 250) stated that unless an ecological or functional aspect is included within the phylogenetic method, the course of the evolution cannot be recognized. We must identify, they claim, that particular mechanism which, in the respective habitat, amounted to a selection advantage. Whereas I do not concur with an apodictic generalization of such statements I do think that the essence in the criticism of the cited authors deserves appreciation (and an English translation).

As a result of evolution there is often apparent a liberation from certain restrictions (see von Wahlert 1968 p. 118). The enormous ecological success of nematodes as compared with their relatives, gastrotrichs and kinorhynchs, can be considered under such a perspective. Gastrotrichs move forward by means of ciliary currents, and are generally restricted to interstitial spaces in sediments. Kinorhynchs, on the other hand, generally are only borers in sediments. Nematodes, however, can achieve both, they can move freely in interstitial habitats and bore in dense media. They accomplish this with their typical undulatory propulsion. This serpent-like locomotion is brought about by contraction of longitudinal muscles without assistance of a chordoid organ or circular muscles.

Let us consider what happens when longitudinal muscles enclosed in a tubular epidermal layer contract themselves: the tube swells strongly, and bends perhaps a little bit. By having the epidermal layer covered with a pressure-resistant but nevertheless longitudinally flexible cuticle the nematodes, without considerable swellings, accomplish their typical and apparently ecologically successful locomotion.

The evolution of such a locomotory apparatus in nematodes had a channeling effect on the evolution of several organ systems. This is because muscular contraction leads to a considerable turgor pressure increase in the pseudocoelom, the cuticle being pressure-resistant in radial direction. Harris & Crofton (1957) first measured and discussed such hydrostatic pressure and pressure waves in *Ascaris*, but in this and subsequent publications (Inglis 1964, Crofton 1966, 1971) perhaps too much stress is given to the generation and maintenance of the hydrostatic pressure acting as an antagonist to muscular contraction (see Riemann 1976). Primarily, in my opinion, the cuticle must be pressure-resistant to avoid swelling of the nematode body. Nevertheless I follow the admirable causal deductions of the functional interdependencies of the nematode structures as given by these English authors.

The cuticle must be flexible longitudinally and strong radially (Inglis 1964). This task is solved by different cuticle constructions: 1. The functional demand is most efficiently and elegantly (Inglis) met by an obliquely crossed arrangement of two or three layers of collagen-like fibres (Fig. 1B). Such structures we find widespread in both nematode subclasses Adenophorea and Secernentea (see the compilation on cuticle structures in Bird 1971). It bears a striking resemblance to the construction of a modern automobile tire. − 2. The relatively rigid, sclerotized, sometimes in a complicate manner interdigitating cuticle rings in Desmodorida and Chromadorida, which apparently have no diagonal crossed fibre system, according to Inglis fulfill the same requirements of pressure resistance as the fibre system, but at the expense of some loss of flexibility (Fig. 5). In functional respect the cuticle of Chromadorida looks like a compromise, an early attempt (see Inglis 1962 p. 217), and may be more primitive than the fibre system cuticle. − 3. Another type of cuticle occurs among Secernentea; it has a striated basal layer (Fig. 1A). This is possibly a modification of the fibre layers, perhaps, as ontogeny reveals, their predecessor (Johnson, Gundy & Thomson 1970) and may be regarded as a strengthening device of the cuticle as well (Wright 1968).

Johnson et al. (1970, p. 57) state that in the literature there is only one nematode genus reported which has neither fibre layer (in certain Oxyuroidea fibres are restricted to lateral regions: Anya 1966, cited by Wright 1968) nor striated layer, and, I want to add, no cuticle with sclerotized elements. This interesting exception is the plant root-attacking *Trichodorus* (see Hirumi, Chen, Lee & Maramorosch 1968; Raski, Jones & Roggen 1969). This genus has a complicated multilayered cuticle, but in the photographs published by these authors I cannot detect particular supporting structures. *Trichodorus* belongs to the Diphtherophorina Coomans & Loof, 1970, and it should be mentioned that the earth nematode *Diphtherophora* itself has a peculiar cuticle which, in some species, is loosely fitting, forming mem-

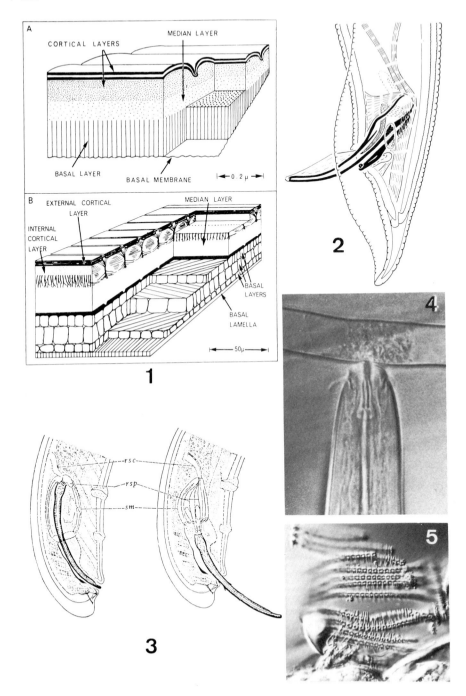

brane-like folds and shifting from side to side as the body bends (Thorne 1939, p. 155, compare Wyss 1971 with photographs of living *Trichodorus*, see also Riemann 1972 in context with Onchulinae). Such a cuticle seemingly does not contribute much to the particular locomotion system of a nematode and hence may be considered as primitive.

The movements of nematodes are accompanied by internal hydrostatic pressure waves (see Seymour 1973), as I have mentioned above, in addition perhaps to a basic tonus turgor pressure (see Chapman 1958, p. 360, Crofton 1966, p. 31). This generation of pressure is in close functional relationship to the general construction plan of nematodes. Crofton (1971, p. 111) remarks: "Probably in no other group is the relationship between form and function so clearly seen"; and Inglis (1964, p. 494) writes: "If the nematodes were derived from some gastrotrich-like form as seems reasonable, by elongation of the body in association with an increased dominance of the longitudinal muscle system and a concomitant increase of the hydrostatic pressure the resulting forms would all, as a matter of mechanical necessity, have the combinations of characters considered diagnostic of the Nematoda".

Such a result of mechanical necessities, for instance, are the pump pharynx of nematodes (see Roggen 1973) and the valves at the ends of the intestine. With the exception of one *Eudorylaimus* species (Zmoray & Guttekova 1969) the intestine lacks cilia which could transport the food. Instead of cilia the pump pharynx takes over the function of transporting the food, thereby acting against the pseudocoelom pressure which tends to compress the intestine (Harris & Crofton 1957, Crofton 1971). Assistance in moving the gut contents is given by somatic muscles (Seymour 1973), and possibly sometimes by circum-enteric muscles (Lee & Anya 1968). Valves and dilator muscles are regulating the direction of the food stream and the defecation (Seymour 1973, Seymour & Doncaster 1972).

In using the word "pseudocoelom" I follow the nematologists' tradition. But there exists serious doubt about the justification of this term. Maggenti, in succession to Remane, wrote (1971, p. 70) that the term "pseudocoelom" has no precise meaning. Attention is called to Remane's (1963, p. 249) statement: "In nematodes the muscle cells of the body are arranged like an epithelium, especially in the platymyarian free-living forms. This outer lining of the body cavity is more similar to a typical coelomic wall than in many coelomates". Meaningful in such comparative morphological considerations as an inner lining of the body cavity I think

---

Fig. 1. Diagrams of nematode cuticle layers. (A) larva of a Secernentea (infective larva); (B) adult *Ascaris* (after Bird 1971). – Fig. 2. Spicular apparatus of *Hoplolaimus* (Tylenchida, after Coomans 1962). - Fig. 3. Spicular apparatus of *Trichodorus* with retracted and protruded spicule positions in the same specimen. (rsc) retractor muscles of spicular capsule; (rsp) retractor muscles of spicule; (sm) protractor muscles of spicule, spicular capsule (after Siddiqi 1973). – Fig. 4. *Trichodorus* sucking on a plant root hair, median view. Note the pale club-shaped structure outside the amphid (courtesy of Dr. U. Wyss; Wyss 1971). – Fig. 5. *Pomponema reductum* Warwick 1970, Chromadorida, from the North Sea. Squash preparation demonstrating interdigitating cuticle rings (original).

is the observation of traces of circum-enteric muscles in some nematodes (see Chitwood & Chitwood 1950, p. 103; Lee & Anya 1968).

The genital tubes in nematodes likewise need valve equipments to overcome problems with the high hydrostatic pressure of the pseudocoelom (see Harris & Crofton 1957, p. 128). Referring to this matter Inglis (1964, p. 494) writes: "I should add that the spicules are probably also a development resulting from the necessity for opening the vulva in the female, and the gubernaculum in many forms appears to be at least as well adapted to opening the male cloaca as it is for guiding spicules, the function which is usually attributed to it".

The spicular apparatus is well known from every taxonomical description; let us look whether we can make causal phylogenetical inferences from their different forms. Basically, most of the spicules and their attaching muscles of free-living and plant-parasiting nematodes are similar (Fig. 2). They are elongate sclerotized structures, obviously outfitted with sensory elements (Lee 1973; Clark, Shepherd & Kempton 1973), and moved by ribbon-like protractor and retractor muscles which attach to the head of the spicules (Chitwood & Chitwood 1950, Coomans 1962). But we know a small group of tripyloid (s.l.) nematodes which have another and seemingly more complicated type of protractor muscle. This protractor is represented by a peculiar capsule of muscle masses surrounding the spicule head but not (!) attached to it. Recently, Siddiqi (1973) studied this in detail in the genus *Trichodorus* (Fig. 3), and came to the conclusion that the protrusion of the spicules seems to be brought about by the squeezing effect caused by the contraction of the capsule which is attached distally to the cloacal wall. Chitwood & Chitwood (1950, p. 119) described the same structures from *Tripyla*, the abundant freshwater and terrestrial genus, and *Triplonchium*, and I found such in *Diphtherophora, Onchulus* and *Stenonchulus* (the latter, a male from the Azores, kindly made available by Dr. Sturhan, Münster), and in the strange Onchulinae genus with a kinorhynch-like head, *Kinonchulus* (Riemann 1972).

The mechanism of protrusion of this type of spicule, I suspect, cannot be brought in accordance with the assumption of a high turgor pressure within the pseudocoelom. Every pressure wave in the course of locomotion would then squeeze the protractor capsule, thereby protruding the spicules. Now we should remember that *Trichodorus* is the unique nematode in the cuticle of which no supporting structures have been found. This may indicate that the cuticle is not particularly pressure resistant, hence a high turgor pressure cannot be generated, a condition which is reflected by the strange spicular apparatus. So I come to the conclusion: If really the turgor pressure-resistance of the nematode cuticle is a measure for evolutionary progress then the terrestrial plant-sucking nematode *Trichodorus* is primitive. Regarding their peculiar similar spicular apparatus, I think that the limnetic genus *Tripyla* and the Onchulinae also belong to the grouping of the most primitive nematodes.

C. The primitive nematodes

This selection of terrestrial and limnetic nematodes as primitive ones sounds unexpected, but the foregoing arguments are not the first to propose members of the Diphtherophorina and Tripyloidea as primitive nematodes. Using embryological evidence Gerlach (1966), in succession to earlier investigations of De Coninck, made the assumption that the distant arrangement of the head sensory organs in three circles in Tripyloidea is a primitive feature. In the more modern taxa the posterior-most circle is fused with the intermediate one. Furthermore, Gerlach considered ventral cervical papillae in males as primitive, and I want to remind that in Tripyloidea and Trichodoroidea such cervical papillae occur. From electron optical investigations Hirumi, Raski & Jones (1971) made the assumption that *Trichodorus* muscle cells may be considered as a very primitive type of somatic muscle cells in nematodes (in this context it should not be forgotten that Stekhoven & De Coninck 1933, p. 139 noted that *Diphtherophora*, according to its meromyarian condition, should not be placed among the enoplids but among the rhabditids. The Diphtherophorina today are considered as Dorylaimida, this placement needs a revision: see Coomans 1971 p. 96).

Another discussion point may be added. Diphtherophorina have interesting amphids with a remarkably conspicuous cilia-bearing proximal part, as many taxonomical descriptions show (ultrastructural investigation: Raski, Jones & Roggen 1969. For general bibliography on amphids see Storch & Riemann 1973; Baldwin & Hirschmann 1975). In *Trichodorus*, a pale club-shaped body is hanging outside the distal pocket-like excavation (see De Man 1884, p. 161, pl. 24, fig. 103c). It is not an artefact but can be seen also in living animals (Fig. 4). This configuration has such striking resemblance to ciliary pits or piston pits of gastrotrichs (see also Riemann 1972a) that we should seriously compare these structures in electron-microscopical dimensions.

Generally, such comparative ultrastructural studies are necessary to elevate the concept of the primitive nematode from the speculative level as presented here to a more substantiated one.

More difficult, however, than the establishment of a primitive stadia group of nematodes is the further analysis of the phylogeny, and in this context I refer the reader to the recent dendrograms of nematode phylogeny given by Maggenti (1963, 1970, 1971), Gerlach (1966) and Andrassy (1974). We must ask for the one character which marks the first apomorphic step in the evolution. In the generally accepted system (see De Coninck 1965) we divide the class Nematoda into the two subclasses Adenophorea (syn. Aphasmidia) and Secernentea (syn. Phasmidia) as taxa of equal rank. Does this really reflect phylogeny? "Examination of the definitions of Phasmidia and Aphasmidia given by the Chitwood's show exceptions and qualifications to every item except the presence or absence of phasmids" (Hyman 1959, p. 745). These phasmids, minute lateral pouches on the tail, characterize the taxon Secernentea (Phasmidia) as monophyletic (single exception among Adenophorea perhaps Desmoscolecida Geeffiellinae, see Schrage & Gerlach 1975, p. 38) without further definition. If Secernentea in a phylogenetic dendrogram are opposed to all other nematodes (Adenophorea) as taxa of equal rank then all features which are shared by some members of both groups must be either primitive, or convergent acquisitions. With this perspective in mind we should, in the future, examine, for example, the similarity in the marginal oesophageal tubes in Secernentea and some Adenophorea (Araeolaimida, Comesomatidae and few Monhysteridae) and the occurrence of a rhabditoid bursa in *Monhystrium* (Monhysterida). Attention is called to the Secernentea-like excretion system of *Anonchus* (Araeolaimida, see Chitwood & Chitwood 1950, p. 133). In addition, it should be noted that not in all Secernentea the excretion canals are located in the lateral chord but are, in some Tylenchida, apparently situated within the pseudocoelom (Allen 1960, p. 139) as in *Anonchus*. As long as a conclusive proof of the either primitive or convergent derived status of these organs is lacking we cannot exclude the possibility that the Secernentea are derived from Adenophorea at a younger, that is lower, category level.

There exists a controversy on the question whether the Hennigean sister-group approach is consistent with the ranking of taxa in a written classification (see Brundin 1972, p. 110, 118. Nelson 1969, p. 531–534 has exemplified an approximation of the congruence of a phylogenetic dendrogram and a written classification in vertebrates). I think systematists cannot escape the necessity to aim at an approximation of the phylogenetic dendrogram and the written classification. Of course, the establishment of the monophyly of each named taxon is a prerequisite. It is the nematologist's main task in phylogenetic work today.

Gadea (1972) recognized "two fundamental groups" within the Nematoda for which he coined the terms Enoplimorpha and Rhabditimorpha, or enoplidean and chromadoridean trunk. According to Gadea, the Enoplia are opposed to all other Adenophorea and some Secernentea. Andrassy (1974, p. 46) in his dendrogram opposed the Enoplia (or Penetrantia sensu Andrassy) to all other nematodes, as does Maggenti (1963, 1970, 1971). The Secernentea are then regarded as younger descendants of the remaining taxon Adenophorea Chromadoria (or Torquentia sensu Andrassy). Again, the monophyly of the Chromadoria and their assumed equal-ranking counterpart (sister-group), the Enoplia, must be discussed. Enoplia are characterized by pocket-like, not circular or spiral, amphids, but exceptions exist in Chromadoria (*Anaplectus, Chronogaster, Anomonema*). Chromadoria (and Secernentea) have a rather constant pattern in the position of oesophageal gland openings (dorsal gland opening near the mouth, both subventral openings near the nerve ring), in contrast to a bewildering variety in Enoplia. These characters are very useful in separating taxa of middle and lower categories, but their value in establishing the first apomorphic step in the nematodes remains to be ascertained.

We nematologists are fascinated and challenged by the task of recognizing the primitive nematode, but when reconstructing the phylogeny we must change our perspective to answer the question of which nematode bears the oldest derived character.

Acknowledgements

This contribution was promoted by a generous travel fund given by Deutsche Forschungsgemeinschaft. Thanks are extended to the members of the International Association of Meiobenthologists for their stimulating discussions at the Bermuda Biological Station.

Bibliography

Allen, M. W.: Alimentary canal, excretory and nervous systems. In: J. N. Sasser and W. R. Jenkins (eds.), Nematology. Fundamentals and recent advances with emphasis on plant parasitic and soil forms. Chapel Hill; Univ. North Carolina Press, 136–139 (1960).

Andrassy, I.: Über den Mundstachel der Tylenchiden (Nematologische Notizen, 9). Acta Zool. Acad. Sci. hung. 8, 241–249 (1962).

–: A nematodák evolúcioja és rendszerezése. Magy. tudom. Akad. biol. Osztál Közl. 17, 13–58 (1974).

Baldwin, J. G. & H. Hirschmann: Fine structure of cephalic sense organs in Heterodera glycines males. J. Nematol. 7, 40–49 (1975).

Bird, A. F.: The structure of nematodes. New York/London; Academic Press, 318 pp. (1971).

Boaden, P. J. S.: Anaerobiosis, meiofauna and early metazoan evolution. Zoologica Scripta 4, 21–24 (1975).

–: Thiobiotic facts and fancies (aspects of the distribution and evolution of anaerobic meiofauna). Mikrofauna Meeresboden 61, 45–63 (1977).

Brundin, L.: Evolution, causal biology, and classification. Zoologica Scripta 1, 107–120 (1972).

Chapman, G.: The hydrostatic skeleton in the invertebrates. Biol. Rev. 33, 338–371 (1958).

Chitwood, B. G.: A revised classification of the nematodes. Papers in Helminthology published in commemoration of the 30 year Jubileum of K. J. Skrjabin. Moskau, 69–79 (1937).

– & M. B. Chitwood: An introduction to nematology. Section 1, anatomy. 2. ed. Baltimore; Monumental Printing Co., 213 pp. (1950).

Clark, S. A., A. M. Shepherd and A. Kempton: Spicule structure in some Heterodera spp. Nematologica 19, 242–247 (1973).

De Coninck, L. A.: Systematique des nématodes. Généralités et sous-classe des Adenophorea. In: P.-P. Grassé, Traité de Zoologie 4; 2. Paris; Masson, 412–432, 586–681 (1965).

Coomans, A.: The spicular muscles in males of the subfamily Hoplolaiminae (Tylenchida, Nematoda). Biol. Jaarb. Dodonaea (Gent), 313–315 (1962).

–: Morphology and nematode systematics. J. Parasitol. 57 (4) Sect. II, Part 2, 95–99 (1971).

Crofton, H.D.: Nematodes. London; Hutchinson, 160 pp. (1966).

–: Form, function, and behavior. In: B. M. Zuckerman, W. F. Mai and R. A. Rohde (eds.), Plant parasitic nematodes, Vol. 1. New York/London; Academic Press, 83–113 (1971).

Fenchel, T. M. & R. J. Riedl: The sulfide system: a new biotic community underneath the oxidized layer of marine sand bottoms. Mar. Biol. 7, 255–268 (1970).

Franzen, J. L., D. Mollenhauer, W. F. Gutmann & D. S. Peters: Was ist Phylogenetik? Natur u. Museum 103, 238–242 (1973).

Gadea, E.: Some aspects on phylogeny of nematoda. Abstracts of the 11th Intern. Symposium on Nematology, Reading 3 to 8 September 1972 (Europ. Soc. of Nematologists), p. 23 (1972).

Gerlach, S. A.: Bemerkungen zur Phylogenie der Nematoden. Mitt. biol. BundAnst. Ld. u. Forstw. 118, 25–39 (1966).

– & F. Riemann: Über Zahnbildungen in der Mundhöhle von Nematoda Monhysteridae: Monhystrella und Hofmaenneria. Nematologica 17, 285–294 (1971).

Gutmann, W. F.: Der Konstruktionsplan der Cranioten: ein phylogenetisches Modell und seine methodisch-theoretischen Konstituentien. Cour. Forsch.-Inst. Senckenberg (Frankfurt) **3**, 36 pp. (1973).

Harris, J. E. & H. D. Crofton: Structure and function in the nematodes: Internal pressure and cuticular structure in *Ascaris*. J. Exp. Biol. **34**, 116–130 (1957).

Hennig, W.: Phylogenetic Systematics. Urbana/Chicago/London; University of Illinois Press, 263 pp. (1966).

Hirumi, H., T. A. Chen, K. J. Lee & K. Maramorosch: Ultrastructure of the feeding apparatus of the nematode *Trichodorus christiei*. J. Ultrastruct. Res. **24**, 434–453 (1968).

Hirumi, H., D. J. Raski & N. O. Jones: Primitive muscle cells of nematodes: morphological aspects of platymyarian and shallow coelomyarian muscles in two plant parasitic nematodes, *Trichodorus christiei* and *Longidorus elongatus*. J. Ultrastruct. Res. **34**, 517–543 (1971).

Hull, D. L.: Certainty and circularity in evolutionary taxonomy. Evolution **21**, 174–189 (1967).

Hyman, L. H.: The Invertebrates, Vol. 5, Smaller coelomate groups. New York/London/Toronto; McGraw-Hill, 783 pp. (1959).

Illies, J.: Zur modernen Systematik. Ein Vergleich der Methoden von Hennig und Remane. Zool. Beiträge N.F. **13**, 521–528 (1967).

Inglis, W.G.: Marine nematodes from Banyuls-sur-mer: with a review of the genus *Eurystomina*. Bull. Br. Mus. nat. Hist. (Zool.) **8**, 209–287 (1962).

–: The structure of the nematode cuticle. Proc. Zool. Soc. London **143**, 465–502 (1964).

Johnson, P. W., S. D. van Gundy & W. W. Thomson: Cuticle ultrastructure of *Hemicycliophora arenaria, Aphelenchus avenae, Hirschmanniella gracilis* and *Hirschmanniella belli*. J. Nematol. **2**, 42–58 (1970).

Lee, D. L.: Evidence for a sensory function for the copulatory spicules of nematodes. J. Zool. Lond. **169**, 281–285 (1973).

– & A. O. Anya: Studies on the movement, the cytology and the associated micro-organisms of the intestine of *Aspiculuris tetraptera* (Nematoda). J. Zool. Lond. **156**, 9–14 (1968).

Lorenzen, S.: Jugendstadien von *Desmoscolex*-Arten (Nematoda, Desmoscolecidae) und deren Bedeutung für die Taxonomie; Mar. Biol. **10**, 343–345 (1971).

Maggenti, A. R.: Comparative morphology in nemic phylogeny. In: E. C. Dougherty (ed.), The lower metazoa. Berkeley; Univ. of California Press, 273–282 (1963).

–: System analysis and nematode phylogeny. J. Nematol. **2**, 7–15 (1970).

–: Nemic relationship and the origins of plant parasitic nematodes. In: B. M. Zuckerman, W. F. Mai and R. A. Rohde (eds.), Plant parasitic nematodes, Vol. 1. New York/London; Academic Press, 65–81 (1971).

De Man, J. G.: Die frei in der reinen Erde und im süßen Wasser lebenden Nematoden der niederländischen Fauna. Eine systematisch-faunistische Monographie. Leiden; Brill, 206 pp. (1884, 2nd abbrev. Ed. 1919, 176 pp.).

Micoletzky, H.: Die freilebenden Erdnematoden. Arch. Naturgesch. 87 A, 1–650 (1922).

Nelson, G. J.: Gill arches and the phylogeny of fishes, with notes on the classification of vertebrates. Bull. Amer. Mus. Nat. Hist. **141**, 475–552 (1969).

Osche, G.: Der dreihöckerige Schwanz, ein ursprüngliches Merkmal im Bauplan der Nematoden. Zool. Anz. **154**, 136–148 (1955).

–: Grundzüge der allgemeinen Phylogenetik. In: L. v. Bertalanffy, Handbuch der Biologie, Band 3. – Frankfurt; Athenaion, 817–906 (1965/1966).

–: Ursprung, Alter, Form und Verbreitung des Parasitimus bei Nematoden. Mitt. biol. Bund Anst. Ld und Forstw. 118, 6–24 (1966).

–: Das Homologisieren als eine grundlegende Methode der Phylogenetik. Aufsätze u. Red. senckenb. naturf. Ges. **24**, 155–165 (1973).

Ott, J. A.: Twelve new species of nematodes from an intertidal sandflat in North Carolina. Int. Rev. ges. Hydrobiol. **57**, 463–496 (1972).

–: Determination of fauna boundaries of nematodes in an intertidal sand flat. Int. Rev. ges. Hydrobiol. **57**, 645–663 (1972a).

Paramonov, A. A.: Plant-parasitic nematodes, vol. 1. Origin of nematodes, ecological and morphological characteristics of plant nematodes, principles of taxonomy. Translated from Russian by S. Nemchonok. Jerusalem; Israel Program for Scientific Translations, 390 pp. (date of issue of Russian original: 1962) (1968).

Peters, D. S.: Homologie – ein Wort und viele Begriffe. Aufsätze u. Red. senckenb. naturf. Ges. **24**, 173–175 (1973).

– & W. F. Gutmann: Über die Lesrichtung von Merkmals- und Konstruktionsreihen. Z. zool. Syst. Evolut.-forsch. **9**, 237–263 (1971).

– & W. F. Gutmann: Modellvorstellungen als Hauptelement phylogenetischer Methodik. Aufsätze u. Red. senckenb. naturf. Ges. **24**, 26–38 (1973).

Raski, D. J., N. O. Jones & D. R. Roggen: On the morphology and ultrastructure of the esophageal region of *Trichodorus allius* Jensen. Proc. helminth. Soc. Wash. **36**, 106–118 (1969).

Rauther, M.: Morphologie und Verwandtschaftsbeziehungen der Nematoden und einiger ihnen nahe gestellter Vermalien. Ergebn. u. Fortschritte der Zool. **1**, 491–596 (1909).

Remane, A.: Die Grundlagen des natürlichen Systems, der vergleichenden Anatomie und der Phylogenetik. Leipzig; Akad. Verlagsges. Geest u. Portig, 400 pp. (1952).

–: The systematic position and phylogeny of the pseudocelomates. In: E. C. Dougherty (ed.), The lower metazoa. Berkeley; Univ. of California Press, 247–255 (1963).

Riemann, F.: Nematoden aus dem Strandanwurf. Beitrag zum natürlichen System freilebender Nematoden. Veröff. Inst. Meeresforsch. Bremerh. **11**, 25–36 (1968).

–: Das Kiemenlückensystem von Krebsen als Lebensraum der Meiofauna, mit Beschreibung freilebender Nematoden aus karibischen amphibisch lebenden Decapoden. Veröff. Inst. Meeresforsch. Bremerh. **12**, 413–428 (1970).

–: *Kinonchulus sattleri* n.g.n.sp. (Enoplida Tripyloidea), an aberrant freeliving nematode from the lower Amazonas. Veröff. Inst. Meeresforsch. Bremerh. **13**, 317-326 (1972).

–: Corpus gelatum und ciliäre Strukturen als lichtmikroskopisch sichtbare Bauelemente des Seitenorgans freilebender Nematoden. Z. Morph. Tiere **72**, 46–76 (1972a).

–: Meeresnematoden (Chromadorida) mit lateralen Flossensäumen (Alae) und dorsoventraler Abplattung. Zool. Jb. (Syst.), **103**, 290–308 (1976).

Roggen, D. R.: Functional morphology of the nematode pharynx: I. Theory of the soft-walled cylindrical pharynx. Nematologica **19**, 349–365 (1973).

Schlee, D.: Die Rekonstruktion der Phylogenese mit Hennigs's Prinzip. Aufsätze u. Red. senckenberg. naturforsch. Ges. **20**, 1–62 (1971).

Schrage, M. & S. A. Gerlach: Über Greeffiellinae (Nematoda, Desmoscolecida). Veröff. Inst. Meeresforsch. Bremerh. **15**, 37–64 (1975).

Schram, F. R.: Pseudocoelomates and a nemertine from the Illinois Pennsylvanian. J. Paleontol. **47**, 985–989 (1973).

Seymour, M. K.: Motion and the skeleton in small nematodes. Nematologica **19**, 43–48 (1973).

– & C. C. Doncaster: Defaecation behaviour of *Aphelenchoides balstophthorus*. Nematologica **18**, 463–468 (1972).

Siddiqi, M. R.: Systematics of the genus *Trichodorus* Cobb, 1913 (Nematoda: Dorylamida), with descriptions of three new species. Nematologica **19**, 259–278 (1973).

Stekhoven, J. H. Schuurmans & L. A. De Coninck: Morphologische Fragen zur Systematik der freilebenden Nematoden. Verh. dt. zool. Ges. **35**, 138–143 (1933).

Storch, V. & F. Riemann: Zur Ultrastruktur der Seitenorgane (Amphiden) des limnischen Nematoden *Tobrilus aberrans* (W. Schneider, 1925) (Nematoda, Enoplida). Z. Morph. Tiere **74**, 163–170 (1973).

Teuchert, G.: Zur Fortpflanzung und Entwicklung der Macrodasyoidea (Gastrotricha). Z. Morph. Tiere **63**, 343–418 (1968).

Thorne, G.: A monograph of the nematodes of the superfamily Dorylaimoidea. Capita zool. **8** (5), 1–261 (1939).

Wahlert, G. von: *Latimeria* und die Geschichte der Wirbeltiere. Eine evolutionsbiologische Untersuchung. Stuttgart; G. Fischer, 125 pp. (1968).

Wieser, W.: Beiträge zur Kenntnis der Nematoden submariner Höhlen. Ergebnisse der österreichischen Tyrrhenia-Expedition 1952, Teil II. Öst. zool. Z. **5**, 172–230 (1954).

– J. Ott, F. Schiemer & E. Gnaiger: An ecophysiological study of some meiofauna species inhabiting a sandy beach at Bermuda. Mar. Biol. **26**, 235–248 (1974).

Wright, K. A.- The fine structure of the cuticle and interchordal hypodermis of the parasitic nematodes, *Capillaria hepatica* and *Trichuris myocastoris*. Can. J. Zool. **46**, 173–180 (1968).

Wyss, U.: Der Mechanismus der Nahrungsaufnahme bei *Trichodorus similis*. Nematologica **17**, 508–518 (1971).

Zimmermann, W.: Methoden der Evolutionswissenschaft (=Phylogenetik). In: G. Heberer (ed.), Die Evolution der Organismen, Vol. 1, 3rd ed. Stuttgart; G. Fischer, 61–160 (1967).

Zmoray, I. & A. Guttekova: Ecological conditions for occurrence of cilia in intestines of nematodes. Biológia (Bratislava) **24**, 97–112 (1969).

| Mikrofauna Meeresboden | 61 | 231−251 | 1977 |

W. Sterrer & P. Ax (Eds.). The Meiofauna Species in Time and Space. Workshop Symposium, Bermuda Biological Station, 1975

Zoogeography and Speciation in Marine Gastrotricha

by

Edward E. Ruppert

Department of Zoology, University of North Carolina, Chapel Hill, North Carolina 27514, USA

Abstract

Worldwide and local distributional patterns of species of marine Gastrotricha are examined. Species groups are cosmopolitan in distribution, and perhaps 10 % of species are true cosmopolitan forms. Support is given to the "plate tectonics" model of Sterrer (1973), but evidence is also presented for local isolation-speciation events. The sympatric occurrences of several congeneric species in one beach zone are examined. It is shown that their in-beach distributional patterns are distinct and that within the total geographic range of their occurrences, sympatry may be rare. The occurrences of elongated cuticular spines, small body sizes, parthenogenesis and the storage of spermatozoa in "bursal organs" are interpreted as adaptations to dispersal. The Gastrotricha are compared to other colonizers of transitional, but renewable, habitats.

A. Introduction

Sediment living meiofaunal species in general are found wherever there is water and the development of interstitial spaces of sufficient sizes, the latter being determined primarily by the force and pattern of water movement (Riedl 1975, Boaden 1968). Local habitats will persist over short periods of time (minutes to years) due to fluctuations in water movement (storm erosion, siltation etc.) and these effects can eliminate many of the fauna in an area either by removing the habitat or by making it unsuitable for their existence (Williams 1972, Gadow & Reineck 1969).

In contrast to the comparatively short life of many local geomorphological features (inlets, bays, barrier islands, beaches, bars etc.) is the long term persistence of the marine interstitial environment through geological time. This fact, in part, explains the occurrence of many old groups of organisms in the environment (Riedl 1971).

But, as seawater and sand are ancient, so local habitats are recent, and there is the necessity of recognizing some means of colonizing suitable local habitats as they arise. Some means of local dispersal in the absence of pelagic larvae must exist in the meiofauna. Additionally, some means of adjusting to variations in local environmental conditions either at the genetic or organismic level must be assumed.

Remane recognized that the systematic and ecological similarities of meiofauna were great over a geographic range (Remane 1952) and this viewpoint has been confirmed by many investigators (Ax & Schmidt 1973, Rieger 1971 etc.). Sterrer (1973) observed that among Gnathostomulida, the two most structurally similar species in many genera occurred on opposite sides of the Atlantic ocean. This observation gave rise to the hypothesis that species of meiofauna in the Atlantic basin in the former warm, shallow seas of the Mesozoic were split apart by continental drift, gene flow was interrupted, and divergence occurred between the species. The result is the occurrence of "trans-allopatric sister-species" observed by Sterrer (1973). Also included in the model is the prediction that speciation does not occur along continuous coastlines due to the "swamping effect" of gene flow and, conversely, that longshore dispersal must be important in maintaining that gene flow along continuous coastlines.

But we are continually presented with records of known morphological types, frequently species (although admittedly, these may be "lumped") from areas that are remote and disjunct to the Atlantic basin (Rao & Ganapati 1968, Saito 1937, Wieser 1957, Ax & Ehlers 1973, Hummon 1972). These occurrences force us to conclude once again that there is a wide constancy of form over geographic distances and that we still lack a complete understanding of the phenomena that produced these patterns.

There are two hypotheses that follow from our current knowledge and must also be considered in relation to the Sterrer model: 1) that the worldwide distribution of species, species groups and genera is the result of enormous powers of contemporary dispersal, or, 2) that the observations are a result not of wholesale dispersal, but of an absence of morphological change; that the fauna are, as *Limulus* is, evolutionary relicts that persist unchanged through extended periods of time (see Gooch 1975, and note his explanation for the persistence of *Limulus*!). Certainly these hypotheses need not be mutually exclusive, and that possibility brings into focus the complexity of the general problem.

The objectives of this paper are: 1) to examine the transatlantic occurrences of species and species groups of marine Gastrotricha, 2) to examine the local occurrences of sympatric species of Gastrotricha[1] 3) to examine structural features in the Gastrotricha that may, in part, explain the distributional patterns, and 4) to interpret these facts in the framework of the above two hypotheses in an effort to bring understanding to speciation and zoogeography in the Gastrotricha.

B. Methods

Sampling and extraction methods as well as locality descriptions are given in Ruppert (1972, 1976) and Crezée (1976). The collection site at Pine Knoll Shores, N.C. is a fine sand flat with a well developed RPD layer. The flat, approximately 60m wide at low tide, receives no wave

---

[1]Sympatric species is used here in the sense of species occurring in the same area, i.e. beach zone, but not necessarily the same sample.

action, and is characterized by a thin surface layer of flocculent organic material and regular patches of diatoms. The gastrotrich collections on this flat (*Turbanella ocellata, Neodasys* sp.) were made at mid tide level at 0–2 cm.

The organisms were observed alive under phase optics and drawn with a camera lucida. Photomicrographs of living organisms were made on Kodak panatomic X (ASA32) film and developed in Microdol X. Photomicrographs of resin embedded specimens (see Ruppert & Rieger, in prep.) were made on Kodak SO–410 (ASA ca. 100) film and developed in Kodak D-19 developer.

Measurements of tentacle lengths in *Tetranchyroderma papii, T. bunti* and *T.* cf. *swedmarki* were made on unsqueezed, fixed specimens.

C. Transatlantic occurrences of Gastrotricha

The occurrence of species of Gastrotricha that are nearly identical structurally on both sides of the Atlantic, while not astoundingly high, is nevertheless significant at our current level of observational technology. Perhaps 10 % of the species must be given the same species name on both sides of the Atlantic (and Pacific! see Wieser 1957, Hummon 1972) and it should not be surprising that most of the remaining species, i.e. those that are not identical, have very similar counterparts on opposite sides of the Atlantic.

Virtually identical species that I have observed and documented from Arcachon, France and from Bogue Banks, North Carolina include: *Ichthydium hummoni, Halichaetonotus aculifer* (Fig. 1A-D) and *Heteroxenotrichula squamosa* (Ruppert 1976). West Atlantic species that in structure agree nearly perfectly with descriptions of european species include: *Tetranchyroderma papii, Pseudostomella roscovita* (Ruppert 1970), *Heteroxenotrichula pygmaea* and *Draculiciteria tesselata* (Ruppert 1976). Finally, there are groups of very similar species that extend across and along both sides of the Atlantic (and other seas!) that can be designated as species groups. It is this latter pattern that I believe to be dominant in the marine Gastrotricha.

The similarity of form over distance is paralleled by a correspondence of position within beaches of these disjunct forms. *Ichthydium hummoni* (Figs. 1C,D, 14F) occurred in the zone of damp sand at the tide level of extreme high water springs (EHWS) on the Plage d'Eyrac in Arcachon, and it occurred in the zone of damp sand at the tide level of mean high water mean (MHWM) on Bogue Inlet Beach in North Carolina. Similarly, *Cephalodasys littoralis* (Fig. 1E,F) occurred on both beaches in the zone of damp sand at the tide level of mean high water mean (MHWM). A group that has been considered in detail, the Xenotrichulidae (Ruppert 1976), showed morphological and ecological correspondences across three beaches, two in France and one in North Carolina (Figs. 7,8,9). These distributional correspondences are not accidental but, in general, reflect the common adaptive radiations of these groups, or more precisely, they record the historical development of form in correlation with environment. And, of course, from such correspondences we infer that the biotic and abiotic interactions of the species are similar, i.e. they are ecological equivalents.

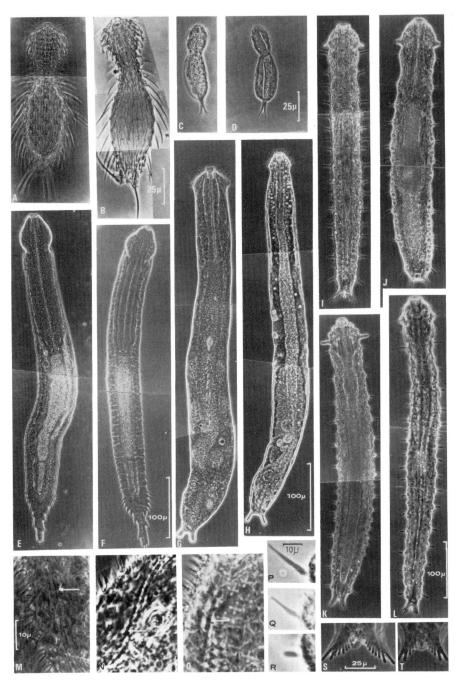

Even though species groups are worldwide in distribution, it is often the case in the Gastrotricha that no clear decision can be made, on the basis of structure, whether the two most similar species in a group are "trans-allopatric" or "cis-

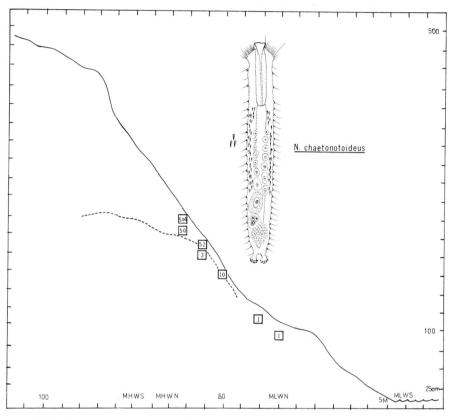

Fig. 2. Distribution of *N. chaetonotoideus* on Plage d'Eyrac, Arcachon, France, December 1971 – January 1972; densities in numbers of individuals per 20cc sediment.

Fig. 1. (A) *H. aculifer*, living specimen, Arcachon, France; (B) *H. aculifer*, formalin fixed whole-mount, Bogue Banks; N. C.; (C) *I. hummoni*, living specimen, Arcachon, France; (D) *I. hummoni*, living specimen, Bogue Banks, N. C.; (E) *C. littoralis*, living specimen, Arcachon, France; (F) *C. littoralis*, living specimen, Bogue Banks, N. C.; (G) *N. chaetonotoideus*, living specimen, Arcachon, France; (H) *N. sp.*, living specimen, Bogue Banks, N. C.; (I) *T. cornuta*, living sub-adult specimen, Arcachon, France; (J) *T. cf. mustela*, living adult specimen, Arcachon, France; (K) *T.* cf. *mustela*, living adult specimen, Chesapeake Bay, Virginia; (L) *T.* cf. *mustela*, living adult specimen, Bogue Banks, N. C., (M) *T.* cf. *swedmarki*, living specimen, sperm and testis; (N) *T. papii*, living specimen, sperm and testis; (O) *T. bunti*, living specimen, sperm and testis; (P) *T. papii*, resin embedded adult, head tentacle; (Q) *T. bunti*, resin embedded adult, head tentacle; (R) *T.* cf. *swedmarki*, resin embedded adult, head tentacle; (S) *T. cornuta*, living subadult, caudal adhesive organ; (T) *T.* cf. *mustela*, living adult, caudal adhesive organ.

allopatric", i.e. whether they occur on opposite sides of the Atlantic or in locally isolated habitats. *Neodasys chaetonotoideus* (Remane 1936; Fig. 1G) is a species that is easily recognized but, as yet, not described from the west Atlantic coast. A second species, *N. uchidai* (definitely not a juvenile of *N. chaetonotoideus*) was also described from Europe, where it appears not to occur with *N. chaetonotoideus* (Remane 1961). It is distinct from the type species in its possession of longitudinal rows of large, red-pigmented Y-cells (Remane 1936, 1961; Rieger et al. 1974; Teuchert 1974), of two pairs of small anterior adhesive papillae and a short caudal peduncle. Its sperm structure agrees with that of *N. chaetonotoideus* (Remane 1936, 1961). A third species in the genus, *N.* sp. (see Ruppert 1976), occurs commonly along the west Atlantic coast. This species lacks a peduncle as an adult, has longitudinal rows of large red-pigmented Y-cells, lacks anterior adhesive papillae and has spermatozoa that are dart-shaped rather than commaform (Figs. 1H, 3,4). Interesting is that *N. uchidai* occurs in shelf sediments off North Carolina and Georgia. The results to date are that *N. chaetonotoideus* is distributed in the east

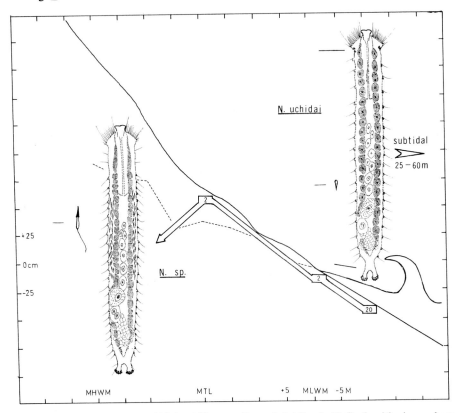

Fig. 3. Distributions of *N. uchidai* and *N.* sp. on Bogue Inlet Beach, N. C.; densities in numbers of individuals per 20cc sediment.

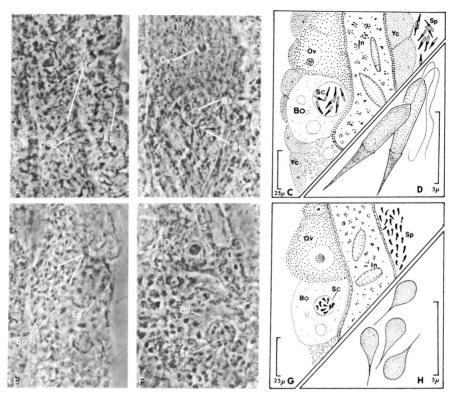

Fig. 4. A, *N.* sp., testis and spermatozoa; B, *N.* sp., sperm storage organ; C, *N.* sp., drawing of caudal body region; D, *N.* sp., spermatozoa; E–H, *N. chaetonotoideus*; E, testis; F, sperm storage organ; G, drawing of caudal body region; H, spermatozoa.

Atlantic, *N.* sp. is known only from the west Atlantic and *N. uchidai* is amphi-Atlantic in distribution. Furthermore, it is impossible to decide whether any pair of these species is more closely related than any other pair.

Similarly, in the *Heteroxenotrichula subterranea* species group (Ruppert 1976; Figs. 7–9, 13A, B, C), *H. subterranea* and *H. affinis* are known from the east Atlantic coast (but not known to co-occur) and *H. transatlantica* is reported from west Atlantic beaches. On the basis of their morphology, it is again not possible to align any two species of this group more closely than any other pair.

These two examples demonstrate that there is not a clear correlation between geographic distance and the degree of morphological divergence, i.e. the two most closely related species of any group need not be on opposite sides of the Atlantic. Rather, it may be reasonable to consider that speciation has occurred in locally isolated populations as well as in populations isolated by vast geographic distances and seemingly insurmountable barriers. It is probable that the results of the former

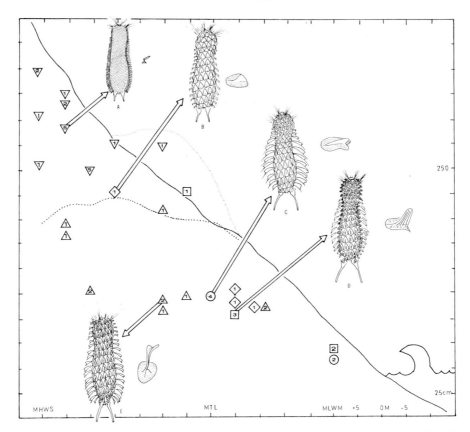

Fig. 5. Distributions of species of *Halichaetonotus* and *Aspidiophorus* on Cap Ferret Beach, France; January 1972; densities in numbers of individuals per 20cc sediment; A, *A. marinus*; B, *H.* sp. *D*; C. *H. aculifer*; D, *H. paradoxus*; E, *H.* sp. *E*.

phenomenon will be superimposed on the general pattern resulting from divergence due to isolation from continental drift. It can be predicted that several species of one species group will occur in one geographic region, and that with even a moderate amount of longshore dispersal many emigration-isolation-speciation-immigration episodes might lead to sympatry, however temporary, in many species groups.

D. Sympatric occurrences of Gastrotricha

Species of the genus *Turbanella* are morphologically homogeneous, eg. there are present lateral and dorsolateral longitudinal rows of adhesive tubes, the head cilia form a complete band, there is a bifid caudal end with a fan-shaped arrangement of adhesive tubes, there is a reflexed male reproductive system etc. Additionally, three principal groups of species can be recognized: 1) *Turbanella cornuta,*

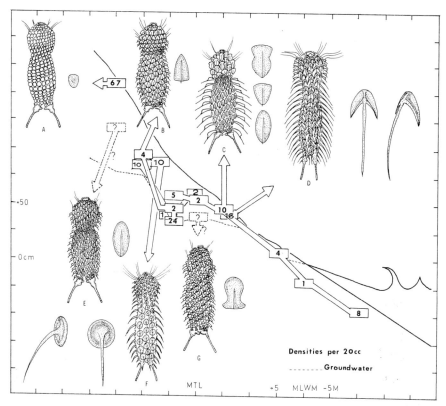

Fig. 6. Distributions of species of *Halichaetonotus, Lepidodermella* and *Chaetonotus* on Bogue Inlet Beach, N. C.; August 1971; A, *L.* sp; B, *H.* sp. *A*; C, *H. aculifer*; D, *C.* sp. *A*; E, *H.* sp. *B*; F, *C.* sp. *B*; G, *H.* sp. *C.*

T. mustela, T. petiti, T. plana, all forms with well developed head appendages (Fig. 1I, J, K, L; 2) *Turbanella hyalina, T. lutheri, T. ocellata, T. subterranea* etc., lacking head appendages (Fig. 13G–J; 3) *Turbanella ambronensis, T. italica (T. veneziana?),* forms with a large folded mouth, rounded head lobes, paired frontal and caudal glands with protruding gland necks and a well developed pair of ventro-lateral tubules (for an alternative grouping of species of this genus see Remane 1953). These species groups occur along both sides of the Atlantic and are known on the Pacific coast of the United States (Wieser 1957, Hummon 1972; Figs. 1I–L, 13A–J).

In the classic comparison of *T. cornuta* and *T. hyalina* (Remane 1925, 1933, 1936), these two similar species were shown to have different habitat preferences, *T. cornuta* being associated with coarse "Amphioxus" sand and *T. hyalina* with fine sand. It may be hypothesized that these two forms speciated into these two distinct habitats. This example may serve as a general model for local speciational

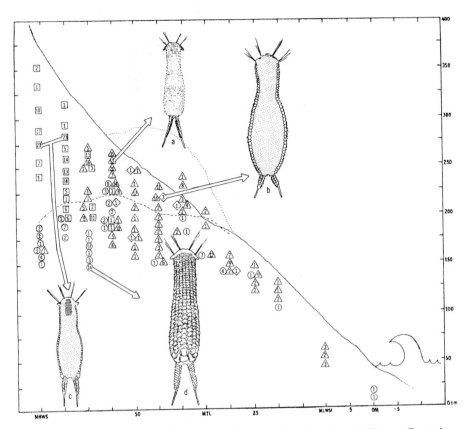

Fig. 7. Distributions of species of Xenotrichulidae on Cap Ferret Beach, France, December 1971 — January 1972; densities in numbers of individuals per 20cc sediment; A, *H. arcassonensis*; B, *X. intermedia*; C, *H. subterranea*; D, *H. wilkei* (after Ruppert 1976).

events, although it is not being proposed that this is the only pattern by which species of Gastrotricha arise. Furthermore, from what we know of their habitat preferences and their morphologies, we can state that if they are found to occur sympatrically, then: 1) that situation might be rare within the total range of their distributions, and 2) that their distributions in the area where they are sympatric might nearly coincide due to the great similarities in their structure.

Five species of the genus *Turbanella* occurred on the Plage d'Eyrac (1971; see Fig. 10). *T. cornuta* and *T. hyalina* were collected in surface sediment in the lower part of the beach up to mid tide level (MTL). *T.* cf. *mustela* and *T. ambronensis* were collected in the zone of damp sand from mean high water neaps (MHWN) to mean high water springs (MHWS) and *T. subterranea* was collected in the zone of damp sand at the tide level of extreme high water springs (EHWS). *T. cornuta* and *T.* cf. *mustela*, though superficially similar in form, exhibit constant

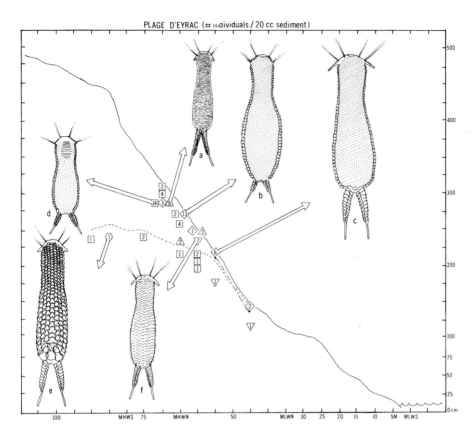

Fig. 8. Distributions of species of Xenotrichulidae on Plage d'Eyrac, Arcachon, France, December 1971 – January 1972; densities in numbers of individuals per 20cc sediment; A, *H. arcassonensis*; B, *X. intermedia*; C, *X. velox*; D, *H. affinis;* E, *H. wilkei*; F, *H. squamosa* (from Ruppert 1976).

differences in adult body lengths and numbers of adhesive tubes especially on the caudal adhesive organ (Fig. 1S, T). Furthermore, I placed living *T.* cf. *mustela* in distilled water in a laboratory watch glass and, after swelling initially, it survived for two days. When this simple experiment was repeated with individuals of *T. cornuta,* they died in minutes.

There is at least one group of structurally similar species in the Thaumastodermatidae. This group consists of the species *Tetranchyroderma hystrix, T. swedmarki, T. antennatum, T. bunti* and *T. papii.* All these species have elongated body forms (250–400 μm), a pair of head tentacles, a pair of short cirri (see Remane 1936) on the dorsofrontal margin of the head, a distinct pair of caudal pedicles each with two distal adhesive tubes and a more dorsal cirrus etc.

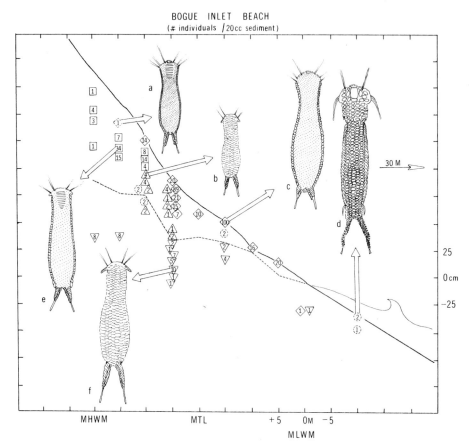

Fig. 9. Distributions of species of Xenotrichulidae on Bogue Inlet Beach, N. C., August-September 1971; densities in numbers of individuals per 20cc sediment; A, *X. lineata*; B, *H. pygmaea*; C, *X. carolinensis*; D, *D. tesselata*; E, *H. transatlantica*; F, *H. squamosa* (from Ruppert 1976).

Three species of this species group occur sympatrically on Bogue Inlet Beach in North Carolina (Fig. 11B, C, D) but have distinct distribution patterns. *T. papii* can be collected in the shallow subtidal portions of the beach where it overlaps somewhat the distribution of *T.* cf. *swedmarki*. The latter species reaches its highest density in surface sediment between the tide levels of mean low water neaps (MLWN) and mid tide level (MTL), but it ranges up the beach where it also overlaps the zone occupied by *T. bunti*. *T. bunti* occurs in surface sediment between MTL and mean high water neaps (MHWN; see Nixon 1976 for a thorough investigation of the static and dynamic distribution patterns of *T. bunti*).

Tetranchyroderma papii, T. bunti and *T.* cf. *swedmarki* are easily confused when observed under a dissecting microscope, but they exhibit clear and constant

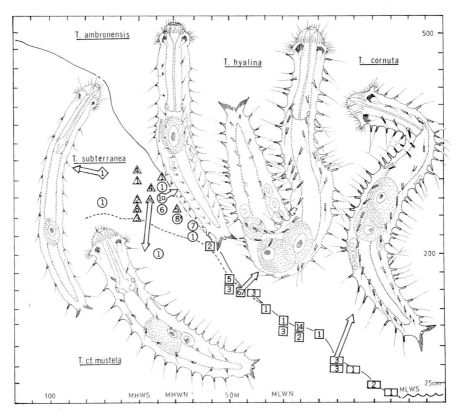

Fig. 10. Distributions of species of the genus *Turbanella* on Plage d'Eyrac, Arcachon, France; November 1971 — January 1972; densities in numbers of individuals per 20cc sediment.

morphological differences when observed at higher magnifications. The most striking dissimilarities occur in the numbers of epidermal glands, the sperm structure, the structure of the cuticular hooks (Fig. 11B—D) and in the shapes and sizes of the head tentacles. Detailed measurements made of tentacle lengths in these three species from over 175 adult specimens demonstrated that the intraspecific variation of this character is so small as to be invariant, while interspecific comparisons of this structure suggest that the differences observed are highly significant (Figs. 13D—F, 12, 1P—R).

There is virtually no intergradation of species characters in this interesting complex of three sympatric forms. Their uniqueness is also reflected in the structure of their spermatozoa. *Tetranchyroderma* cf. *swedmarki* occupies the zone of beach between *T. papii* and *T. bunti* and has a sperm structure markedly different than that of the former two species. The sperm structure of *T. bunti* and *T. papii*, on the other hand, is very similar but their distributions never overlap (at least at low tide; Figs. 1M—O, 11B—D).

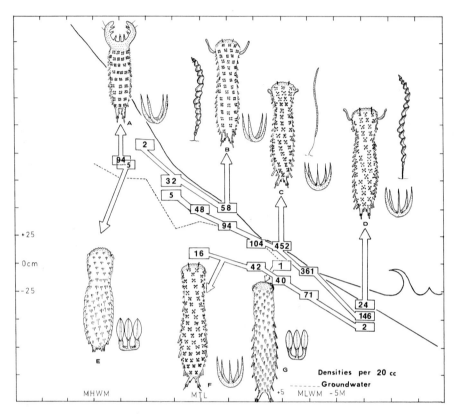

Fig. 11. Distributions of species of Thaumastodermatidae on Bogue Inlet Beach, N. C.; August–
September 1971; A, *P. roscovita*; B, *T. bunti*; C, *T.* cf. *swedmarki*; D, *T. papii*; E, *T.* sp. *A.*;
F. *T.* sp. *B*; G, *T.* sp. *C*.

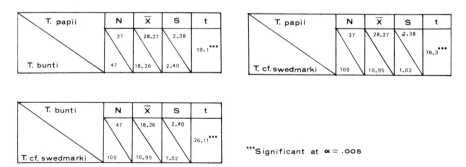

Fig. 12. Statistical comparison of the tentacle lengths of *T. papii, T. bunti* and *T.* cf. *swedmarki*
from Bogue Inlet Beach, N. C.

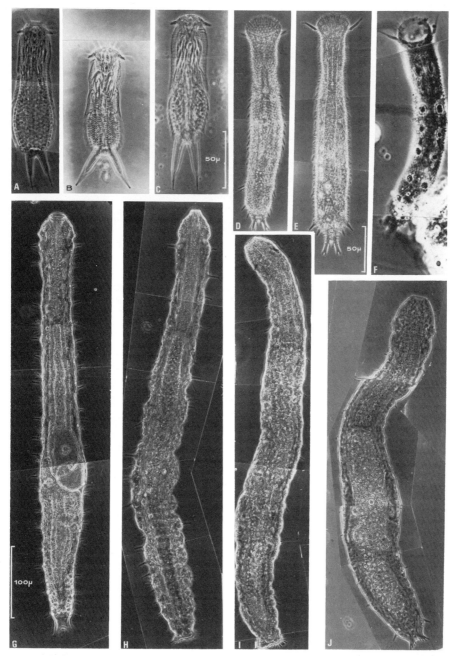

Fig. 13. Species groups in the Gastrotricha: A–C, *H. transatlantica*, Bogue Banks, N. C., *H. affinis*, Plage d'Eyrac, Arcachon, France; *H. subterranea*, Cap Ferret Beach, France; D–F, *T.* cf. *swedmarki, T. papii, T. bunti*, Bogue Banks, N. C.; G–J, *T. hyalina*, Plage d'Eyrac, Arcachon, France; *T.* cf. *hyalina*, White Oak River, N. C.; *T. ocellata*, Pine Knoll Shores, N. C.; *T. subterranea*, Plage d'Eyrac, Arcachon, France. A–E, G–J, living specimens; F, resin embedded adult.

These examples of sympatry in the Turbanellidae and Thaumastodermatidae were chosen to demonstrate that sympatry can be observed frequently in species groups in the Gastrotricha; that the sympatric forms are distinct morphologically; that the specific distributional patterns of sympatric forms are spatially distinct but frequently adjacent; that within the total range of their occurrences, the areas of sympatry may be few; and that it is likely that these forms arose in locally isolated habitats and have become sympatric through parts of their ranges after mutual colonizations of suitable habitats.

E. Adaptations to Colonization

Gastrotricha lack pelagic larvae. If longshore dispersal is important in producing the currently observed distributional patterns and if it is to be effective, then the Gastrotricha must be adapted to disperse themselves or their eggs, while attached to sand grains, or while freely floating in the water column (see Pilkey & Field 1972, Rodolfo, Buss & Pilkey 1971, for dynamics of sediment transport); and it is reasonable to look for these adaptations.

One suborder of the Gastrotricha, viz. the Chaetonotida-Paucitubulatina, has evolved in the direction of small body size, toward the development of a specialized spined or scaled cuticle and toward the adoption of parthenogenesis as a reproductive specialization. Long spines and small body size can aid in flotation (Fig. 14C, D) and parthenogenesis can enhance the possibility that a single individual will initiate a new population (White 1973; Fig. 14E). These trends in the suborder can be considered to culminate in the freshwater families Neogosseidae and Dasydytidae, where a semi-pelagic existence has been adopted and where parthenogenesis is the sole mode of reproduction. Indeed, it may have been this trend to become an adult dispersive propagule in the Chaetonotida-Paucitubulatina that, in part, enabled them to invade the freshwater habitat. In the ubiquitous marine genus *Halichaetonotus, H. aculifer* is not only cosmopolitan in distribution but has the longest ventrolateral spines of all members of this genus (Fig. 14D).

That the evolution of parthenogenesis is of major adaptive value and be important in dispersal is further documented by its independent derivation in two species in different subfamilies of the hermaphroditic family Xenotrichulidae, viz. *Draculiciteria tesselata* and *Heteroxenotrichula pygmaea* (Ruppert 1976). It is probably no mere coincidence that *D. tesselata* and *H. pygmaea* are widely distributed species (Ruppert 1976, Schmidt 1974, Luporini et al. 1971, 1973; Renaud-Mornant 1964, Remane 1934, Hummon 1974). In addition, there is the possibility that in *Ichthydium hummoni*, two types of eggs are produced (Fig. 14F) paralleling the condition of summer and winter egg production in many freshwater micrometazoa. Though this adaptation in *I. hummoni* would appear to be a rare example, it nevertheless illustrates a trend toward selecting for adaptations to unstable environments.

There is an extensive literature (Suomalainen 1950, Rollins 1967, White 1973) discussing the "short-sightedness" of parthenogenesis in abandoning the possibility

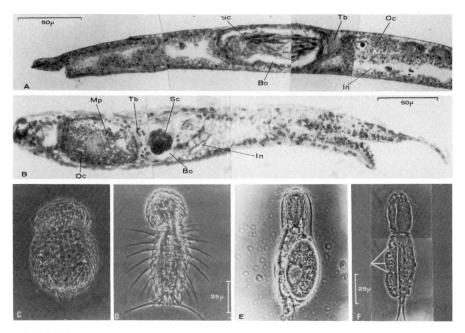

Fig. 14. Adaptations to low density colonizations in Gastrotricha: A, B, median sagittal sections through the caudal end of *Cephalodasys* sp. and through the entire body of *Diplodasys* cf. *ankeli* showing storage of spermatozoa; C, "curled" position of *H.* sp. *C*; D, *H. aculifer* with spines extended, both positions possibly aiding in floatation; E, parthenogenesis in *A. marinus*; F, second egg type (?), possibly resistant eggs in *I. hummoni* (N. C.). Sections stained with Heidenhain's haematoxylin and eosin-Y.

of giving rise to new adaptive lines in favor of retaining a form that is particularly well adapted to existing conditions. It should be considered that the monotonous replications of parthenogenetic species of many Chaetonotida-Paucitubulatina (perhaps also *Urodasys viviparus*) may be a superior means of tracking the monotonous continuance of the sandy environment through geological time. Though we may not expect this group to give rise to new families or even genera, neither do we expect to see them become extinct.

The Macrodasyida, the Chaetonotida-Multitubulatina (*Neodasys*) and some Chaetonotida-Paucitubulatina (many Xenotrichulidae, *Musellifer*, *Polymerurus delamarei* etc.) are hermaphrodites that have developed or retained organs for the receipt and storage of spermatozoa (Remane 1936, Ruppert 1976; Fig. 14A, B). In terms of the "low density" model discussed by Ghiselin (1969) in relation to hermaphroditism, opportunities for fertilization are maximized in hermaphrodites and, with the storage of spermatozoa, a single mating may be sufficient to fertilize all the eggs produced by an individual in its lifetime. This adaptation might not only increase the possibility that all the eggs will be fertilized but it may also ensure that a single adult founder in a new habitat can populate the area. In addition, we have data on the structure of the reproductive organs of the Thaumastodermatidae (Ruppert & Rieger, in prep.) that suggest that self-fertilization, at least facultatively, is structurally feasible in some genera. It is perhaps significant that parthenogenesis, vegetative reproduction, hermaphroditism and self fertilization are widely known as reproductive mechanisms of adaptive value for low density colonizations or for opportunistic species (Ghiselin 1969, Stebbins 1957, Baker 1955, White 1973).

F. Summary

Some observations may now be summarized: 1) there is an enormously wide distribution of species occasionally and species groups frequently (see for example, *Turbanella* cf. *mustela*, Fig. 1J–L); 2) there is little correlation between geographic distance and the degree of morphological divergence between species, though this may reflect an inadequate number of observations. Speciation is likely to be occurring on a local scale; 3) in some cases , the "swamping effect" of gene flow in maintaining form can be flatly rejected, e.g. in the parthenogenetic species *Halichaetonotus aculifer*; rather this form must considered stable; 4) sympatry is commonly observed though the areas of sympatry may be rare in comparison to the total ranges of the species; 5) the occurrence of sympatric species may reflect the dynamic character of the local environment and the ability of the species to undergo longshore dispersal; 6) there is an absence, at least in the *Tetranchyroderma papii, T. bunti, T.* cf. *swedmarki* assemblage, of any intergradation of characters or character displacement, suggesting that these are distinct forms; 7) there may be some adaptations to remaining afloat in the water column; 8) there are reproductive specializations that could optimize founder success though they may simultaneously sacrifice long term variability.

The marine interstitial environment can be viewed as a perpetually renewed but locally transitional habitat, i.e. it is nearly timeless in its general distribution but ephemeral locally. As it is locally originated, maintained and destroyed by water movement, so it is reasonable to assume that colonization is also made possible by that same force. That the species and species groups are worldwide in distribution is a zoogeographic pattern, while not unique to colonizing species, at least characteristic of opportunistic forms (Grassle & Grassle 1974).

Acknowledgements

I gratefully acknowledge my friend Dr. Reinhard Rieger for his stimulating discussions and untiring encouragement as well as his careful reading of the manuscript. I also thank Dr. Pierre Lasserre for his kind invitation to Arcachon in 1971 and I cheerfully acknowledge Drs. C. Cazaux and J. Boisseau for providing me access to the facilities at the Institut de Biologie Marine in Arcachon. This investigation was supported by National Science Foundation grants #Des 74–23761 to the University of North Carolina, and #DES 75–20227 to the Bermuda Biological Station.

Abbreviations

MLWS	mean low water springs
MLWM	mean low water mean
MLWN	mean low water neaps
MTL	mid-tide level
MHWN	mean high water neaps
MHWM	mean high water mean
MHWS	mean high water springs
EHWS	extreme high water springs
Bo	bursal organ
Ep	epidermis
In	intestine
Mp	male pronucleus
Oc	oocyte
Sc	sperm cluster
Sp	spermatozoa
Tb	tissue bridge
Te	testis
Yc	Y-cells

Bibliography

Ax, P. & P. Schmidt: Interstitielle Fauna von Galapagos I. Einführung. Mikrofauna Meeresboden **20**, 1–38 (1973).

– & U. Ehlers: Interstitielle Fauna von Galapagos. III. Promesostominae (Turbellaria, Typhloplanoida). Mikrofauna Meeresboden **23**, 203–216 (1973).

Baker, H. G.: Self-compatibility and establishment after "long-distance" dispersal. Evol. 9(3), 347–349 (1955).

Boaden, P. J. S.: Water movement – a dominant factor in interstitial ecology. Sarsia **34**, 125–136 (1968).

Crezée, M.: Monograph of the Solenofilomorphidae (Turbellaria, Acoela). Int. Rev. ges. Hydrobiol. (In Press) (1976).

Gadow, S. & H. E. Reineck: Ablandiger Sandtransport bei Sturmfluten. Senckenbergiana Maritima **1**, 63–78 (1969).

Gerlach, S.: Gastrotrichen aus dem Küstengrundwasser des Mittelmeeres. Zool. Anz. **150**, 203–211 (1953).

Ghiselin, M. T.: The evolution of hermaphroditism among animals. Quart. Rev. Biol. **44**, 188–208 (1969).

Gooch, J. L.: Mechanisms of evolution and population genetics. In: Marine Ecology, O. Kinne, ed. Vol. **2**(1), 373–374 (1975).

Grassle, J. F. & J. P. Grassle: Opportunistic life histories and genetic systems in marine benthic polychaetes. J. Mar. Res. **32**(2), 253–284 (1974).

Hummon, W. D.: Dispersion of gastrotricha in a marine beach of the San Juan Archipelago, Washington, Mar. Biol. **16**, 349–355 (1972).

–: S$_H$': a similarity index based on shared species diversity, used to assess temporal and spatial relations among intertidal marine Gastrotricha. Oecologia (Berl.) **17**, 203–220 (1974).

Luporini, P., G. Magagnini & P. Tongiorgi: Contribution à la connaissance des Gastrotriches des côtes de Toscane. Cah. Biol. Mar. **12**, 433–455 (1971).

–: Chaetonotoid Gastrotrichs of the Tuscan Coast. Boll. Zool. **40**, 31–40 (1973).

Nixon, D.: Distribution and dynamics of a sandy beach Gastrotrich on Bogue Banks, North Carolina. Int. Revue ges. Hydrobiol. (In Press) (1976).

Pilkey, O. H. & M. E. Field: Onshore transportation of Continental shelf sediment: Atlantic southeastern U. S. In: Shelf Sediment Transport. Swift, Duane and Pilkey, eds. Dowden, Hutchinson and Ross, Inc., Stroudsburg, Pa. (1972).

Rao, G. C. & P. N. Ganapati: Some new interstitial Gastrotrichs from the beach sands of Waltair coast. Proc. Ind. Acad. Sci. **67**(2), B, 35–53 (1968).

Remane, A.: Neue aberrante Gastrotrichen II. *Turbanella cornuta* nov. spec. und *T. hyalina* Schultze 1853. Zool. Anz. **64**, 309–314 (1925).

–: Verteilung und Organisation der benthonischen Mikrofauna der Kieler Bucht. Wiss. Meeresunters. Abt. Kiel **21**, 161–221 (1933).

–: Die Gastrotrichen des Küstengrundwassers von Schilksee. Schr. naturw. Schlesw.-Holst. **20**, 475–478 (1934).

–: Gastrotrichen. Bronn's Klassen u. Ordn. Tierreichs 4(2), Buch 1, Teil 2, Lfrg. 1–2, 242 pp. ·(1936).

–: Die Besiedlung des Sandbodens im Meere und die Bedeutung der Lebensformtypen für die Ökologie. Zool. Anz. (Suppl.) **16**, 327–359 (1952).

–: Ein neues Gastrotrich aus dem Pazifik, *Turbanella palaciosi* nov. spec. Zool. Anz. **151** (9/10), 272–276 (1953).

–: *Neodasys uchidai* n. sp., eine zweite *Neodasys*-Art (Gastrotricha, Chaetonotoidea). Kieler Meeresf. **17** (1), 85–88 (1961).

Renaud-Mornant, J.: Note sur la faune interstitielle du Bassin d'Arcachon et description d'un Gastrotriche nouveau. Cah. Biol. Mar. **5**, 111–123 (1964).

Riedl, R.: Energy exchange at the bottom/water interface. Thalassia Jugoslavica **7** (1), 329–339 (1971).

–: Water Movement. Chapter 5 in Marine Ecology, O. Kinne, ed., Vol. 1(1) (1975).

Rieger, R. M.: *Bradynectes sterreri* gen. nov., spec. nov., eine neue psammobionte Macrostomide (Turbellaria). Zool. Jb., Syst. **98**, 205–235 (1971).

–: E. Ruppert, G. E. Rieger & C. Schoepfer-Sterrer: On the fine structure of gastrotrichs with description of *Chordodasys antennatus* sp. n. Zool. Scripta **3**, 219–237 (1974).

Ruppert, E. E.: On the genus *Pseudostomella* Swedmark, 1956, with descriptions of *P. plumosa* nov. spec., *P. cataphracta* nov. spec., and a form of *P. roscovita* Swedmark, 1956 from the west Atlantic coast. Cah. Biol. Mar. **11**, 121–143 (1970).

–: An efficient, quantitative method for sampling the meiobenthos. Limnol. Oceanogr. **17** (4), 629–631 (1972).

–: Monograph of the Xenotrichulidae (Gastrotricha, Chaetonotida); their systematics, phylogeny and biogeography. Int. Rev. ges. Hydrobiol. (submitted) (1976).

–: & R. M. Rieger: The reproductive system of Gastrotrichs. I. The structure of the reproductive organs of Thaumastodermatidae (In prep.).

– & R. M. Rieger: A method for preserving, concentrating and observing entire samples of meiofauna in Epon-Araldite flat molded embeddments (In prep.).

Rodolfo, K. S., B. A. Buss & O. H. Pilkey: Suspended sediment increase due to hurricane Gerda in continental shelf waters off Cape Lookout, N. C. J. Sed. Petr. **41** (4) , 1121–1125 (1971).

Rollins, R. C.: The evolutionary fate of inbreeders and nonsexuals. Amer. Nat. **101** (920), 343–351 (1967).

Saito, I.: Neue und bekannte Gastrotrichen der Umgebung von Hiroshima (Japan). J. Sci. Hiroshima Univ. Ser. B., Div. **1**, 245–265 (1937).

Schmidt, P.: Interstitielle Fauna von Galapagos. IV. Gastrotricha. Mikrofauna Meeresboden **26**, 497–570 (1974).

Stebbins, G. L.: Self fertilization and population variability in the higher plants. Amer. Nat. **91** (861), 337–354 (1957).

Sterrer, W.: Plate tectonics as a mechanism for dispersal and speciation in interstitial sand fauna. Neth. J. Sea Res. **7**, 200–222 (1973).

Suomalainen, E.: Parthenogenesis in animals. Adv. Genetics. **3**, 193–253 (1950).

Teuchert, G.: Aufbau und Feinstruktur der Muskelsysteme von *Turbanella cornuta* Remane (Gastrotricha, Macrodasyoidea). Mikrofauna Meeresboden **39**, 223–246 (1974).

Thane-Fenchel, A.: Interstitial Gastrotrichs in some south Florida beaches. Ophelia **7** (2), 113–138 (1970).

White, M. J. D.: Animal cytology and evolution. Cambridge University Press, 961 pp. (1973).

Wieser, W.: Gastrotricha Macrodasyoidea from the intertidal of Puget Sound. Trans. Amer. Micros. Soc. **76** (4), 372–381 (1957).

Williams, R.: The abundance and biomass of the interstitial fauna of a graded series of shell-gravels in relation to available space. J. Anim. Ecol. **41**, 623–646 (1972).

Wilke, U.: Mediterrane Gastrotrichen. Zool. Jahrb. Syst. **82**, 497–550 (1954).

Mikrofauna Meeresboden	61	253–262	1977

W. Sterrer & P. Ax (Eds.). The Meiofauna Species in Time and Space. Workshop Symposium, Bermuda Biological Station, 1975

Jaw Length as a Tool for Population Analysis in Gnathostomulida[1]

by

Wolfgang Sterrer

Bermuda Biological Station, St. George's West, 1–15, Bermuda

Abstract

Analysis of jaw length in 25 samples (containing 371 specimens) of *Tenuignathia rikerae* from Bermuda revealed that 1) individual samples are very homogeneous, with a standard deviation usually not exceeding 5% of the mean; 2) samples from localities only a few hundred meters apart are significantly different from each other in the majority (58%) of the cases, and 3) samples from the same locality taken over a four-year span are not significantly different from each other. Whereas no explanation is offered as to the genetic or phenetic origin of the differences, it is noted that only little mixing seems to take place between closely adjacent phena, and over the timespan observed, which would strengthen the hypothesis that dispersal in interstitial sand fauna is very restricted.

A. Introduction

Ever since the phylum Gnathostomulida (Ax 1956; Riedl 1969; Sterrer 1972) became known in its rather unexpected abundance and diversity, their jaw morphology (together with that of the basal plate) has been used as the primary character for species identification (Sterrer 1968). Apart from the complexity of these hard structures, this was mostly due to the fact that they did not seem to change with the age or sexual maturity of a specimen, and that their variability, both within a population and between considerable geographic distances, was surprisingly small (Sterrer & Farris 1975). These observations applied particularly to the jaw length, and led to the adoption of a very narrow definition of jaw length for species diagnosis (Sterrer 1972).

When working on the description of *Tenuignathia rikerae* Sterrer, 1976, I noticed an unusually wide range of jaw lengths not only between distant localities but even between samples taken only a few hundred meters apart. The present paper, while unable to offer an explanation for this inconsistency, is the first statistical analysis of jaw length variability in Gnathostomulida, and may thus contribute to a better evaluation of this character for taxonomy, zoogeography and speciation studies.

[1] Contribution No. 689 from the Bermuda Biological Station for Research, Inc.

B. Material and Methods

Qualitative sediment samples of 5–7 l volume were collected by hand, taken back to the lab. in buckets, and treated with MgCl$_2$ space technique (Sterrer 1971). Extracted live specimens were then placed under a coverslip, anesthesized again with MgCl$_2$, and squeezed until the mouthparts were optimally visible. Whereas the degree of squeezing may considerably distort the measurements of "soft" characters (such as body length and width, etc.), it has a negligible effect on the dimensions of the "hard" mouth parts. The jaw length was measured to the nearest μm, along the inner edge of that half of the jaws which appeared the longer of the two.

In the following, "sample" means all the specimens of a particular species that were extracted from one bucket full of sand. "Phenon" (sensu Mayr 1969) denotes a group of samples (usually from the same locality) that are not significantly different from each other. "Locality" is the place where samples were taken; the accuracy of repeated sampling was probably within maximally 20 m. A total of 25 samples taken over a four-year span (1972–1975) in the north-eastern part of Bermuda (fig. 1), were processed for statistical analysis. They yielded 371 specimens of *Tenuignathia rikerae*, as well as 102 specimens of *Gnathostomula tuckeri* Farris, 1977, and 52 specimens of *Problognathia minima* Sterrer & Farris, 1975. In addition, 9 specimens of *Tenuignathia rikerae* were collected near Morehead City (N. Carolina), and 2 near Big Pine Key (Florida). Throughout the present study, emphasis was on the Bermudian samples of *T. rikerae*; the other species were used for comparative purposes when necessary.

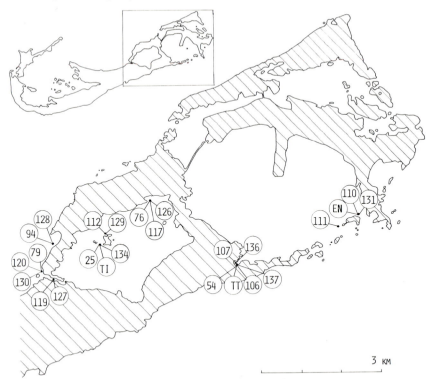

3 KM

Fig. 1. Sampling localities in Bermuda.

The simple statistical methods included Student's T-test (two-tailed) and linear regression analysis. The 99% (p = 0.001) level of confidence was used for all tests of significance. In the two-sample comparisons (Table III), samples from the same locality were first tested against each other; if they were not significantly different (which was the case in all except # 131 and 110), they were then lumped for subsequent tests against other localities.

C. Problems, hypotheses and results

An analysis of earlier data on jaw length particularly of representatives of the genera *Haplognathia, Pterognathia* (Sterrer 1968) and *Austrognatharia* (Sterrer 1970) shows a remarkably narrow range of variability within the same locality (Table I). Even when samples were taken several months to more than one year apart, the standard deviation (s) of jaw length reached maximally 12% of the mean ($\overline{x}$) but generally did not exceed 5%. The few specimens collected in Porta-ferry (N. Ireland), about 1.500 km from the main collecting locality (Kristineberg, Swedish west coast), showed jaw length data which, although slightly different, seemed to generally fall within the range of the specimens from the main collecting site. Furthermore, Müller & Ax (1971), reporting seven species (of *Haplognathia* and *Pterognathia* originally found by me in Sweden and N. Ireland) from Sylt (North Sea), give jaw lengths well within the range of the original description. It could be assumed, therefore, that jaw length in Gnathostomulida was a stable character within a considerable geographic range. This meant, on the other hand, that I had no doubt that a sample of 7 specimens of *Gnathostomaria* sp. collected in N. Carolina (U.S. east coast), with a mean jaw length of 23.7μm, should be described as a new species and thus differentiated from the Mediterranean species *G. lutheri* Ax, 1956 (jaw length: $\overline{x}$ = 30.1 μm). Similarly, it seemed justified, in the case of *Nanognathia exigua* Sterrer, 1973, to at least distinguish between a Florida (or type) form ($\overline{x}$ = 15.25 μm, s = 1.25 μm; n = 20) and a North Carolina from ($\overline{x}$ = 12.5 μm, s = 0.71 μm; n= 2).

When first recording data of the new species and genus *Tenuignathia rikerae* (which lacks a basal plate), the jaw length of N. Carolina specimens ranged from 24 to 28 μm, and of Florida specimens from 27 to 28 μm. The first Bermuda specimens (from Harrington Sound) measured 27 to 30 μm. Then I came across specimens from another Bermuda locality, Tucker's Town, which measured 28 to 37 μm. This was far beyond the known range of variability, and suggested the present study.

Analysis of the Bermudian samples of *T. rikerae* by means of two-tailed t-test produced three immediate results:
1) There is very little variability within a sample — the average standard deviation being only 5.4 % of the mean (but see sample #131).
2) There are no significant differences (on the 99 % level of probability) between samples taken at the same locality but at different times (but see sample #131), and

Wolfgang Sterrer

Table I. Earlier data on jaw length in some species of Gnathostomulida.

Species	Reference	Locality	n	Mean ($\bar{x}$)	Standard Deviation (s)	Standard Dev. in % of the mean
Haplognathia simplex	Sterrer 1968	Kristineberg	35	30.09	.95	3.2
Haplognathia simplex	Sterrer 1968	Portaferry	3	30.0	1.73	5.7
Haplognathia simplex	Müller & Ax 1971	Sylt	30	(30)		
H. filum	Sterrer 1968	K	15	19.93	.46	2.3
H. filum	Müller & Ax 1971	Sylt	3	(19)		
H. gubbarnorum ("small form")	Sterrer 1968	K	20	20.50	.89	4.3
H. gubbarnorum ("small form")	Sterrer 1968	P	1	23		
H. gubbarnorum ("small form")	Müller & Ax 1971	Sylt	1	20		
H. lunulifera	Sterrer 1968	K	8	20.37	.74	3.6
H. lunulifera	Sterrer 1968	P	1	19		
H. rosea	Sterrer 1968	K	15	17.47	2.1	12.0
H. rosea	Sterrer 1968	P	1	(21)		
H. rosea	Müller & Ax 1971	Sylt	20	(17−18.5)		
Pterognathia swedmarki	Sterrer 1968	K	34	26.85	1.18	4.4
Pterognathia swedmarki	Sterrer 1968	P	1	29		
Pterognathia swedmarki	Müller & Ax 1971	Sylt	20	(26)		
Pt. atrox	Sterrer 1968	K	7	19.57	.79	4.0
Pt. atrox	Sterrer 1968	P	2	20.5	.71	3.4
Pt. meixneri	Sterrer 1968	K	6	25.33	.52	2.0
Pt. sorex	Sterrer 1968	K	5	17.80	.84	4.7
Pt. sorex	Müller & Ax 1971	Sylt	1	(18)		
Haplognathia rosacea	Sterrer 1970	Wrightsville, N. C.	11	18.54	.82	4.4
H. lyra	Sterrer 1970	W	9	16.22	.97	5.9
Austrognatharia kirsteueri	Sterrer 1970	W	6	20.0	1.26	6.3

3) There are significant differences between most samples from different localities. The following working hypotheses were formulated:

1) That jaw length, contrary to previous evidence, is in fact correlated with body length, age, or sexual maturity, and therefore only of limited use for variability studies.
2) That jaw length changes with the seasons, regardless of body size, age, etc.
3) That there is a cline, or a series of disjunct populations, over a geographic distance along the Bermuda shoreline, indicating incipient speciation. This seemed particularly appealing in view of the fact that Harrington Sound (where several sampling localities are situated) is a very isolated though fully marine water body with only one shallow opening to the ocean.
4) That the differences in jaw length are a phenotypic response to a particular parameter of the respective (micro-)environment, and thus not primarily an expression of ongoing speciation.

It was relatively easy to eliminate the first hypothesis. Fig. 2, based on two different samples, shows that there is no significant correlation between jaw length and either the body size, or the presence of testes, a malestylet, a bursa, or an

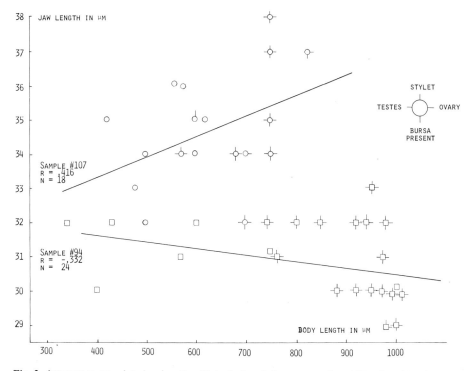

Fig. 2. Attempt to correlate jaw length with body length in two samples of *Tenuignathia rikerae* from Bermuda. There is no correlation on the 99% significance level.

Table II. Samples, grain size analysis and jaw length data of some species of Gnathostomulida in Bermuda, 1972–1975.

Sample #	Date coll.	Locality	Depth in m	>4	4–2	2–1	1–.5	500–250	250–125	125–63	63–37	<37	Tenuignathia rikerae n	x̄	S	S in % of x̄	Gnathostomula tuckeri n	x̄	S	S in % of x̄	Problognathia minima n	x̄	S	S in % of x̄
25	28/4/72	Harrington Sound W of Trunk Island	2–3										6	27.83	1.17	4.3					5	22.90	1.67	7.3
TI	9/73		2–3										17	28.29	1.38	4.9					6	22.66	.52	2.3
134	10/12/75		3	1.1	1.6	13.0	34.0	40.5	8.8	.8	.3	.2	2	29.50	2.12	7.3	15	16.46	.52	3.1				
112	29/8/75	Harrington Sound N of Trunk Island	1	3.8	2.2	3.6	12.9	39.6	35.3	2.4	.3	.2	20	31.50	1.64	5.3								
129	19/11/75		1	4.6	.7	1.3	12.0	43.8	34.1	2.4	.5	.3	19	30.78	1.58	5.3	13	16.15	.80	5.0				
76	4/9/73	Harrington Sound Church Bay	3–4	.2	1.3	2.2	6.4	36.9	47.9	4.2	.6	.2	7	28.93	2.03	7.0								
117	3/9/75		2										7	30.43	1.90	6.3								
126	19/11/75		4	1.4	.8	2.1	5.6	27.7	53.9	7.0	.7	.2	8	28.87	.99	3.4	9	15.22	.97	6.5				
119	3/9/75	in Flatts Inlet S of boat channel	1–2	.1	.6	.9	5.8	25.4	63.3	2.7	.1	.1	21	30.57	2.01	6.7								
127	19/11/75		1	.1	.1	.5	2.4	25.9	69.7	1.8	.1	.1	22	29.72	1.35	4.5	5	15.80	.84	5.2				
79	4/9/73	N of entrance to Flatts Inlet	3	2.4	3.0	9.4	18.4	28.7	36.7	.6	.1	.1	14	31.57	1.96	6.3								
120	3/9/75		2–3										21	31.71	1.38	4.3								
130	19/11/75		3	.7	3.9	9.3	11.1	19.1	53.6	2.0	0	0	16	32.93	1.69	5.1	2	15.50	.71	4.7				
94	26/7/75	Shelly Bay, West side	4	1.9	5.3	25.8	27.4	18.4	19.6	1.2	.2	.1	26	30.92	1.16	3.7								
128	19/11/76		6	1.3	4.6	26.8	31.1	22.3	13.4	.7	.1	.2	11	31.54	1.37	4.3	17	16.82	.73	4.3				
111	21/8/75	Castle Harbour, W of Nonsuch Is.	3	0	.1	1.8	12.9	46.9	36.1	1.6	.1	.1	13	29.69	1.18	3.9					7	23.14	1.07	4.6
EN	21/8/75	Castle Harbour, E of Nonsuch Island	2										4	31.25	1.26	4.1								
110			2										7	30.00	2.00	6.7								
131	29/11/75		2	.8	.6	1.9	7.6	48.8	18.9	1.1	.2	.1	21	35.57	3.64	10.1	16	15.25	.93	6.2				
54	15/12/72	Castle Harbour, Tucker's Town Cove, above sandbar	0–1										17	33.18	1.94	5.9					16	23.59	1.02	4.4
TT	9/73		0–1										7	33.14	2.55	7.7					4	24.50	1.00	4.2
106	21/8/75		0–1										21	33.57	1.78	5.2					6	22.83	.41	1.8
137	17/12/75		0–1	.1	.4		7.9	51.9	38.6	.6	.2	.2	20	33.60	1.46	4.3	12	17.08	.67	3.9				
107	21/8/75	Castle Harbour, Tucker's Town Cove, below sandbar	1–2										20	34.70	1.56	4.5								
136	7/12/75		1–2	0	.1	.9	13.2	54.8	30.3	.3	.1	.1	20	33.90	1.29	3.8	13	16.53	.66	4.1				

ovary. Fig. 2 also shows that the mean body length of adults is smaller in sample #107, and larger in sample #94; body length seems inverse to jaw length, a possible correlation which deserves further analysis in the future.

While data are not available over a whole year's period (as would have been desirable), the great consistency between samples taken at the same locality but in up to three different months and as many different years (e. g. #54, TT, 106 and 137) suggests that there is no significant change within these time spans. This is paralleled by the consistency in grain size distribution (e. g. Table II, #112 and #129, or #94 and #128), indicating that environmental conditions did not change significantly either.

Table III. Two-tailed, two-sample t-test between jaw length samples taken in Bermuda, 1972–1975 (see Table II). "+" stands for "significant difference", "–" for "no significant difference", on a significance level of 99%. The upper right half of the square applies to *Tenuignathia rikerae*, the lower half to *Gnathostomula tuckeri* and (in brackets) *Problognathia minima*.

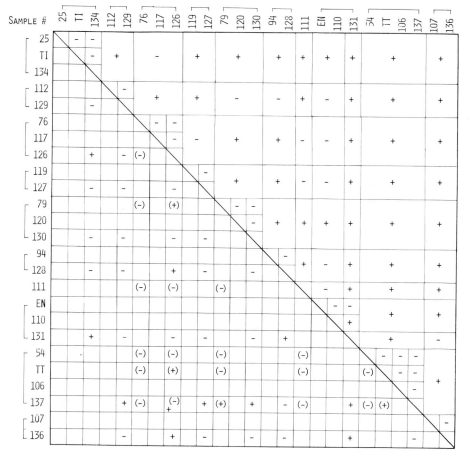

The third step, an attempt to arrange samples according to mean jaw length (from shortest to longest) and to then correlate this series with the geographic distances of their respective localities, failed miserably. Not only is there no correlation between jaw length differences and geographic distance, but some of the most striking and best documented differences occur between samples only a few hundred meters apart. This is evident when comparing sample #25 with #112 (both in Harrington Sound, but on different sides of Trunk Island; (linear distance 200 m). Consequently, there is no pattern suggestive of a cline, or a series of disjunct though related populations.

Although this finding does not rule out genetic differences between samples as such, the next step was to find clues as to a possible phenetic nature of the differences. The only environmental parameters available were water depth and granulometry of the sample. Since the tidal range in Bermuda is up to 1.5 m, and most samples were collected in the intertidal or shallow subtidal, water depth was useless as a parameter. To test grain size distribution against jaw length, on the other hand, seemed justified, either because of the possibility of a direct correlation (e.g. jaws might have to be longer in order to scrape food off larger sand particles), or because the grain size composition of a sample can be regarded as an indicator for other parameters (e. g. wave energy, chemistry of interstitial water, etc.). Linear regression analysis, however, failed to reveal a significant correlation between jaw length and any of the size classes of sand grains.

Results were equally negative when trying to correlate the mean jaw lengths of *Tenuignathia rikerae* with those of *Gnathostomula tuckeri* or *Problognathia minima* from the same samples. This might have given an indication as to the selectiveness of an environmental influence with regard to jaw length.

Two-sample comparison between jaw length data in the three species studied revealed that *T. rikerae* has by far the widest range of differences (out of 76 pairs of samples tested, 43 – or 58 % – are significantly different from each other), whereas *G. tuckeri* (11 out of 34; or 32 %) and *P. minima* (4 out of 21; or 19 %) are more homogeneous over the same sampling range. The fact, however, that there are highly significant differences also in those species deserves attention.

D. Conclusions

It is clear that the present material does not give an answer to the question: incipient speciation versus phenotypic variation. While the lack of a geographic pattern (such as a cline) seems to contradict an obvious speciation hypothesis, it cannot be ruled out that the samples did in fact come from genetically different populations. Other techniques (such as chromatography assays for gene frequency, or cross-breeding and rearing in the lab.) will have to be used to clarify this problem.

The three positive results of the present study, however, deserve consideration. The first is the simple fact that, probably in all Gnathostomulida known so far, jaw length is independent of age, is readily measured, only slightly variable within a sample, and therefore a useful tool for population analysis. The second result is the occurrence of highly significant differences between samples only short horizontal distances apart. The third is the consistency, over long periods of time, between samples taken from the same locality. Together, these results might be interpreted in the following way. Highly localized factors (genetic or environmental) produce equally localized phena. These factors do not change significantly over a period of (four) years, or otherwise the samples taken at the same localities would reflect this (the consistency in the grain size distribution between samples from the same locality may attest to the consistency of environmental factors). Mixing of adjacent (but significantly different) phena must be rare at least within the lifetime of a specimen, or otherwise no differences could be recorded, nor would the variability within a sample be so small. (The case of sample #131, which was not only significantly different from another sample recorded at the same locality, but also had an unusually high standard deviation, suggests such mixing; in fact the sampling locality is exposed to the south, and sampling took place after a period of predominantly southerly winds).

In conclusion, it is interesting to note that a study whose initial hypothesis was one of ongoing, rapid speciation (Harrington Sound is only 2–3,000 years old as a marine environment) has produced as its main result the fact that mixing of phena is rare even over very short distances. This would seem to support my earlier hypothesis (Sterrer 1973b) of limited dispersal, and again characterize interstitial fauna as the most sedentary of all marine faunas.

Acknowledgements

I am indebted to my friends R. Rieger (Chapel Hill) and R. Farris (McMinnville) for stimulating discussion. My work was supported by NSF Grant DES 7520227.

Bibliography

Ax, P.: Die Gnathostomulida, eine rätselhafte Wurmgruppe aus dem Meeressand. Akad. d. Wiss. u. d. Literatur. Abhandl. d. Math.-nat. Kl. 8, 534–560 (1956).

Farris, R. A.: Three new species of Gnathostomulida from the West Atlantic. Int. Rev. ges. Hydrobiol. (1977) (in press).

Mayr, E.: Principles of Systematic Zoology, McGraw-Hill, New York, 428 pp. (1969).

Müller, U. & P. Ax.: Gnathostomulida von der Nordseeinsel Sylt mit Beobachtungen zur Lebensweise und Entwicklung von Gnathostomula paradoxa Ax. Mikrofauna Meeresboden 9, 1–41 (1971).

Riedl, R. J.: Gnathostomulida from America. Science 163, 445–452 (1969).

Sterrer, W.: Beiträge zur Kenntnis der Gnathostomulida. I. Anatomie und Morphologie des Genus Pterognathia Sterrer. Ark. Zool. 22 (1), 1–125 (1968).

—: On some species of *Austrognatharia, Pterognathia* and *Haplognathia* nov. gen. from the North Carolina coast (Gnathostomulida). Int. Revue ges. Hydrobiol. **55** (3), 371–385 (1970).

—: Gnathostomulida: problems and procedures. In: Proceedings of the First Int. Conference on Meiofauna, N.C. Hulings, ed. Smithson. Contr. Zool. 76, 9–15 (1971).

—: Systematics and evolution within the Gnathostomulida. Syst. Zool. **21**, 151–173 (1972).

—: On *Nanognathia,* a new gnathostomulid genus from the east coast of the United States. Int. Rev. ges. Hydrobiol. **58**, 105–115 (1973a).

—: Plate tectonies as a mechanism for dispersal and speciation in interstitial sand fauna. Netherlands J. Sea Res. 7, 200–222 (1973b).

—: *Tenuignathia rikerae* nov. gen. nov. spec., a new gnathostomulid from the West Atlantic. Int. Rev. ges. Hydrobiol. **61**, 249–259 (1976).

— & R. A. Farris.: *Problognathia minima* nov. gen. nov. spec., representative of a new family of Gnathostomulida, Problognathiidae n. fam. from Bermuda. Trans. Amer. Micros. Soc. **94**, 357–367 (1975).

| Mikrofauna Meeresboden | 61 | 263–270 | 1977 |

W. Sterrer & P. Ax (Eds.). The Meiofauna Species in Time and Space. Workshop Symposium, Bermuda Biological Station, 1975

Life Histories of Marine Nematodes. Influence of Temperature and Salinity on the Reproductive Potential of Chromadorina germanica Bütschli[1]

by

John H. Tietjen and John J. Lee

Department of Biology and Institute of Marine and Atmospheric Science, City College of New York, New York, New York 10031, U.S.A.

Abstract

Chromadorina germanica Bütschli, a free-living marine nematode common in the aufwuchs assemblages of marine macrophytes located in North Sea Harbor, Southampton, New York was isolated from *Enteromorpha intestinalis* and established in laboratory culture. Experiments were conducted to evaluate the influence of temperature and salinity on the reproductive potential of the organism. Reproduction was quickest at $26^o/oo$ salinity and over the temperature span of 24–28 °C. Reproduction was slightly lower at $39^o/oo$ and severely depressed at $13^o/oo$. Upper lethal temperature varied with salinity; at $26^o/oo$ it was 35 °C. Generation time at $26^o/oo$, calculated from the net reproductive rate per generation and instantaneous rate of maximum increase, was 12–15 days between 20 and 30 °C. Generation time increased significantly below 20 °C and above 30 °C; at 34 °C (one degree below the upper lethal limit) it was calculated to be nearly 100 days.

Values of the reproductive potential calculated from laboratory cultures agreed well with those obtained from the study of natural populations. Assuming a salinity of $26^o/oo$, the number of generations of *C. germanica* produced annually under temperature conditions prevailing in North Sea Harbor is estimated at 13. This compares favorably with the numbers of generations estimated for other species with similar reproductive rates.

A. Introduction

Within the past decade, information on various aspects of the life histories of marine nematodes gained from the successful laboratory culturing of these organisms has significantly increased. A number of studies have focused on the effects of temperature on the life cycles of marine nematodes (Tietjen et al. 1970; Gerlach & Schrage 1971, 1972; Tietjen & Lee 1972; Hopper et al. 1973); however, there are only two species, *Rhabditis marina* Bastian and *Monhystera denticulata* Timm for which experimental data on the effects of both temperature and salinity on the life cycle are available (Tietjen et al. 1970; Tietjen & Lee 1972).

---

[1]Contribution No. 98 from the Institute of Marine and Atmospheric Science.

If the importance of free-living nematodes in the benthic habitat is to become known, a knowledge of the number of generations produced annually by these organisms will be necessary. For longer lived species, it may be possible to estimate the number of generations produced annually in the benthic habitat from field studies. However, for shorter lived species, this is difficult due to the fact that the generations of such species often overlap. For shorter lived species, it may prove necessary to estimate the annual production from life cycle information generated from laboratory studies. Such laboratory experiments, however, should be conducted over the annual range of particular variables normally encountered by the organism (temperature, salinity, etc.) and not only at optimum conditions. From such experiments, a realistic estimate of annual production might prove possible.

The aim of the present paper is to present information on the influences of temperature and salinity on the reproductive potential of *Chromadorina germanica* Bütschli, a species of nematode common in shoal benthic habitats, and to compare the response of this organism to variations in temperature and salinity with other marine nematodes for which such information is available.

B. Methods

Chromadorina germanica was isolated from the aufwuchs present on *Enteromorpha intestinalis* located in North Sea Harbor, Southampton, New York. The nematode is quite abundant in this area during mid summer, and may account for as much as 85 % of the total nematode fauna present in the aufwuchs assemblages of the marine macrophytes in the area in July and August.

Isolation and maintainance. Our standard isolation technique was employed in isolating *C. germanica* into continuous culture; this technique is detailed elsewhere (Tietjen & Lee 1972, 1973) and will not be described here. Continuous culture of the organism was achieved on the diatom *Cylindrotheca closterium* and the chlorophyte *Chlorococcum* sp; information on the feeding habits of the nematode has been published (Tietjen & Lee, 1977). *Chromadorina germanica* has a definite preference for algae over bacteria, ingesting, on a daily average, a weight of algae three to six orders of magnitude higher than bacteria. Furthermore, some recent experiments have shown that the nematode is unable to maintain itself on a diet of bacteria alone. In the temperature and salinity experiments conducted, *C. germanica* was maintained in liquid Erdschreiber medium, using *Chlorococcum* sp. and *Cylindrotheca closterium* as foods.

Temperature and salinity experiments. Experiments were conducted using an aluminum thermal gradient block. Details and specifications of the block are given by Thomas et al. (1963). Briefly, the block, constructed of aluminum alloy and measuring 100 cm by 30.5 cm, has a thermal gradient established through it by circulating cold water in a channel at one end and warm water in a similar channel at the other end. The block is insulated with styrofoam plastic. A total of 78 holes, each with an inside diameter of 2.1 cm were drilled into the block to accomodate 20 x 125 mm screw-cap test tubes; there are 13 rows of holes with six holes in each row. The block is especially designed so that temperature over the entire end of the block is uniform. In the experiments performed, we were able to examine the population growth of *C. germanica* at 2 °C intervals over a temperature span of 2–42 °C . The salinities

at which experiments were conducted were 6.5 $^o/_{oo}$, 13 $^o/_{oo}$, 26 $^o/_{oo}$, 39 $^o/_{oo}$ and 52 $^o/_{oo}$.
The block was maintained on its side in front of six cool white fluorescent lamps with a 16
hour diurnal light period. The lamps were so arranged that the total energy reaching the test
tubes containing the nematodes and algae was approximately equal.

The 20 x 125 mm borosilicate screw-cap test tubes were inoculated with 50–100 adult
nematodes and placed in the block. The tubes were examined the following day to check
nematode viability. The initial population level in each tube was noted, and the population
levels in the tubes was then monitored at seven day intervals.

The influences of salinity and temperature on the populations of nematodes in the tubes
were evaluated by examining the reproductive potential of the nematodes at the different
temperatures and salinities tested. If an animal population is allowed to increase without
limit, the increase of the population at any moment is, of course, proportional to the size
of the population at that moment. This is represented by the well-known expressions,

$$dN/dT = (b - d) N = rN$$

or in integrated form:

$$N_t = N_o \, e^{rt}$$

where N_o and N_t are the population sizes at time zero and time t respectively e the base of
natural logarithms and r the intrinsic rate of natural increase. r is equal to

$$r = \frac{1}{t} \ln \frac{N_t}{N_o}$$

and its measurement allows the comparison of the reproductive potential of unrelated groups
of organisms, different species, or representatives of the same species grown under different
environmental conditions.

Since the experiments were conducted in an environment in which food was non-limiting,
the reproductive potential observed represented the animals' maximum rate of natural increase,
or r_{max}.

C. Results

At the extreme salinities tested (6.5 and 52 $^o/_{oo}$), 100 % mortality of the
worms was observed at all temperatures. For the other salinities r_{max}, plotted
as a function of temperature, is given in Figure 1. At all three salinities given in
Figure 1, r_{max} occurs in the temperature range of 24–28 °C. At 13 and 39 $^o/_{oo}$,
the reproductive potential of C. germanica at temperatures below 17.5 °C was
too low to measure with the technique employed in this experiment; at 13 and
39 $^o/_{oo}$ the numbers of nematodes in the tubes remained relatively constant
or declined slightly at temperatures between 17.5 ° and 11 °C. Below 11 °C,
the nematodes died within a few days. Population growth was most restricted
at 13 $^o/_{oo}$; at this salinity (1) the range of temperature tolerated by C. germanica
was the narrowest observed (19–30 °C); (2) the upper lethal temperature (31 °C)
was the lowest for all three salinities considered and (3) at any given temperature
within its tolerance range, the reproductive potential of the organism was the
lowest for all three salinities considered. Population growth was better at 39 $^o/_{oo}$
than at 13 $^o/_{oo}$; at 39 $^o/_{oo}$ the upper lethal limit was higher (33 °C) and population growth within the temperature tolerance range of the organism was considerably greater than at 13 $^o/_{oo}$.

John H. Tietjen and John J. Lee

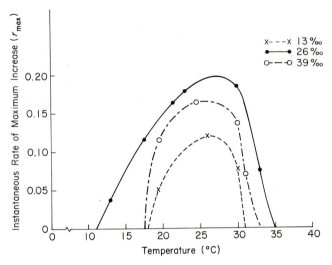

Fig. 1. Instantaneous rate of maximum increase of *Chromadorina germanica* at 13, 26 and 39 $^0/_{00}$ over a 10–35 °C temperature span.

Optimum salinity was 26 $^0/_{00}$. At this salinity the upper lethal temperature was at its highest (35 °C), and increases in population levels were observed at temperatures as low as 11.5 °C. At 26 $^0/_{00}$ some survival of *C. germanica* was observed at temperatures as low as 7 °C, but the population levels were severely depressed. In addition to being able to tolerate a broader temperature range at 26 $^0/_{00}$ the rate of population growth, within the 20–30 °C optimum, was fastest at this salinity.

From the *r* values obtained from these experiments, it was possible to calculate the generation time of *C. germanica* as

$$T = \frac{\ln R}{r}$$

where T is the generation time, *r* the instantaneous rate of natural increase and R the net reproductive rate per generation (N_{t+1}/N_t). The generation times so calculated are given in Figure 2.

At all three salinities the generation time decreased with increasing temperature until 25 °C, above which it increased. As the lethal temperatures are approached, generation time lengthened significantly. Between 20° and 30 °C, the generation time of nematodes grown at 13 $^0/_{00}$ was approximately twice as long as that of worms grown at 26 and 39 $^0/_{00}$. Outside the 20–30 °C range, but still within the range of temperature tolerance, the generation time of worms grown at 26 $^0/_{00}$ was considerably shorter (by a factor of 2 or 3) than those grown at 39 $^0/_{00}$. Finally, at the optimum salinity of 26 $^0/_{00}$, the estimated generation time for worms grown between 20° and 30 °C was relatively short (12–15 days); as the extremes of temperature at which the populations were still able to maintain them-

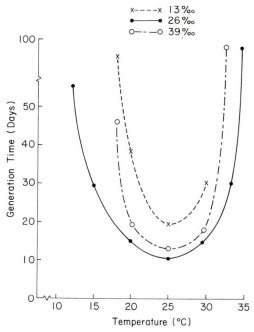

Fig. 2. Estimated generation time (days) of *Chromadorina germanica* at 13, 26 and 39 $^o/_{oo}$ over a 10–35 °C temperature span.

selves were approached, the generation times increased to 50 days or more. This phenomenon was especially interesting in the region of higher temperatures, where the calculated generation time approached 100 days at 34 °C, one degree below the observed lethal limit.

D. Discussion

The data indicate that the reproductive potential of *Chromadorina germanica* may be significantly influenced by both temperature and salinity. Within the optimum temperature range (20–30 °C) reproduction was significantly lower for worms grown at 13 $^o/_{oo}$ than it was for worms grown at 26 and 39 $^o/_{oo}$. Over the same temperature span, the calculated generation time of worms grown at 13 $^o/_{oo}$ was approximately double that of the worms grown at 26 and 39 $^o/_{oo}$, whereas there was little difference between the calculated generation times of worms grown at the latter two salinities. These results were remarkably similar to those obtained for *Monhystera denticulata*, another species of nematode isolated from the same locality. In *M. denticulata*, while the difference between the observed generation times (at 15 ° and 25 °C) of worms maintained at 26 and 39 $^o/_{oo}$ was greater than those calculated at these salinities for *C. germanica*, the differences in generation times were not as great as those observed between 13 and 26 %o. In other words, it appears that the generation times of both nematodes may be more significantly lengthened by exposure to lower salinities than to higher ones.

The depression of reproductive potential (and hence increase in the generation time) of *C. germanica* at decreasing temperatures below 20 °C was likewise similar to what was observed for *M. denticulata*. Low temperatures are known to prolong the generation times of several marine nematodes (Gerlach & Schrage 1971, 1972; Tietjen & Lee 1972; Hopper et al. 1973). However, as stated by Hopper et al., the potential effects of increased temperature on marine nematodes has not received attention. The reproductive potential of *C. germanica* was significantly influenced by elevated temperatures; generation times were significantly lengthened at elevated temperatures just below the lethal limit. In this respect the results obtained for *C. germanica* were in excellent agreement with those obtained by Hopper et al. (1973), who found that the life cycles of the six nematodes they studied tended to lengthen as the upper lethal limits were reached. They also noted that, for each species studied, most rapid growth occurred over the 24–31 °C range and that lethal limits were in the 33–35 °C range. Ambient temperature of their stock cultures was 24 °C, nearly identical to ours (25 °C).

Chromadorina germanica attains its peak abundance in North Sea Harbor in mid-summer (July and August); average water temperature and salinity at this time are 24 °C and 25 %o, respectively. However, in the area of the sampling station, temperatures as high as 33 °C are common at low tide in July and August, and salinity levels may also be elevated (Matera and Lee 1972). It is quite possible, therefore, that a combination of factors, namely the ability of the nematode to reproduce quite rapidly at elevated temperatures and salinities between 26 and 39%o, coupled with the availability of food, enables *C. germanica* to overwhelmingly dominate the fauna at this time.

Both *C. germanica* and *Monhystera denticulata* appear to be better adapted for life in the middle and upper portions of their salinity range than in the lower portions. The physiological explanation for this is unknown, although Croll & Viglierchio (1969) have shown that *Deontostoma californicum*, collected from an intertidal area, could osmoregulate very well in hypertonic solutions, but was unable to maintain osmotic equilibrium in hypotonic solutions below 0.6 M NaCl. Salt and water balance have not been studied in marine nematodes with the exception of the above work by Croll & Viglierchio. One may speculate, however, that the reduced reproductive potentials observed at lower salinities for *C. germanica* and *M. denticulata* in our studies, and the similar phenomenon which appears to exist for *Monhystera disjuncta* (von Thun 1968; Gerlach & Schrage 1971), may partially be the result of a need to spend a high proportion of energy on osmoregulation at reduced salinities. This, of course, might result in a lower quantity of energy available for expenditure on reproduction. Further studies are obviously called for.

The generation times for *C. germanica* calculated in this study are in rather good agreement with those that have been observed for other marine nematodes grown in the laboratory (Hopper & Meyers 1966; Gerlach & Schrage 1971, 1972; Tietjen & Lee 1972, 1973; Hopper et al. 1973). Heip (1972), however, found that for the harpacticoid *Tachidius discipes* the values of *r* generated from laboratory studies were three times those suggested by study of the natural populations. It is a well-known fact that estimates of the reproductive potential of any organism reared in the laboratory will likely be high if such estimates come from cultures grown only at optimum conditions. We have seen that the reproductive potential of *C. germanica* may be significantly affected by variations in temperature and salinity. At 26 ^{o}/oo, the mean value of *r* for *C. germanica* over the 20–30 °C temperature span in our laboratory studies was 0.12. If one calculates *r* for *Chromadora macrolaimoides* using our published data (Tietjen & Lee 1973), we find that this compares quite favorably with the mean value for *C. germanica*. At 25 °C and 26 ^{o}/oo salinity, the value of *r* for *C. macrolaimoides* was 0.15. The values of *r* obtained from studying the natural populations of *C. germanica* in North Sea Harbor over the 20–30 °C temperature span in July and August ranged between 0.098 and 0.167; the mean value was 0.121. The agreement between field-generated and laboratory-generated values is rather good, and we are confident that, at least for *Chromadorina germanica* and *Monhystera denticulata*, our estimates of the annual number of generations produced by these species are probably not much in error.

From the data on generation time, it is possible to calculate the number of generations of *C. germanica* produced annually under temperature conditions prevailing in North Sea Harbor (Table 1). The estimated number of generations produced per year is 13, a number which compares favorably with the estimated annual production of *Monhystera denticulata* in North Sea Harbor (Tietjen & Lee 1972) and *M. disjuncta* in Helgoland Bay (Gerlach & Schrage 1971), two species whose reproductive rates are similar to *C. germanica*.

As with *M. denticulata*, allowance should be made for the fact that egg deposition occurs over a period of several days; this probably reduces the number of generations that can be produced. However, *Chromadorina germanica*, like *Monhy-*

Table 1. Number of generations of *Chromadorina germanica* produced per year in North Sea Harbor. Extrapolation from data obtained in experiments conducted at 2–42 °C, to values of temperature prevailing in North Sea Harbor.

Period and temperature (°C)		Number of generations
January–April	120 days at 0.5–5	less than 0.5
May–June	60 days at 5–20	3.0
June–September	90 days at 20–28	7.0
October–November	60 days at 20–5	3.0
December–January	30 days at 5–0.5	less than 0.5

stera denticulata, nevertheless belongs to that group of nematodes having short life histories.

Acknowledgements

The authors wish to acknowledge the laboratory assistance provided by Mrs. Judith Garrison, Mr. Carmine Mastropaolo and Mr. Michael Grofik in this research.
This work was supported by National Science Foundation Grant GA 33388.

Bibliography

Groll, N. A. & D. R. Viglierchio: Osmoregulation and the uptake of ions in a marine nematode. Proc. Helm. Soc. Wash. **36**, 1–9 (1969).
Gerlach, S. A. & M. Schrage: Life cycles in marine meiobenthos. Experiments at various temperatures with *Monhystera disjuncta* and *Theristus pertenuis* (Nematoda). Mar. Biol. **9**, 274–280 (1971).
–: Life cycles at low temperatures in some free-living marine Nematodes. Veröff. Inst. Meeresforsch. Bremerh. **14**, 5–11 (1972).
Heip, C.: The reproductive potential of copepods in brackish water. Mar. Biol. **12**, 219–221 (1972).
Hopper, B. E., Fell, J. W. & R. C. Cefalu: Effect of temperature on life cycles of nematodes associated with the mangrove (*Rhizophora mangle*) detrital system. Mar. Biol. **23**, 293–296 (1973).
– & S. P. Meyers: Aspects of the life cycles of marine nematodes. Helgoländer. wiss. Meeresunters. **13**, 444–449 (1966).
Matera, N. J. & J. J. Lee: Environmental factors affecting the standing crop of foraminifera in sublittoral and psammolittoral communities of a Long Island salt marsh. Mar. Biol. **17**, 89–103 (1972).
Thomas, W. H., Scotten, H. L. & J. S. Bradshaw: Thermal gradient incubators for small aquatic organisms. Limnl. Oceanogr. **8**, 357–360 (1963).
Thun, W. von: Autökologische Untersuchungen an freilebenden Nematoden des Brackwassers. Thesis, Kiel University (1968).
Tietjen, J. H. & J. J. Lee: Life cycles of marine nematodes. Influence of temperature and salinity on the development of *Monhystera denticulata* Timm. Oecologia **10**, 167–172 (1972).
–: Life history and feeding habits of the marine nematode, *Chromadora macrolaimoides* Steiner. Oecologia **12**, 303–314 (1973).
–: Feeding behavior of marine nematodes. In: B. C. Coull, ed.: Ecology of marine benthos. Univ. South Carolina Press, Columbia, pp 22–36 (1977).
Tietjen, J. H., Lee, J. J., Rullman, J., Greengart, A. & J. Trompeter: Gnotobiotic culture and physiological ecology of the marine nematode, *Rhabditis marina* Bastian. Limnol. Oceanogr. **15**, 535–543 (1970).

| Mikrofauna Meeresboden | 61 | 271—286 | 1977 |

W. Sterrer & P. Ax (Eds.). The Meiofauna Species in Time and Space. Workshop Symposium, Bermuda Biological Station, 1975.

Ultrastructure and Systematics:
an Example from Turbellarian Adhesive Organs

by
Seth Tyler

University of Maine at Orono, Dept. of Zoology, Murray Hall, Orono, Maine 04473 U. S. A.

Abstract

The stability of ultrastructural characters at various levels of a taxonomic hierarchy is investigated with a comparative study of adhesive organs in turbellarians.

Differences between adhesive organs of three species within a single genus, *Macrostomum*, are not great and relate mostly to dimensions of component parts. Ultrastructural characters are also relatively stable among genera of a single well-delimited family, the Dolichomacrostomidae. More significant differences emerge at the level of the order, for instance in a comparison of adhesive organs of the macrostomids with those of the rhabdocoel *Carcharodorhynchus*. The stability of ultrastructural characters at lower levels of the taxonomic hierarchy shows that the variability encountered at higher levels is significant and that the characters can be applied validly to systematics.

A. Introduction

As for most organisms of small size, the present taxonomic system for the Turbellaria rests almost exclusively on the use of characters that are discernible at the light microscopic level. In particular, the histology and arrangement of such structures as the genital apparatus and pharynx provide the most important characters for subdividing the turbellarian taxa. However, there is some indication from several recent electron microscopical studies of turbellarians that ultrastructural characters, i.e. characters observed through the use of electron microscopy, can be systematically significant as well. Ultrastructural studies, for instance, of rhabdites (Reisinger & Kelbetz 1967; Reisinger 1969), of epidermis (Bedini & Papi 1974), and of sperm morphology (Hendelberg 1975) have revealed characters that are potentially applicable to questions of phylogeny and systematics of the Turbellaria.

There is clearly a need for more characters in turbellarian systematics. Turbellarians are relatively simple animals and offer few characters on which to base a taxonomic system. In terms of the characters that are available from light microscopic studies, the relationships of turbellarian taxa to one another are not clear. Ax (1963) and Karling (1975) have proposed phylogenetic schemes relating the turbellarian orders, and though these schemes agree in several respects, there are significant differences in interpretations of what characters are primitive or advanced and as to what phylogenetic relations certain orders have to one another. Hopefully, such controversies can be resolved with the use of new characters such as can be obtained with electron microscopy.

I have investigated the value of ultrastructural characters for such questions of phylogenetics and systematics with a comparative study of the ultrastructure of a single organ system in the Turbellaria, namely adhesive organs. Even though there is considerable variety to be found in the ultrastructure of adhesive organs, there are characters in these organs which are stable within given systematic ranges and which therefore show promise of being systematically useful characters. The variation in structure of adhesive organs between genera of the order Macrostomida and between the Macrostomida and other turbellarian orders has been dealt with in another publication (Tyler 1976). In the present paper, the variation between species of a single genus, *Macrostomum*, and between genera in a single well-defined family, the Dolichomacrostomidae, is dealt with in more detail, and, for comparative purposes, a description is given of adhesive organ ultrastructure in a higher, non-macrostomid turbellarian, the rhabdocoel *Carcharodorhynchus*.

B. Materials and methods

Three undescribed species of *Macrostomum* have been studied here; two of these, *M.* sp. I and *M.* sp. II, are closely related to *M. hystricinum* Beklemischev, 1951, and the third, *M.* sp. III, resembles *M. evelinae* Marcus, 1946 in that its stylet has a lateral spike. These three species are the same three that Rieger (1977) deals with in the publication in this volume (p. 212), and the distinctive features of these species' stylets can be seen in his figures. The species of *Carcharodorhynchus* used here is likewise an undescribed species designated *Carcharodorhynchus* sp. I by Rieger & Doe (1975). Three species of dolichomacrostomids, *Myozonaria jenneri* Rieger & Tyler, 1974, *Paramyozonaria simplex* Rieger & Tyler, 1974, *P. rosacea* (Rieger, in prep.) and *Cylindromacrostomum mediterraneum* (Ax, 1955), and a new macrostomid with asexual reproduction (Genus I) which occupies a systematic position intermediate between the Dolichomacrostomidae, Macrostomidae, and Microstomidae (see Rieger & Tyler 1974) are also studied. Collecting sites for all of these species is given in Table 1.

The animals were extracted from sand samples using a magnesium chloride anesthetization technique described by Sterrer (1968). For electron microscopy, the animals were fixed in 2.5 % phosphate-buffered glutaraldehyde, post-fixed in phosphate-buffered 2 % osmium tetroxide, and embedded in a mixture of Epon and Araldite. Sections cut from the embedment blocks were stained with uranyl acetate and lead citrate and examined with a Zeiss EM9S2. Reconstructions of adhesive organs in *Macrostomum* sp. I and *Myozonaria jenneri* were made with the use of serial sections mounted on slot grids with a composite carbon/formvar film for support. Other species' adhesive organs were reconstructed from unsupported sections mounted in carefully recorded order on standard mesh grids.

Table 1. Material

Species	Location of Collecting Sites	Number of Specimens Examined by EM	Number of Adhesive Organs Viewed by EM
Macrostomum sp. I	US Atlantic coast, N. C., Bogue Bank, sound side, 300 m from Bogue Inlet, LTL, fine sand with detritus	2	130
Macrostomum sp. II	US Atlantic coast, Morehead, N. C. Radio Island, LTL, fine sand with detritus	3	60
Macrostomum sp. III	US Atlantic coast, N. C., Bogue Bank, sound side, 300 m from Bogue Inlet, HTL, fine sand with detritus	2	30
Paramyozonaria simplex (Rieger & Tyler 1974)	US Atlantic coast, N. C., New River, 1.5 miles from inlet, south shore, LTL, medium to fine sand	1	30
Paramyozonaria rosacea (Rieger in prep.)	US Atlantic coast, N. C., New River, 1.5 miles from inlet, south shore, LTL, medium to fine sand	1	5
Myozonaria jenneri (Rieger & Tyler 1974)	US Atlantic coast, Bahia Honda Key, Florida, subtidal (1–2 m)	2	30
Cylindromacrostomum mediterraneum (Ax 1955)	Adriatic, near Rovinj, Yugoslavia, shallow subtidal, fine sand	1	5
Genus I	US Atlantic Coast, N. C., New River, 1.5 miles from inlet, south shore, LTL, medium to fine sand	1	10
Carcharodorhynchus sp. I (see Rieger & Doe 1975)	US Atlantic coast, Emerald Isle, N. C., Surfer Beach, shallow subtidal	1	2

C. Observations

Representatives of the turbellarian orders Macrostomida, Haplopharyngida, Polycladida, Rhabdocoela, Proseriata, and Tricladida have adhesive organs with basically the same structure and composed of the same cell types. These are glandular

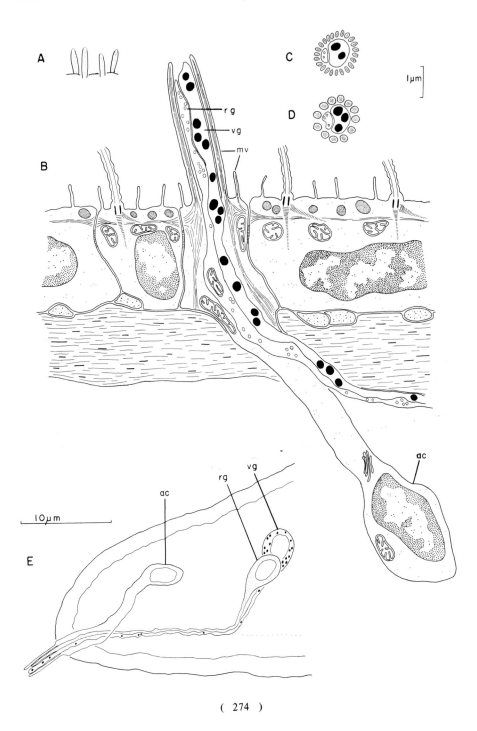

organs, each consisting of two gland cell types and a modified epidermal cell type through which the necks of the glands project to the epidermal surface. The gland cells lie below the epidermis of the animal and their necks penetrate the modified epidermal cell through one or more channels in the cell. Where the gland necks emerge on the epidermal surface, the epidermal cell surrounds them with a collar of microvilli, tonofilament-reinforced microvilli arranged in a fence-like array around one or both of the gland types' necks. The two gland cell types can be distinguished by differences in the morphology of their secretion granules. One of the gland types has relatively large, dense granules and is called the "viscid gland" on the basis of evidence that it secretes the adhesive substance by which the adhesive organ attaches to surfaces; and the other gland type has relatively smaller and less-dense granules and is called the "releasing gland" on the basis of evidence that it secretes a substance by which adhering adhesive organs are released from surfaces. The modified epidermal cell has a well-developed cell web and is evidently the tension-bearing element of the adhesive organ; it ist accordingly called the "anchor cell." (See Tyler 1976 for further explanation of this terminology.)

Macrostomum spp.

Adhesive organs of all three *Macrostomum* species investigated have one of each of the three cell types, i.e., one viscid gland and one releasing gland whose necks project together through one anchor cell (Fig. 1). The anchor cell enwraps the distal parts of the gland necks with a relatively long tubiform collar of microvilli, and this collar with its enclosed gland necks is readily visible in the living animal as an "adhesive papilla" (Fig. 1A). Papillae are about 5 μm long and 1.5 μm wide and lie in a wide arc along the ventral margin of the animal's spatulate tail plate. The cell bodies of the gland and anchor cells lie below the epidermis, deep within the parenchyma of the animal. The anchor cells therefore are insunk epidermal cells – i.e., the portion of the cell bearing the nucleus lies below the muscle layer – while all other epidermal cells in the *Macrostomum* species are intra-epithelial (Fig. 1B). The part of the anchor cell that lies within the epidermis is relatively small and narrow, and it completely enwraps the gland necks through the thickness of the epidermis and muscle layer. At its outer surface, it bears only

Fig. 1. *Macrostomum* spp. A) Appearance of adhesive papillae in a living animal (freehand). B) Reconstruction of adhesive organ in *Macrostomum* sp. I; the anchor cell's position in the body wall and the distal parts of the gland necks penetrating the anchor cell are shown; gland cell bodies lie to the right (anterior) out of the figure. C)–D) Arrangement of microvilli and gland necks as seen in transverse section of papillae in *Macrostomum* sp. II and *M.* sp. III (C) and *M.* sp. I (D). E) Reconstruction of a single adhesive organ at the posterior body tip in *M.* sp. I, showing the relationship of anchor cell (ac), releasing gland (rg), and viscid gland (vg) to each other. (B–D at same magnification).

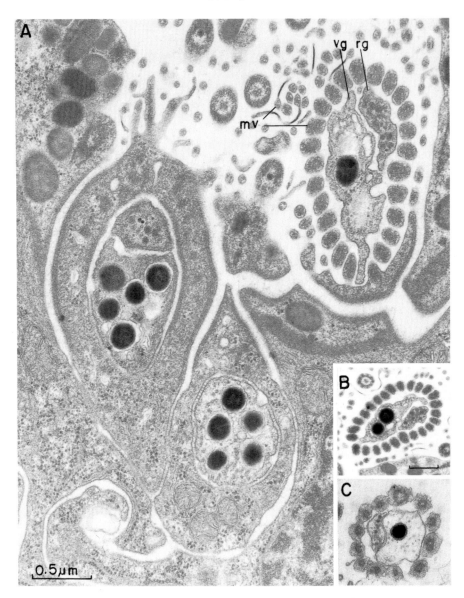

Fig. 2. *Macrostomum* spp., electron micrographs. A) Three adhesive organs of *M.* sp. III in transverse section. The organ in the upper right is sectioned at a level above the epidermal surface so that the microvilli of its papilla are visible; the other two are sectioned at a level below the bases of the microvilli and show how the gland necks penetrate through the center of the intraepithelial part of the anchor cell. The lower-most organ lacks a releasing gland neck; this is the only such organ encountered in all of the *Macrostomum* species studied. B)–C) Papillae of *M.* sp. III (B) and *M* sp. I (C); scale is 0.5 μm for both figures.

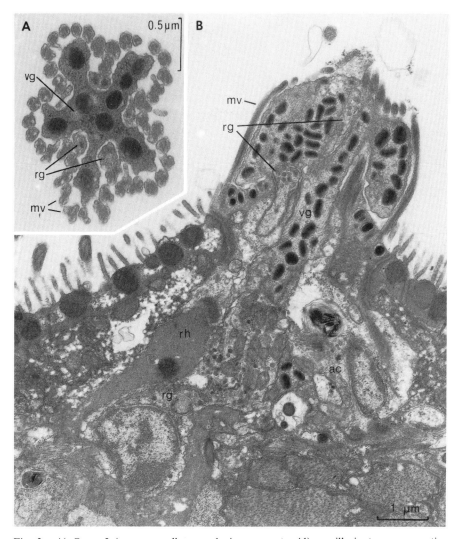

Fig. 3. A) Genus I (new asexually reproducing macrostomid); papilla in transverse section showing star-shaped folding of viscid gland with branches of the releasing gland in the grooves between each fold. B) *Paramyozonaria rosacea*; adhesive organ in longitudinal section; portions of the releasing gland nucleus and anchor cell nucleus are visible near "rg" and "ac" respectively.

microvilli, including the tonofilament-reinforced microvilli of the tubiform collar and a few smaller microvilli standing to the outside of this collar, but the anchor cell does not bear cilia as do other epidermal cells in the animals. The body of each anchor cell lies somewhat anterior to the papilla it bears (Fig. 1E), and the bodies of its associated gland cells lie yet further anterior in the body of the animal.

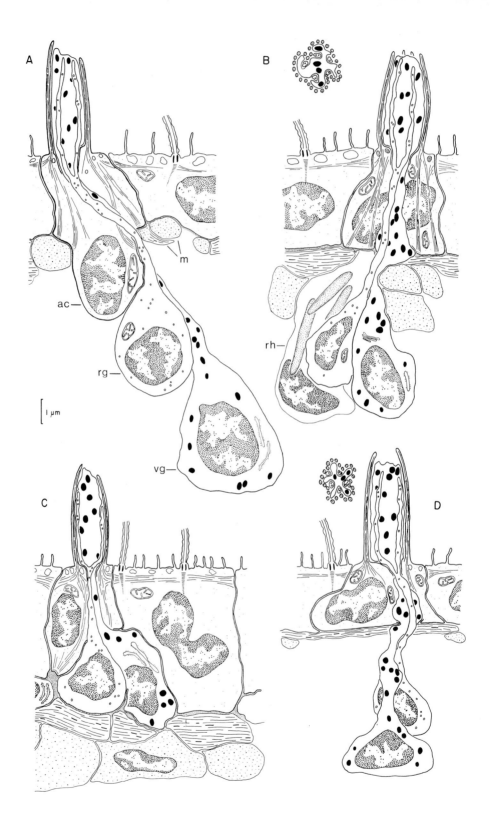

A

m

ac

rg

1 μm

vg

B

rh

C

D

Differences between the three *Macrostomum* species in the ultrastructure of their adhesive organs are not great. The most obvious difference is in the number and shape of microvilli in the collar around the gland necks. In *Macrostomum* sp. I, there are 9–13 microvilli in each papilla, each circular in transverse section and quite thick for a microvillus (0.2–0.3 μm), some being as thick as the cilia of neighboring epidermal cells (Figs. 1D, 2C). Papillae of *M.* sp. II and *M.* sp. III have more microvilli (22–26) and these microvilli are thinner, averaging 0.15 μm in diameter. In *M.* sp. III the microvilli tend to be oval in transverse section (0.1 x 0.2 μm) and stand in the papilla with their broader sides in contact with one another (Figs. 1C, 2A). Aside from this difference in shape and number of microvilli in the adhesive papillae, the only difference in adhesive organs of the three species are differences in dimensions of the organs' parts. The papillae in *M.* sp. I are slightly wider than those of *M.* spp. II and III (1.6–1.7 μm diameter in *M.* sp. I as compared to 1.0–1.5 μm in *M.* sp. II and 1.3–1.6 μm in *M.* sp. III), and the necks of the gland cells and anchor cells of *M.* sp. I (which is the largest of the three species) seem to be longer than those in *M.* spp. II and III. (Complete reconstructions have been made only of *M.* sp. I's organs, however). The anchor cell bodies of many adhesive organs in *M.* sp. III lie directly below the animal's muscle layer, in contact with the muscles, while in spp. I and II the anchor cell bodies are sunken further below the muscle layer, at least 12 μm from the intraepithelial part of the anchor cell in *M.* sp. I.

Myozonaria jenneri, Paramyozonaria simplex, P. rosacea, Cylindromacrostomum mediterraneum, and the new macrostomid with asexual reproduction.

As in *Macrostomum*, the dolichomacrostomids and the new macrostomid with asexual reproduction have adhesive organs with one of each of the three cell types; one viscid gland, one releasing gland, and one anchor cell. These species can be distinguished from *Macrostomum*, however, in a number of characters relating to the shape and position of the anchor cell and the morphology of the gland cell necks.

The most distinctive feature shared by adhesive organs in the dolichomacrostomids and in the new asexually reproducing macrostomid is that the papilla of each organ is folded into longitudinal ridges so that, in transverse section, the papilla appears star-shaped (Fig. 3A, 4B, D; see also figures of another dolichomacrostomid, *Paromalostomum* in Tyler 1976). It is the folding of the tip of the viscid gland where it emerges into the papilla that gives the papilla its star shape; the tip of the releasing gland is not folded but instead branches at the base of the papilla and sends a fingerlike extension along the grooves between the viscid gland's folds.

Fig. 4. Diagram of adhesive organs in A) *Myozonaria jenneri*, B) *Paramyozonaria simplex*, C) *Cylindromacrostomum mediterraneum*, and D) Genus I (asexually reproducing macrostomid). Insets in B) and D) are of papillae in transverse section.

These five species differ from one another in the position of their anchor cells and gland bodies (Fig. 4). Anchor cells in *Cylindromacrostomum mediterraneum*, the two *Paramyozonaria* species, and the asexually reproducing macrostomid are intraepithelial, i.e., the nuclei lie within the epidermis at the level of nuclei in other epidermal cells of these animals (Fig. 4B–D), while anchor cells in *Myozonaria jenneri* are slightly sunken into the muscle layer (Fig. 4A). The cytoplasm of all species' anchor cells is dense with tonofilaments, and these attach by hemidesmosomes to an intercellular matrix at the anchor cells' bases in all species except the new asexually reproducing macrostomid which has no discernible intercelluar matrix.

The gland cell bodies of the adhesive organs lie below the muscle layer in *Myozonaria jenneri* and the new asexually reproducing macrostomid; in both species of *Paramyozonaria*, the glands tend to lie within the muscle layer and are especially close to the epidermis in *P. rosacea* (Fig. 3B). *Cylindromacrostomum mediterraneum* is unique in that the adhesive organ glands lie above the muscle layer (Fig. 4C). In fact, all glands with ducts to the exterior in this species lie above the muscle layer and separate the muscles from parts of the overlying epidermis.

Rhabdite glands can be found associated with adhesive organs in the two *Paramyozonaria* species, even in *P. rosacea* whose rhabdite glands are modified and produce only a few rhabdites, and in the new asexually reproducing macrostomid.

Carcharodorhynchus sp. I

In contrast to the numerous small adhesive organs of the macrostomids, adhesive organs in the rhabdocoel *Carcharodorhynchus* are in the form of broad fields with necks of glands penetrating to the epidermal surface over a relatively large area. By light microscopy these adhesive organs are seen as continuous belts or girdles in the animal's epithelium which completely encircle the body; but despite this macroscopic difference from the small discrete organs of macrostomids, electron microscopy reveals that *Carcharodorhynchus*'s adhesive organs have the same cell types — viscid and releasing glands and a modified epidermal component — as are seen in macrostomid organs (Fig. 5). There are significant differences, however, between *Carcharodorhynchus* and the macrostomids in the organization of these cell types within adhesive organs.

Whereas gland necks in macrostomid organs emerge together in a common collar of microvilli, each gland neck in adhesive organs of *Carcharodorhynchus* has its own separate pore on the epidermal surface, and the viscid and releasing glands thus emerge separately from one another. Only the viscid gland necks are surrounded by a collar of microvilli where they emerge on the epidermal surface; necks of releasing glands emerge relatively free of microvilli (Fig. 5A, C). Although the whole epidermal surface of the adhesive organ bears microvilli, it is only those microvilli surrounding the viscid gland necks which have a core of tonofilaments comparable to the cores in microvilli of macrostomid adhesive organs. This core

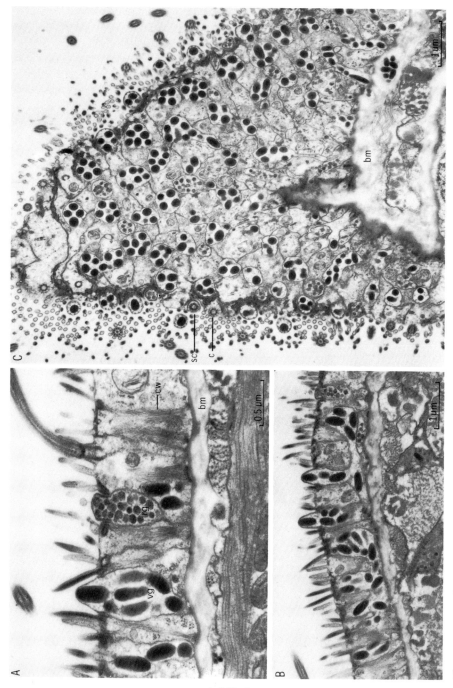

Figure 5. *Carcharodorhynchus* sp. I. A) Adhesive field in longitudinal section; the tips of a viscid gland and releasing gland neck, as well as a cilium are visible. B) Branching viscid gland necks. C) Adhesive field in tangential section.

is densest towards one side of each microvillus, the side that stands closest to the gland neck. Tonofilaments extend basally from these cores as fibers of the cell web and attach to the basement membrane by hemidesmosomes. There are 9–12 of the cored microvilli around each viscid gland neck.

Another significant feature of adhesive organs in *Carcharodorhynchus* that distinguish it from adhesive organs of *Macrostomum* is that necks of glands branch before reaching the epidermal surface (Fig. 5B). Branching is especially evident in viscid glands: up to six branches can be seen to arise from a single stem neck within the epidermis. Each of these branches penetrates to the epidermal surface through its own separate pore.

The general epidermis of the body in *Carcharodorhynchus* is synctial, and accordingly the adhesive field does not have discrete anchor cells; rather the necks of the glands project through restricted regions of the epidermal syncytium. (Other higher turbellarians, however, even other rhabdocoels, do have truly cellular anchor cells [Tyler 1976]). Except for the presence of cored microvilli and of a more highly developed cell web, the epidermis of *Carcharodorhynchus'* adhesive organs is little differentiated from other epidermis of the animal. It is ciliated and is penetrated by cilium-bearing sensory processes just as is other unmodified epidermis.

D. Discussion

Ultrastructural characters in adhesive organs of turbellarians are apparently quite stable within a genus. At least in the species of *Macrostomum* examined here, the variation in structure between species in the genus is not great, nor is there a significant difference between the two *Paramyozonaria* species. What differences do emerge between species do not necessarily coincide with other recognized species differences. For example, *Macrostomum* sp. II is more like *M.* sp. III in terms of morphology of the collar in its adhesive papillae while it is more like *M.* sp. I in morphology of its stylet.

Collar structure in *M.* sp. III is similar to that of another macrostomid genus, *Psammomacrostomum* (see Tyler 1976) and it is clear that of all the macrostomids studied by EM so far, *Macrostomum* is most closely related to *Psammomacrostomum*. Both genera have narrow elongate papillae with insunk anchor cells. Differences between these genera can be seen in the position of the gland and anchor cell bodies (in *Psammomacrostomum* these cells lie in discrete groups quite distant from their respective papillae) and in the absence of ultrarhabdites in *Macrostomum's* anchor cells.

Variation in adhesive organ characters between genera of the family Dolichomacrostomidae falls within a range which logically could be expected for intrafamilial variation. There are slight differences in such characters as the position of gland and anchor cell bodies, but all dolichomacrostomids investigated so far are alike in that their releasing gland branches within the papilla and in that the viscid gland tip is folded into a star shape. The family is evidently a well delimited unit in terms of adhesive organ characters, clearly distinct from other macrostomid families. The new asexually reproducing macrostomid is closely allied with the Dolichomacrostomidae by similarities in its adhesive organs. No character could be found in adhesive organs that could be used to distinguish the dolichomacrostomid subfamilies, Dolichomacrostominae and Karlingiinae, from one another.

More significant differences emerge between other genera of macrostomids (see Tyler 1976). *Microstomum*, for instance, though superficially like *Macrostomum* in having prominent elongate papillae, is distinguished from *Macrostomum* by the branching of the releasing gland neck in its papillae, branching somewhat like that in the dolichomacrostomids. Its organs are unlike those of the dolichomacrostomids, however, in having fully insunk anchor cells and in lacking the star-shaped folding of the viscid gland. *Myozona* is more radically different yet in that it has cilia on its anchor cells and two viscid glands in each adhesive organ unlike any other macrostomid. In this regard, *Myozona* provides a link between the Macrostomida and another turbellarian order, the Haplopharyngida. *Haplopharynx*, the only genus in this order, also has cilia on its anchor cells and two, or as many as three, viscid glands in each adhesive organ. Unlike *Myozona*, it has intraepithelial anchor cells.

The form of adhesive organs in higher turbellarian orders is significantly different from that of macrostomids even though the organs consist of the same cell types. For example, the rhabdocoel *Carcharodorhynchus,* as shown here, has its viscid and releasing glands emerging through separate pores, with collars around only the viscid gland neck tips, and with gland necks that branch below the epidermal surface. These same characters can be seen in other higher turbellarians including representatives of the Proseriata, Tricladida, Polycladida, and other representatives of the Rhabdocoela (see Tyler 1976). Therefore, the Macrostomida and Haplopharyngida with their viscid and releasing gland necks emerging in a common collar of microvilli, and with simple gland necks that do not branch below the epidermal surface, stand as a distinct unit, clearly separable from the higher turbellarians.

Though there is this clear dichotomy between the higher turbellarians on the one hand and the Macrostomida and Haplopharyngida on the other, it is highly probable that adhesive organs in all of these turbellarians are homologous and that these animals share a common ancestor which also had this type of adhesive organ. The fact that there are other types of adhesive systems in the turbellarian orders Acoela, Nemertodermatida, and Lecithoepitheliata (see Tyler 1973, 1976) strengthens the likelihood of this homology since it shows that the type of adhesive system encountered in the macrostomids and higher turbellarians is not the only means by which adhesion could be accomplished. On the basis of this assumed homology, a scheme of the evolution of adhesive organs is presented in Figure 6.

Characters in these organs do have a bearing on considerations of phylogenetic relationships in the Turbellaria. The stability of these characters at lower levels of

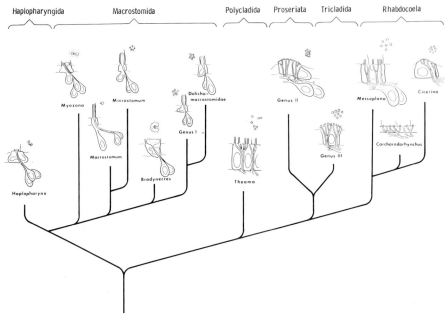

Fig. 6. Scheme of the evolution of adhesive organs in the Turbellaria, based for the most part on ultrastructural characters of the organs. Consideration is given, as well, to recognized relationships deduced from light microscopic characters, especially for the higher turbellarians. Adhesive organs in representative genera are diagramed. Genus I is the new macrostomid with asexual reproduction; Genus II is an otoplanid proseriate; Genus III is a procerodid triclad. (After Tyler 1976.).

the taxonomic hierarchy, for example within the genus *Macrostomum* and within the family Dolichomacrostomidae, show that the variability seen at higher levels of the hierarchy, e.g. between the Macrostomida and neoophoran orders, or even between some genera of the Macrostomida, is systematically significant. The evolutionary scheme depicted in Figure 6 has a number of implications for turbellarian systematics. First of all, it shows that there is a direct link between the higher turbellarian orders and the Macrostomida by virtue of homology of the two groups' adhesive organs; such a connection has been lacking until now (Ax 1963). It also shows that the family Macrostomidae is a rather heterogeneous grouping of genera. Whereas the Dolichomacrostomidae is a well-defined group in terms of adhesive organ characters, the Macrostomidae includes forms with widely dissimilar adhesive organs (especially *Myozona* and *Bradynectes*, see Tyler 1976); this heterogeneity has been indicated previously on the basis of some light microscopical characters (Rieger 1971). Furthermore, adhesive organ characters point to a close relationship of the Haplopharyngida with the Macrostomida and a close relationship of the Polycladida with the Neoophoran orders. (For further discussion of these relationships see Tyler 1976).

Even though characters of adhesive organs are ultrastructural characters, observable only with the use of electron microscopy, they are as useful as any other characters in elucidating phylogenetic relationships. They are small, but they are sufficiently complex and sufficiently stable within given systematic ranges that they can be used validly in systematics. New characters are clearly needed for such a systematically complex group as the Turbellaria. It is primarily through the discovery of new characters that the taxonomic system for any group of organisms can be improved, and not through simply re-evaluating characters which are already known. Electron microscopy can reveal new characters, characters which can be systematically valid, and it should be useful for defining new systems in the future.

Acknowledgements

I thank Dr. R. M. Rieger, of the University of North Carolina, for his inspirational guidance throughout the course of this study. I thank him and David Doe also for use of sections of *Carcharodorhynchus* sp. I. Financial support of this research was provided by grants from the National Science Foundation (GB 42211), from the University of North Carolina at Chapel Hill University Research Council (VF 349) and the Austrian Science Foundation (2021). This manuscript was prepared while the author was a Killam postdoctoral fellow at the Department of Anatomy, Dalhousie University.

List of abbreviations

ac — anchor cell
bm — basement membrane
c   — cilium
cw — cell web
m   — muscle
mv — microvilli
rg — releasing gland
rh — rhabdite gland
sc — sensory cilium
vg — viscid gland

Bibliography

Ax, P.: Monographie der Otoplanidae (Turbellaria). Morphologie und Systematik. Akad. Wiss. Lit. Mainz. Abh. Math. Naturw. Kl. 1955. Nr. 13. 499–796 (1956).
—: Relationships and phylogeny of the Turbellaria. In: The Lower Metazoa. E. C. Dougherty, ed. Univ. California Press, Berkeley, Calif. pp 191–224 (1963).
Bedini, C., & F. Papi.: Fine structure of the turbellarian epidermis. In: Biology of the Turbellaria. N. W. Riser and M. P. Morse, eds. McGraw-Hill, New York, N. Y. pp 108–147 (1974).
Beklemishev, W.N.: The species of the genus *Macrostomum* (Turbellaria, Rhabdocoela) of the Soviet Union. [In Russian]. Bull. Soc. Nat. Moscow (Biol.) 56: 31–40 (1951).
Hendelberg, J.: Spermiogenesis, sperm morphology, and biology of fertilization in the Turbellaria. In Biology of the Turbellaria. N. W. Riser and M. P. Morse, eds. McGraw-Hill, New York, N. Y. pp 148–164 (1974).

Karling, T. G.: On the anatomy and affinities of the turbellarian orders. In: Biology of the Turbellaria. N. W. Riser and M. P. Morse, eds. McGraw-Hill, New York, N. Y. pp 1–16 (1974).

Marcus, E.: Sobre Turbellaria limnicos brasileiros. Bol. Fac. Filos. Ciênc. Let., Univ. São Paulo, Ser. Zool. 11, 5–254 (1946).

Reisinger, E.: Ultrastrukturforschung und Evolution. Ber. Physik. Med. Gesellsch. Würzburg 77, 1–43 (1969).

— & S. Kelbetz: Feinbau und Entladungsmechanismus der Rhabditen. Z. wiss. Mikrosk. 65, 472–508 (1964).

Rieger, R. M.: *Bradynectes sterreri* gen. nov., spec. nov., eine neue psammobionte Macrostomide (Turbellaria). Zool. Jahrb. Syst. Oekol. Tiere 98, 205–235 (1971).

—: The relationship of character variability and morphological complexity in copulatory structures of Turbellaria-Macrostomida and -Haplopharyngida. Mikrofauna Meeresboden 61, 197–216 (1977).

—: Die Turbellarienfamilie Dolichomacrostomidae Rieger. III. Teil. Dolichomacrostominae II. (in preparation).

— & D. Doe: The proboscis armature of Turbellaria – Kalyptorhynchia, a derivative of the basement membrane? Zool. Scripta 4, 25–32 (1975).

— & S. Tyler: A new glandular sensory organ in interstitial Macrostomida. I. Ultrastructure. Mikrofauna Meeresboden, 42, 1–41 (1974).

Sterrer, W.: Beitrag zur Kenntnis der Gnathostomulida. I. Anatomie und Morphologie des Genus *Pterognathia* Sterrer. Ark. Zool. 22, 1–125 (1968).

Tyler, S.: An adhesive function for modified cilia in an interstitial turbellarian. Acta Zoologica 54, 139–151 (1973).

—: Comparative ultrastructure of adhesive systems in the Turbellaria. Zoomorphologie 84, 1–76 (1976).

Mikrofauna Meeresboden	61	287–302	1977

W. Sterrer & P. Ax (Eds.). The Meiofauna Species in Time and Space. Workshop Symposium, Bermuda Biological Station, 1975

The Geographical Distribution of Interstitial Polychaetes

by

Wilfried Westheide

II. Zoologisches Institut der Universität, Berliner Str. 28, D-3400 Göttingen
Fed. Rep. Germany

Abstract

The interstitial polychaete fauna of the world's surf-beaten, pure sand beaches is remarkable uniform. This uniformity is expressed in terms of the spectra of species and genera, the number of species, the percentage of the total number of individuals, the number of dominants, and individual distribution patterns. It supports the hypothesis that widely distributed genera, even species, of the interstitial fauna are very old and were distributed by the extensive movement of the continental plates. Amphiatlantic distribution patterns of species and "trans-allopatric sister-species" within the polychaetes provide further support. On the other hand, the situation in the beaches of the Galapagos Islands shows, that interstitial polychaetes can cross open oceans within a relatively short period of time. Thus, the great faunistic uniformity of this interstitial group and its beach community might be the result of both a common historical origin and a more or less continuous gene flow between geographically separated regions.

A. Introduction

The worldwide distribution of many polychaete taxa was already apparent from the early investigations of this marine animal group. Quatrefages (1865) was among the first to recognize this peculiarity, but stated that it characterized only genera and higher taxa. Fauvel (1925) and other authors demonstrated that a large number of polychaete species are also distributed worldwide and that separate zoological provinces do not exist for many of these animals (Hempelmann 1931; Friedrich 1937). For example, Hartmann-Schröder (1971) included 43 out of the 426 species of polychaetes in the fauna of Germany and the bordering seas among the truly cosmopolitan and widely distributed species. Almost as great is the portion of bipolar forms, whose distribution is usually explained in terms of an equatorial submergence connection (Ekman 1935, Rullier 1958). Out of 409 polychaetes from South Africa, 142 belong to the widely distributed species, 43 are considered circumtropical (Hartmann-Schröder & Hartmann 1974). In general, the wide distribution of benthic polychaetes can be accounted for by free-swimming larval stages (Fauvel 1925, and others). The proportion of widely distributed species is especially high, however, in the interstitial habitat (Westheide 1971). Since these polychaetes, as a rule, do not possess free-swimming developmental stages and are not suited for active migration within the water, other reasons must be sought to explain their distribution spectrum.

Sterrer (1973), in a broadly based hypothesis, has presented an historical explanation for this wide distribution pattern which can also be observed in other interstitial taxa. According to this hypothesis the widely distributed species are supposed to be very old, having existed prior to the separation of an old supercontinent, and being distributed by the extensive movement of the continental plates. Rao (1972) reached the same conclusions.

This paper will present further evidence for this hypothesis using faunistic and simple ecological data on polychaetes. There are two reasons why the best support herefore is offered by species[1] from surf-beaten pure sand beaches ("stark-, mittel- und schwachlotische Sandhänge" in the sense of Schmidt 1968). Qualitative and quantitative investigations are available for this biotope in different geographic areas, and submergence connections can be excluded for these species. Do the findings support so-called passive contemporary dispersal (e.g. by air, birds, floating objects) or plate tectonics as mechanisms for species distribution?

B. Ecological data

The number of polychaete species within a beach is generally higher than that of tardigrades, mollusc or acari, but substantially lower than that of turbellarians, nematodes, and copepods. In the seven beaches from which reliable data are available, the number of interstitial polychaete species is similar (Table 1). There is no discernible decrease in species diversity with latitude. The somewhat lower species diversity reported from Tromsø (Norway) and Hood (Galapagos) is correlated with the unique conditions found on these two beaches. (A higher number of investigations, however, of continental tropical beaches in particular would be most desirable.)

The percentage contributed from the polychaetes to the total number of individuals in a beach is usually slight, lying as a rule below 5 % in a collection of a whole beach transect (method see Schmidt 1968) (Table 2). Only in single samples are higher percentages sometimes found. This is explained by the fact that the different species do not appear evenly over the total area of the beaches, but

---

[1] Species spectrum and family membership of interstitial polychaetes has been summarized in Hartmann-Schröder (1964), Laubier (1967c) and Westheide (1971). The archiannelids are here included. Their separation from the polychaetes in the narrower sense, as practiced earlier, is not justified. Though the definite position of the families Polygordiidae, Saccocirridae, Protodrilidae, Nerillidae and Dinophilidae within the polychaete system is still in question, they are unequivocally, though aberrant, polychaetes (Hermans 1969; Orrhage 1974). Most of the species of the 5 families are to be placed in the interstitial fauna (Jouin 1971); a great many are characteristic of surf beaches.

Table 1. Number of interstitial polychaete species (including archiannelids) in different beaches

Location	Number of species	Source
Tromsø, Northern Norway (Station 1)	6	Schmidt 1972
List/Sylt, German Bight (Litoralstation)	11	Westheide 1966; Schmidt 1969
Arcachon, Atlantic coast, France (Camp Americaine)	13	Renaud-Debyser 1963; Westheide 1972/73
Amilcar, Mediterranean Tunisia	11	Westheide 1972b
Hood, Galapagos (Bahia Gardner)	7	Schmidt, unpublished
Barrington, Galapagos	11	Schmidt, unpublished

Table 2. Percentage values contributed by the polychaetes to the total fauna

Locality	Percentage	Source
Tromsø, Northern Norway	< 1 %	Schmidt 1972
List/Sylt, German Bight	< 5 %	Schmidt 1968
Arcachon, Atlantic coast France (Camp Americaine)	0.5 % (excluding Archiannelida)	Renaud-Debyser 1963
Amilcar, Mediterranean	1.5 to 4.4. %	Westheide, unpublished
Bimini, Bahamas	2 % (excluding Archiannelida)	Renaud-Debyser 1963
Porto Novo, South-East India	1 %	McIntyre 1968
Galapagos, different beaches	0.4 to 4.9 %	Schmidt, unpublished

colonize certain microhabitats, in which they even accumulate during reproduction, during climatically dependent withdrawal movements, etc. (Schmidt 1969; Westheide 1972b).

Dominance represents another important ecological characteristic. One or two species represent more than 50 % of all polychaete individuals; three to five species represent more than 90 % of the polychaete fauna (Table 3). Thus about half of all species present in a beach compose less than 10 % of the total number.

The dominant role of some species becomes even more clear when one investigates their distribution within the biotope. Only in a few cases the areas of two dominant species overlap extensively. As a rule they exclude each other, especially if they are closely related or congeneric species (Westheide 1972b). Every distribution pattern is, furthermore, an expression of the physiological tolerances of the individual species (Lasserre & Renaud-Mornant 1973). Each pattern has definite

Table 3. Number of species that represent more than 50 or 90 % of the total polychaete fauna

%	Sylt, Litoralstation	Arcachon, Camp Americaine	Amilcar	Hood, Bahia Gardner	Barrington
50	2	1–2	2	2	1
	Hesionides arenaria arrenaria	*Hesionides arenaria arenaria*	*Hesionides arenaria arenaria*	*Hesionides arenaria pacifica*	*Protodrilus* spec.
	Trilobodrilus axi (Protodriloides chaetifer)	*Trilobodrilus axi (Protodriloides chaetifer)*	*Hesionides gohari*	*Eusyllis homocirrata*	
90	3	3–4	4–5	2	3

horizontal and vertical boundaries which are similar from beach to beach, largely independent of differences in sediment quality, and independent of geographic latitude (Schmidt 1970, 1972). The consistency of this distribution pattern is remarkable and only changes according to certain hydrographic or climatic factors.

Table 4 contains species lists from 4 similar, surf-beaten, pure sand beaches: the Atlantic-Boreal region (List, Sylt, Germany, North Sea); the transition region between Boreal and the Mediterranean-Atlantic (Arcachon, France, Atlantic); the Mediterranean region (Amilcar, Tunisia); the Eastern Pacific subregion of the Tropics (Hood, Galapagos). The genera spectra are very similar. Although the similarity decreases with increasing difference in latitude, the value of the index of similarity (Sørensen 1948) is 0.3 between the Eastern Pacific and the North Sea, 0.4 between the Eastern Pacific and the Mediterranean, and 0.7 between the North Sea and the Mediterranean. The similarity becomes even greater when other littoral mesopsammic environments are included. Every truly interstitial genus in the Galapagos Islands is also represented by at least one species in either the North Sea or the Mediterranean. This includes the following genera: *Pisione, Hesionura,*

Table 4. Species list (Polychaeta) of four different, geographically separated sandy beaches

List/Sylt, German Bight (Litoralstation)	Arcachon, Atlantic coast, France (Camp. Am.)
Hesionides arenaria arenaria Friedrich	*Hesionides arenaria arenaria* Friedrich
Hesionides maxima Westheide	*Hesionides maxima* Westheide
Microphthalmus listensis Westheide	*Hesionides gohari* Hartmann-Schröder
Parapodrilus psammophilus Westheide	*Microphthalmus listensis* Westheide
Stygocapitella subterranea Knöllner	*Petitia amphophthalma* Siewing
Protodriloides chaetifer (Remane)	*Parapodrilus psammophilus* Westheide
Protodrilus adhaerens Jägersten	*Stygocapitella subterranea* Knöllner
Protodrilus spec.	*Protodriloides chaetifer* (Remane)
Diurodrilus minimus Remane	*Protodrilus similis* Jouin (?)
Diurodrilus subterraneus Remane	*Protodrilus* spec. (?)
Trilobodrilus axi Westheide	*Diurodrilus benazzii* Gerlach
	Diurodrilus spec. (?)
	Trilobodrilus axi Westheide

Amilcar, Mediterranean, Tunisia	Hood, Galapagos (Bahia Gardner)
Hesionides arenaria arenaria Friedrich	*Hesionides arenaria pacifica* Westheide
Hesionides maxima Westheide	*Hesionides unilamellata* Westheide
Hesionides gohari Hartmann-Schröder	*Eusyllis homocirrata* Hartmann-Schröder
Hesionura augeneri (Friedrich)	*Polygordius pacificus floreanensis* Schmidt & Westheide
Stygocapitella subterranea Knöllner	*Saccocirrus sonomacus* Martin
Protodriloides spec.	*Protodrilus infundibuliformis* Schmidt & Westheide
Protodrilus similis Jouin	*Protodrilus* spec. 2
Protodrilus spec. 1	
Protodrilus spec. 2 (?)	
Diurodrilus benazzii Gerlach	
Diurodrilus spec.	

Microphthalmus, Hesionides, Typosyllis, Syllides, Brania, Sphaerosyllis, Exogone, Protodorvillea, Schistomeringos, Ophryotrocha, Ctenodrilus, Macrochaeta, Polygordius, Protodrilus, Saccocirrus, Nerilla, Mesonerilla, Nerillidium, Diurodrilus, Dinophilus (Westheide 1974, Schmidt & Westheide 1977, Westheide unpublished). The similarity between tropical India (Rao & Ganapati 1967, 1968) and the Mediterranean is as strong, except that the genus *Goniadides* appears to be strictly tropical in its distribution. On the other hand, there are four genera in the beaches of Sylt (North Sea) which have not been found in the Galapagos Islands (*Parapodrilus, Stygocapitella, Protodriloides, Trilobodrilus*). At least *Trilobodrilus* occurs in the Tropics (Rao 1973).

The index of similarity between the different faunas is smaller at the species level. The agreement between Sylt and Arcachon is relatively large (0.6). Sylt and Amilcar have 3 species in common (0.2). On the Galapagos there is only one subspecies (*Hesionides arenaria pacifica* Westheide) whose nominate subspecies is found on Sylt. In addition to this subspecies there are three other more or less interstitial Mediterranean species, which are identical with Galapagos species: *Syllides edentula* (Claparède), *Brania subterranea* (Hartmann-Schröder) and *Sphaerosyllis hystrix* Claparède. The Indian fauna possesses 9 species in the surf beaches which are probably the same as those in the Mediterranean or in the Boreal-Atlantic region (Rao & Ganapati 1967, 1968).

C. Faunistic data

The values of the index of similarity reflect the different geographical distribution patterns of the individual species. In fact, all transitions between locally limited and globally distributed forms exist. The questionable nature of taxonomical analyses of interstitial forms which are geographically widely separated has been pointed out (Westheide 1971; Sterrer 1973). Faunistic work is made more difficult by the existence of unrecognized sibling species or morphologically similar end populations, which behave with respect to one another as true species even though they are connected through a chain of interbreeding populations (Mayr 1963). Even so, the wide distribution of a number of species within the polychaetes can be considered proven, especially when identical ecological requirements and distribution patterns support the morphologically based conclusions. Slight geographical variations are naturally not excluded herewith. In fact, the complete absence of physiological or morphological variability in such widely distributed species seems improbable. The best proof of species identity would be crossbreeding experiments which, for several reasons, cannot be carried out with most of these species. Experiments of this kind with *Ophryotrocha* have shown that morphologically identical populations are separated by genetic barriers (Åkesson 1977). Such experiments, however, could also prove that crossbreeding between widely separated populations is possible.

Several distribution types can be demonstrated for largely validated species, for the most part inhabitants of the surf-beaten beaches (Figs. 1–2). The discontinuities between localities outside of Europe reflect the lack of collecting activity in many geographical areas. Actually one should imagine the different coast lines as more or less connected population areas of individual species, as shown, for example, by the collections of *Hesionides arenaria* by Hartmann-Schröder (1974) on the coast of Southern Africa.

The distribution of the latter, inclusive of the subspecies *Hesionides arenaria pacifica* from the Galapagos Islands, is the broadest of any interstitial polychaete species (Fig. 1A). It inhabits the Tropics as well the Boreal and Antiboreal region.

Most of the species with worldwide distributions are more or less limited to a broad tropical belt including the Mediterranean-Atlantic region (Fig. 1B).

A Boreal species group stands in contrast to these forms. *Stygocapitella subterranea* is an example (Fig. 2A).

The distribution of a larger number of species is limited regionally. This includes numerous archiannelids and representatives of the genera *Pisione, Hesionura* (Fig. 2B), and *Microphthalmus*. In *Hesionura*, however, the differences between some of the species are so slight that some taxonomists might consider them to be mere geographic variations.

Finally, there are a series of endemisms. Whether the composition of this last category can be maintained is doubtful. At least several island forms definitely belong in this categroy.

D. Discussion

The ecological and faunistic picture of the polychaetes in the surf-beaten beaches supports the hypothesis that the species were widely distributed by the extensive movement of the continental plates.

Thus, the wide distribution of species, already known for a longer time, is validated, supplemented and broadened to include other species.

The amphiatlantic distribution of species or the presence of "trans-allopatric sister-species", considered by Sterrer (1973) as a special indication for this hypothesis, can be supported with several examples. A new subspecies of *Hesionides maxima* was found on the North American Atlantic Coast; three localities of the nominate subspecies are situated on the European Coast and in the Mediterranean (Fig. 3A). A very closely related new form of the monotypic genus *Parapodrilus* (Fig. 3D) was also found on the Atlantic Coast of North America. The small differences between *Microphthalmus listensis* in two American localities and in Europe do not require the creation of a new taxonomic category (Fig. 3C)[1]. The specimens identified as *Microphthalmus sczelkowii* from the Atlantic Coast of North America

---

[1] Footnote (added in proof): actual investigations including new material of *Microphthalmus listensis* from the Atlantic Coast of North America show small morphological differences between the species from Europe and North America; perhaps the populations of *M. l.* on both sides of the North Atlantic belong to two different subspecies.

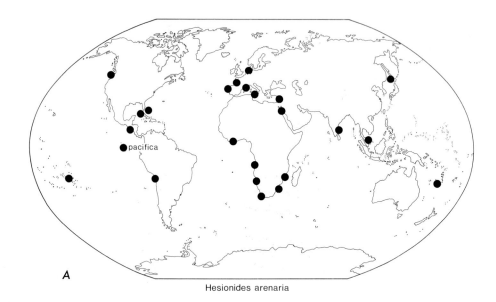

A

Hesionides arenaria

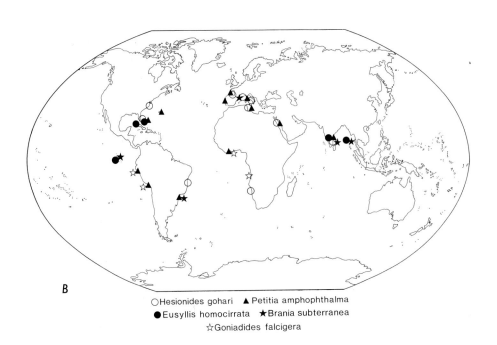

B

○Hesionides gohari ▲ Petitia amphophthalma
● Eusyllis homocirrata ★Brania subterranea
☆Goniadides falcigera

probably lie within the variability of that species. *Microphthalmus aberrans* occurs on both sides of the Atlantic (Fig. 3B). The only two species of the genus *Schroederella* are found on the West and Southwest Coasts of Africa and on the Coast of Massachusetts in North America.

The agreement in the spectrum of genera, in the number of species and dominants, and their percentage contribution to the total fauna, as well as in the similar distribution pattern within the various beaches is best interpreted in terms of a common origin of meiofauna communities.

An essential prerequisite for the hypothesis is the demanded age of the corresponding taxa. Although polychaete fossils are rare, a number of finds from the Cambrian and Ordovician, as well as from later periods, exist. Some of them show remarkable similarities to recent forms (Roger 1959; Walliser 1972). It is thus reasonable to assume that many of the recent interstitial polychaetes existed in their present form prior to the segmentation of an old supercontinent.

Distribution by means of longe-range dispersal (Gerlach 1977) cannot be disproved, however. The Galapagos Islands have never been connected with the continent and the age of these islands is roughly 3 million years (Bailey 1976). Yet, within this relatively short period, almost the entire spectrum of the typical interstitial groups reached the islands (Ax & Schmidt 1973). This includes almost all the characteristic polychaete taxa of the interstitial habitat. A high percentage of these species are not endemic. The ecological and faunistic agreement between Galapagos Island and continental beaches suggests that all forms arrived at nearly the same time after crossing an ocean barrier of at least 1000 km. Almost every niche appears to have been occupied by the species or genus typical for it before extensive radiation of a few species could take place. Signs of especially high speciation, or vicarism of higher taxa, as one would expect if colonization was gradual, are absent in the polychaetes.

That vicarism in surf beaches is possible, within the absence of a characteristic species, has been demonstrated. In Bermuda the genus *Hesionides* is absent in the

---

Fig. 1A. Distribution of *Hesionides arenaria arenaria* Friedrich, 1937, and *H. arenaria pacifica* Westheide, 1974 (Hesionidae). Localities: Westheide (1971), Hartmann-Schröder (1973, 1974b), Renaud-Mornant & Serène (1967), Salvat & Renaud-Mornant (1969), and Hokkaido, Japan (coll. T. Itô, personal communication).
B. Distribution of *Hesionides gohari* Hartmann-Schröder, 1960 (Hesionidae); *Petitia amphophthalma* Siewing, 1956; *Eusyllis homocirrata* Hartmann-Schröder, 1958; *Brania subterranea* (Hartmann-Schröder, 1956) (Syllidae) and *Goniadides falcigera* Hartmann-Schröder, 1962 (Goniadidae). Localities for *Hesionides gohari*: Westheide (1971), Westheide (1974a), Hartmann-Schröder (1974b), Coast of North Carolina (coll. R. Rieger), and Florida, USA. Localities for *Petitia amphophthalma*: Westheide (1971), Hartmann-Schröder (1962), Hartmann-Schröder & Hartmann (1965), Westheide (1974a), and Bermuda. Localities for *Eusyllis homocirrata:* Westheide (1974b), and Andaman Archipelago (coll. G. C. Rao). Localities for *Brania subterranea*: Westheide (1974b), and Andaman Archipelago (coll. G. C. Rao). Localities for *Goniadides falcigera*: Hartmann-Schröder (1974b).

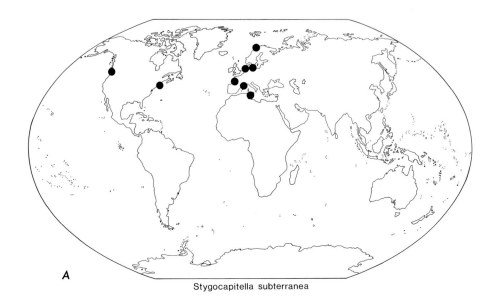

Stygocapitella subterranea

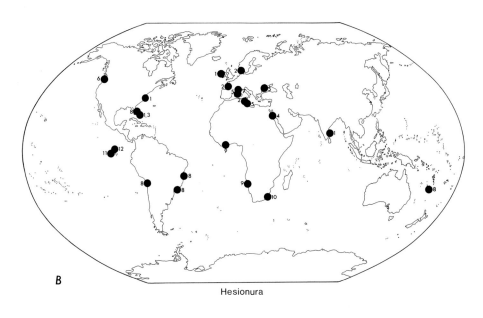

Hesionura

surf zone. It is replaced by an endemic species of *Microphthalmus* which closely resembles the corresponding *Hesionides* species in body size, color, and habitus (Westheide 1973).

E. Conclusions

From these points of discussion the following conclusions can be reached. They are similar to Sterrer's (1973) considerations.

1. At the beginning of the colonization of the surf-beaten beach biotope, an intensive speciation occurred which led to the development of new, optimally adapted taxa. This evolution involved numerous biological and morphological adaptations in the interstitial polychaetes (Westheide 1971). At the same time the biotope was divided among the participating groups forming a stable species distribution pattern and thus a stable community.

2. Evolution of the taxa and development of community pattern thus reached a climax. Only minor speciation occurred after this time (Noodt 1974) because of the constancy of the living conditions in this area. Adaptations to changing factors, such as temperature, probably did not result in significant morphological or ethological changes.

3. At the same time the structure of the community also became constant. The biotope was optimally occupied and thus essentially closed to immigration, since the species already present were superiorly adapted to all potential invaders.

4. Both taxa and interspecific structure were dispersed by the separation of the continental plates, but due to the great stability of the system only minor changes occurred.

5. In addition to this dispersal by plate tectonics, some other dispersal has taken place, not only along shores but also across open oceans. The latter type of dispersal, whatever the mechanism, caused the settlement of the sandy beaches of the Galapagos Islands within a relatively short period of time. Similarly the same type

---

Fig. 2A. Distribution of *Stygocapitella subterranea* Knöllner, 1934 (Parergodrilidae). Localities: Schmidt (1970). United States localities are San Juan Island, Washington (Westheide 1966) and the coast of Massachusetts (N. W. Riser, personal communication).
B. Distribution of species and subspecies of the genus *Hesionura* Hartmann-Schröder, 1958 (Phyllodocidae). 1. *Hesionura elongata* (Southern, 1914). 2. *H. augeneri* (Friedrich, 1927). 3. *H. fragilis* Hartmann-Schröder, 1958. 4. *H. serrata* (Hartmann-Schröder, 1960). 5. *H. coineaui* (Laubier, 1962). 6. *H. coineaui difficilis* (Banse, 1963). 7. *H. mystidoides* (Hartmann-Schröder 1963). 8. *H. laubieri* (Hartmann-Schröder, 1963), all descriptions and localities in Hartmann-Schröder (1963). 9. *H. portmanni* Laubier, 1967 (Laubier 1967a). 10. *H. natalensis* Hartmann-Schröder, 1974 (Hartmann-Schröder 1974b). 11. and 12. Two different *Hesionura* species from the Galapagos Islands, which cannot be described owing to insufficient material. The locality with question mark (?8) refers to several specimens from Florida (coll. T. H. Perkins), which are very close to *H. laubieri*. Further localities can be found in: Day (1973), Laubier (1965, 1967b), Marinov (1963), Rao & Ganapati (1967), Westheide (1972a, 1974a).

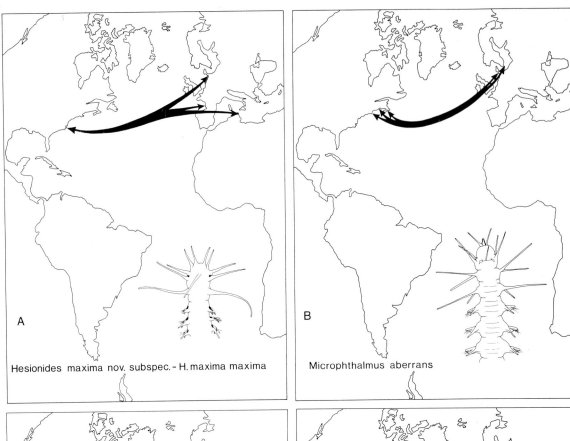

A

Hesionides maxima nov. subspec. – H. maxima maxima

B

Microphthalmus aberrans

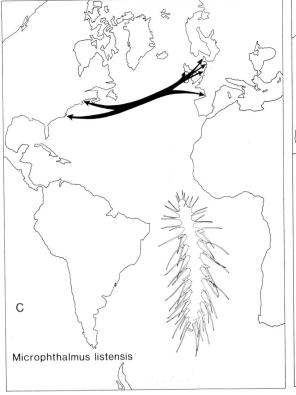

C

Microphthalmus listensis

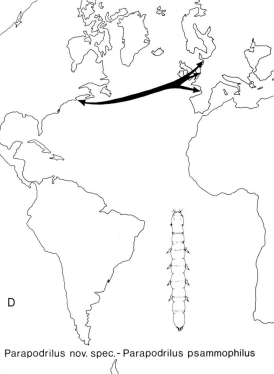

D

Parapodrilus nov. spec. – Parapodrilus psammophilus

of dispersal would provide a more or less continuous gene flow between geographically separated regions containing preexisting communities and species.

6. Thus the great faunistic uniformity of the interstitial polychaetes and their beach community might be the result of b o t h a common historical origin and recent passive dispersal mechanism.

A c k n o w l e d g e m e n t s

I wish to thank Prof. Dr. P. Schmidt for use of some of his unpublished ecoligical data, and Dr. C. Marschall for assistance in translation. The essay profited especially from the comments of Prof. Dr. C. C. Hermans. The presentation of this paper was supported by a grant from the Deutsche Forschungsgemeinschaft.

Bibliography

Åkesson, B.: Crossbreeding and geographic races: Experiments with the polychaete genus *Ophryotrocha*. Mikrofauna Meeresboden **61**, 11–18 (1977).

Ax, P. & P. Schmidt: Interstitielle Fauna von Galapagos. I. Einführung. Mikrofauna Meeresboden **20**, 1–38 (1973).

Fig. 3. Distribution of four interstitial polychaetes in the Atlantic region.

A. *Hesionides maxima* Westheide, 1967, respectively new subspecies (Hesionidae). Localities on the east side of the Atlantic: Sylt, Germany (Westheide 1966, p. 206; Westheide 1967, p. 130). Arcachon, France (Westheide 1972/73, p. 366). Amilcar, Tunisia (Westheide 1972a, p. 456). Locality of the new subspecies on the west side of the Atlantic: North Carolina, USA (coll. R. Rieger).

B. *Microphthalmus aberrans* (Webster & Benedict, 1887) (Hesionidae). Localities on the east side of the Atlantic: Öresund, Baltic Sea (Eliason 1920, p. 16; Eliason 1962, p. 29). Isefjord, Denmark (Rasmussen 1956, p. 49, as *M. sczelkowii*). Sylt, Germany (Westheide 1966, p. 206; Westheide 1967, p. 139). Arkona Bassin, Baltic Sea (Hagberg 1972, p. 1). Localities on the west side of the Atlantic: Eastport, Maine and Provincetown, Massachusetts, USA (Webster & Benedict 1887, p. 713); Maine to Massachusetts, USA (Pettibone 1963, p. 104); West Falmouth, Mass. (Grassle & Grassle 1974, p. 256). Halifax, Nova Scotia, Canada (coll. Levings, id. Pettibone, U.S. Nat. Museum).

C. *Microphthalmus listensis* Westheide, 1967 (Hesionidae). Localities on the east side of the Atlantic: Sylt, Germany (Westheide 1966, p. 206; Westheide 1967, p. 142). Scharhörn, Germany (Hartmann-Schröder & Stripp 1968, p. 11). Estuarine area of the rivers Rhine, Meuse and Scheldt, Netherlands (Wolff 1969, p. 309). Arcachon, France (Westheide 1972/73, p. 367). Firemore, Westcoast Scotland (McIntyre & Murison 1973, p. 106). Skagerrak, near Denmark (Hartmann-Schröder 1974a, p. 189). Localities on the west side of the Atlantic: East Point, Massachusetts, USA (N. W. Riser, personal communication); North Carolina, USA (coll. R. Rieger).

D. *Parapodrilus psammophilus* Westheide, 1965, respectively new species (see text above) (Dorvilleidae). Localities on the east side of the Atlantic: Sylt, Germany (Westheide 1965, p. 207, Schmidt 1969, p. 131). Arcachon, France (Westheide 1972/73, p. 367). Westkapelle, Netherlands (Wolff & Stegenga 1975, p. 87). Locality of the new species on the west side of the Atlantic: East Point, Massachusetts, USA (N. W. Riser, personal communication).

Bailey, K.: Potassium-Argon ages from the Galapagos Islands. Science **192**, 465–466 (1976).

Day, J. H.: New Polychaeta from Beaufort, with a key to all species recorded from North Carolina. NOAA Technical Report NMFS CIRG-375, U. S. Dep. Commerce Publ. 1–140 (1973).

Ekman, S.: Tiergeographie des Meeres. Akademische Verlagsgesellschaft, Leipzig (1935).

Eliason, A.: Biologisch-faunistische Untersuchungen aus dem Öresund. Polychaeta. Lunds Univ. Arsskr., Avd. **2, 16**, 1–103 (1920).

– : Weitere Untersuchungen über die Polychaetenfauna des Öresunds. Lunds Univ. Arsskr., n. s. **58**, 1–98 (1962).

Fauvel, P.: Bionomie et distribution géographique des annélides polychètes. Bull. Soc. Océanogr. Fr. **6**, 307–317 (1925).

Friedrich, H.: Polychaeta. Tierwelt N.- u. Ostsee. **6b**, 1–201 (1938).

Gerlach, S. A.: Means of meiofauna dispersal. Mikrofauna Meeresboden **61**, 89–103 (1977).

Grassle, J. F. & J. P. Grassle: Opportunistic life histories and genetic systems in marine benthic polychaetes. J. mar. Res. **32**, 253–284 (1974).

Hagberg, A.: Notes on the distribution of some polychaetes in the Baltic. Meddn. Havsfiskelaboratoriet Lysekil **128**, 1–3 (1972).

Hartmann-Schröder, G.: Zweiter Beitrag zur Polychaetenfauna von Peru. Kieler Meeresforsch. **18**, 109–147 (1962).

– : Revision der Gattung *Mystides* Theel (Phyllodocidae; Polychaeta Errantia). Mit Bemerkungen zur Systematik der Gattungen *Eteonides* Hartmann-Schröder und *Protomystides* Czerniavsky und mit Beschreibungen zweier neuer Arten aus dem Mittelmeer und einer neuen Art aus Chile. Zool. Anz. **171**, 204–243 (1963).

– : Zum Problem der Anpassung von Polychaeten an das Leben im Küstengrundwasser. Mitt. Hamburg. Zool. Mus. Inst., Kosswig-Festschrift 67–78 (1964).

– : Annelida, Borstenwürmer, Polychaeta. Tierwelt Dtl. **58**, 1–594 (1971).

– : Die Polychaeten der Biospeologischen Expedition nach Kuba 1969. Résultats des expéditions biospéologiques cubano-roumaines à Cuba, Bucuresti, 89–98 (1973).

– : Polychaeten von Expeditionen der „Anton Dohrn" in Nordsee und Skagerak. Veröff. Inst. Meeresforsch. Bremerh. **14**, 169–274 (1974a).

– : Zur Kenntnis des Eulitorals der afrikanischen Westküste zwischen Angola und Kap der Guten Hoffnung und der afrikanischen Ostküste von Südafrika und Mocambique unter besonderer Berücksichtigung der Polychaeten und Ostracoden. Teil II. Die Polychaeten des Untersuchungsgebietes. Mitt. Hamb. Zool. Mus. Inst., Ergbd. **69**, 95–228 (1974b).

– & G. Hartmann: Zur Kenntnis des Sublitorals der chilenischen Küste unter besonderer Berücksichtigung der Polychaeten und Ostracoden. Mitt. Hamb. Zool. Inst., Ergbd. **62**, 1–384 (1965).

– & –: Zur Kenntnis des Eulitorals der afrikanischen Westküste zwischen Angola und Kap der Guten Hoffnung und der afrikanischen Ostküste von Südafrika und Mocambique unter besonderer Berücksichtigung der Polychaeten und Ostracoden. Teil I. Beschreibung der Lebensräume, Ökologie und Zoogeographie. Mitt. Hamb. Zool. Mus. Inst., Ergbd. **69**, 5–94 (1974).

– & K. Stripp: Beiträge zur Polychaetenfauna der Deutschen Bucht. Veröff. Inst. Meeresforsch. Bremerh. **11**, 1–24 (1968).

Hempelmann, F.: Archiannelida und Polychaeta. Handb. Zool. **2**, 2, 1–212 (1931).

Hermans, C.O.: The systematic position of the Archiannelida. Syst. Zool. **18**, 85–102 (1969).

Jouin, C.: Status of the knowledge of the systematics and ecology of Archiannelida. In: N.C. Hulings, ed.: Proceedings of the First International Conference on Meiofauna. Smithson. Contr. Zool. **76**, 47–56 (1971).

Lasserre, P. & J. Renaud-Mornant: Resistence and respiratory physiology of intertidal meiofauna to oxygen-deficiency. Netherl. J. Sea Res. **7**, 290–302 (1973).

Laubier, C.: Sur la présence d'*Etéonides augeneri* dans les sables du Bassin d'Arcachon. Act. Soc. Linn. Bordeaux **102**, 1–3 (1965).

— : Quelques annélides polychètes interstitielles d'une plage de Côte d'Ivoire. Vie Milieu, Ser. A., **18**, 573–594 (1967a).

— : Annélides polychètes interstitielles de Nouvelle-Calédonie. Expéd. franc. Réc. corall. Nouvelle-Calédonie. **2**, 91–101 (1967b).

— : Adaptions chez les annélides polychètes interstitielles. Ann. Biol. **6**, 1–16 (1967c).

Marinov. T.: Über die Polychaetenfauna der Sandbiozönose vor der bulgarischen Schwarzmeerküste. Bull. Inst. Centr. Rech. Sci. Pisc. Pêch., Varna **3**, 61–78 (1963) (In Bulgarian).

Mayr, E.: Animal species and evolution. Belknap Press of Harvard Univ. Press, Cambridge, Mass. (1963).

McIntyre, A.D.: The meiofauna and macrofauna of some tropical beaches. J. Zool. Lond. **15**, 377–392 (1968).

— & D.J. Murison: The meiofauna of a flatfish nursery ground. J. mar. biol. Ass. U.K. **53**, 93–118 (1973).

Noodt, W.: Anpassung an interstitielle Bedingungen: Ein Faktor in der Evolution höherer Taxa der Crustacea? Faun-ökol. Mitt. **4**, 445–452 (1974).

Orrhage, L.: Über die Anatomie, Histologie und Verwandtschaft der Apistobranchidae (Polychaeta sedentaria) nebst Bemerkungen über die systematische Stellung der Archianneliden. Z. Morph. Tiere **79**, 1–45 (1974).

Pettibone, M.H.: Marine polychaete worms of the New England region. 1. Aphroditidae through Trochochaetidae. Bull. U.S. Nat. Mus. **227**, pt. 1, 1–356 (1963).

Quatrefages, A. de: Mémoire sur la distribution géographique des annélides. Nouv. Arch. Mus. Paris, ser. 1, **1**, 1–14 (1865).

Rao, G.Ch.: On the geographical distribution of interstitial fauna of marine beach sand. Proc. Ind. Nat. Sci. Acad. 38, B, 164–178 (1972).

— : *Trilobodrilus indicus* n.sp. (Dinophilidae, Archiannelida) from Andhra coast. Proc. Ind. Acad. Sci., 77, B, 101–108 (1973).

— & P.N. Ganapati: On some interstitial polychaetes from the beach sands of Waltair Coast. Proc. Ind. Acad. Sci. **65**, 10–15 (1967).

— & —: On some archiannelids from the beach sands of Waltair coast. Proc. Ind. Acad. Sci. · 67, B, 24–30 (1968).

Rasmussen, E.: Faunistic and biological notes on marine invertebrates III. The reproduction and larval development of some polychaetes from the Isefjord, with some faunistic notes. Biol. Meddel. Kongl. dansk Vedensk. Selsk. **23**, 1–84 (1956).

Renaud-Debyser, J.: Recherches écologiques sur la faune interstitielle des sables. Bassin d'Arcachon, île de Bimini, Bahamas. Vie Milieu, Suppl. **15**, 1–143 (1963).

Renaud-Mornant, J. & P. Serène: Note sur la microfaune de la côte oriental de la Malaisie. Cah. Pacif. **11**, 51–73 (1967).

Roger, J.: Annélides fossiles. Traité de Zoologie **5**, 1, 687–713 (1959).

Rullier, F.: Répartition géographiques des annélides polychètes. Bull. Lab. marit. Dinard. **43**, 69–71 (1958).

Salvat, B. & J. Renaud-Mornant: Etude écologique du macrobenthos et du meiobenthos d'un fond sableux du Lagon de Mururoa (Tuamotu-Polynésie). Cah. Pacif. **13**, 159–179 (1969).

Schmidt, P.: Die quantitative Verteilung und Populationsdynamik des Mesopsammons am Gezeiten-Sandstrand der Nordseeinsel Sylt. I. Faktorengefüge und biologische Gliederung des Lebensraumes. Int. Revue ges. Hydrobiol. **53**, 723–779 (1968).

— : Die quantitative Verteilung und Populationsdynamik des Mesopsammons am Gezeiten-Sandstrand der Nordsee-Insel Sylt. II. Quantitative Verteilung und Populationsdynamik einzelner Arten. Int. Revue ges. Hydrobiol. **54**, 95–174 (1969).

— : Zonation of the interstitial polychaete *Stygocapitella subterranea* (Stygocapitellidae) in European sandy beaches. Mar. Biol. **7**, 319–323 (1970).

— : Zonierung und jahreszeitliche Fluktuationen der interstitiellen Fauna in Sandstränden des Gebietes von Tromsφ (Norwegen). Mikrofauna Meeresboden **12**, 1–86 (1972).

— & W. Westheide: Interstitielle Fauna von Galapagos. XVII. Polygordiidae, Saccocirridae, Protodrilidae, Nerillidae, Dinophilidae (Polychaeta). Mikrofauna Meeresboden **62**, (in press, 1977).

Sφrensen, T.: A method of establishing groups of equal amplitude in plant society based on similarity of species content. K. danske Vidensk. Selsk. Skr. **5**, 1–34 (1948).

Sterrer, W.: Plate tectonics as a mechanism for dispersal and speciation in interstitial sand fauna. Netherl. J. Sea Res. **7**, 200–222 (1973).

Walliser, O.H.: Die Entwicklung der Wirbellosen im Erdaltertum. In: Grzimeks Tierleben, Suppl.: G. Heberer und H. Wendt, ed. Entwicklungsgeschichte der Lebewesen. 175–230 (1972).

Webster, H. E. & J. E. Benedict: The Annelida Chaetopoda, from Eastport, Maine. Rep. U. S. Commnr Fish. Wash. 1885, pt. **13**, 707–758 (1887).

Westheide, W.: *Parapodrilus psammophilus* nov. gen. nov. spec., eine neue Polychaeten-Gattung aus dem Mesopsammal der Nordsee. Helgoländer wiss. Meeresunters. **12**, 207–213 (1965).

— : Zur Polychaetenfauna des Eulitorals der Nordseeinsel Sylt. Helgoländer wiss. Meeresunters. **13**, 203–209 (1966).

— : Monographie der Gattungen *Hesionides* Friedrich und *Microphthalmus* Mecznikow (Polychaeta, Hesionidae). Ein Beitrag zur Organisation und Biologie psammobionter Polychaeten. Z. Morph. Tiere **61**, 1–159 (1967).

— : Zur Organisation, Biologie und Ökologie des interstitiellen Polychaeten *Hesionides gohari* Hartmann-Schröder (Hesionidae). Mikrofauna Meeresboden **3**, 1–37 (1970).

— : Interstitial Polychaeta (Excluding Archiannelida). In: N. C. Hulings, ed.: Proceedings of the First International Conference on Meiofauna. Smithson. Contr. Zool. **76**, 57–71 (1971).

— : La faune des polychètes et des archiannélides dans les plages sableuses à ressac de la côte méditerranéenne de la Tunesie. Bull. Inst. Océanogr. Pêche, Salammbô, **2**, 449–468 (1972a).

— : Räumliche und zeitliche Differenzierungen im Verteilungsmuster der marinen Interstitialfauna. Ver. dt. Zool. Ges. Helgoland 1971, 23–32 (1972b).

— : Nouvelles récoltes d'annélides interstitielles dans les plages sableuses du Bassin d'Arcachon. Vie Milieu Sér. A, **23**, 365–370 (1972/73).

— : Zwei neue interstitielle *Microphthalmus*-Arten (Polychaeta) von den Bermudas. Mikrofauna Meeresboden **14**, 1–16 (1973).

— : Interstitielle Polychaeten aus brasilianischen Sandstränden. Mikrofauna Meeresboden **31**, 1–16 (1974a).

— : Interstitielle Fauna von Galapagos. XI. Pisionidae, Hesionidae, Pilargidae, Syllidae (Polychaeta). Mikrofauna Meeresboden **44**, 1–146 (1974b).

Wolff, W. J.: Three species of *Microphthalmus* (Polychaeta) new to the Netherlands. Zool. Meded. **43**, 307–311 (1969).

— & H. Stegena: *Hesionura augeneri, Goniadelle bobretzkii*, and *Parapodrilus psammophilus* (Annelida, Polychaeta) new to the Netherlands. Zool. Bijdr. **17**, 82–87 (1975).

| Mikrofauna Meeresboden | 61 | 303–304 | 1977 |

W. Sterrer & P. Ax (Eds.). The Meiofauna Species in Time and Space. Workshop Symposium, Bermuda Biological Station, 1975

The Species Problem in Halammohydridae (Cnidaria)

by

Claus Clausen

Zoological Laboratory, University of Bergen, Norway

Abstract

The paper is a review of the systematics of the hydrozoan family Halammohydridae based on taxonomic and population studies on the taxon. An attempt is also made to establish the phylogenetic relationships within the group.

Special attention is given to the nematocysts. It has been a prevailing view that races of certain *Halammohydra* species exist which differ with respect to the cnidome when compaired with the respective type forms. Inasmuch as distinct nematocyst types are involved, it is assumed that such diverging "races" should be designated new species. This is the case in a supposed local form of *H. intermedia* which has aspiroteles, an astomocnide nematocyst type. When the difference between the forms in question is expressed in the possession of either the one or both of two different size categories of a certain nematocyst type (stenoteles or euryteles), the term subspecies is proposed for such forms. This also indicates that we may be dealing with incipient species. *H. octopodides, H. schulzei, H. vermiformis* and *H. coronata* behave in this way.

With respect to the cnidome there are two distinct species groups in *Halammohydra*, those with both isorhizas and stenoteles (*H. vermiformis, H. andamanensis, H. octopodides, H. schulzei, H. intermedia, H.* sp. (with aspiroteles)), and those with euryteles only (*H. adherens, H. coronata, H. chauhani*). Within each of these groups morphological reductions and specializations occur. These involve body size and shape, and tentacle number. Presence of astomocnides, the supposedly most primitive nematocyst type, is believed to express an original condition.

Bibliography

Clausen, C.: Morphological studies of *Halammohydra* Remane (Hydrozoa). Sarsia **29**, 349–370 (1967).

– : Interstitial Cnidaria: Present status of their systematics and ecology. Smiths. Contr. to Zool. **76**, 1–8 (1971).

Lacassagne, M.: Deux nouveaux types de nématocystes astomocnides chez les Actinulides (Hydrozoaires): les eurytèloides spirotèles et aspirotèles. C. R. Acad. Sc. Paris **266**, 892–894 (1968a).

– : Les anisorhizes atriches, nouveau type de nématocystes stomocnides présents chez deux formes d'*Halammohydra* (Hydrozoaires, Actinulides). C. R. Acad. Sc. **266**, 2090–2092 (1968b).

Rao, G. C.: *Halammohydra cauhani* n. sp. (Hydrozoa) from Andamans, India. Dr. B. S. Chanhan Comm. Vol., 299–303 (1975).

— : *Halammohydra andamanensis* n. sp. Current Science (in press) (1977).

Remane, A.: *Halammohydra*, ein eigenartiges Hydrozoon der Nord- und Ostsee. Z. Morph. Ökol. Tiere 7, 643–678 (1927).

Swedmark, B.: Variation morphologique des différentes populations régionales d'*Halammohydra*. Année biologique **44** (3/4), 183–189 (1957).

— : & G. Teissier: *Halammohydra vermiformis* n. sp. et la famille des Halammohydridae Remane. Bull. Soc. Zool. France **82**, 38–49 (1957).

— : The Actinulida and their evolutionary significance. Symp. Zool. Soc. Lond. **16**, 119–133 (1966).

— : Structure et adaptation d'*Halammohydra adherens*. Cah. Biol. Mar. **8**, 63–74 (1967).

Werner, B.: Die Nesselkapseln der Cnidaria, mit besonderer Berücksichtigung der Hydroida. I. Klassifikation und Bedeutung für die Systematik und Evolution. Helgoländer wiss. Meeresunters. **12** (1–2), 1–39 (1965).

Mikrofauna Meeresboden	61	305	1977

W. Sterrer & P. Ax (Eds.). The Meiofauna Species in Time and Space. Workshop Symposium, Bermuda Biological Station, 1975

Studies on Differentiation in Conspecific Disjunct Oligochaete Populations: I. Experiments on Salinity Preference and Tolerance.

by

Olav Giere

Zool. Inst. und Museum der Universität Hamburg, D-2 Hamburg 13, W. Germany

Abstract

Several eurytopic oligochaete species occur in ecologically most divergent habitats. It was tried to find out whether disjunct populations of these species display differences in their eco-physiological tolerances and preferences, and thus might indicate a possible differentiation also in their respective genetical background.

The salinity preference of specimens from habitats widely different in salinity (beaches from the North Sea, Baltic Sea; oligohaline river inlets) was tested in experimental salinity gradients. Corresponding tolerance experiments supported the investigation. The examined animals belonged to the macrofauna enchytraeid species *Enchytraeus albidus* and *Lumbricillus lineatus*, and to the meiofauna species *Marionina southerni*.

The short time available and the complicated experimental set-up rendered so far only tentative results which require additional studies:

Lumbricillus lineatus: Preference – in general accordance with the respective home habitat (North Sea: a rather wide range between $5-20$ $^0/_{00}$ S; Baltic Sea: appr. 10 $^0/_{00}$ S. Tolerance – a markedly higher range ($0-55$ $^0/_{00}$ S for 7 days).

Enchytraeus albidus: Preference – North Sea: no specially preferred range between 0 and 60 $^0/_{00}$ S. Baltic Sea: appr. $5-10$ $^0/_{00}$ S. Tolerance – all populations, including a terrestrial one from garden soil, survived in 0 to 55 $^0/_{00}$ S.

Marionina southerni: The specimens lived preferably between 5 and 15 $^0/_{00}$ S. Tolerance range was extended up to 30 $^0/_{00}$ S (in 40 $^0/_{00}$ S, LD 50-time: 2 days).

These preliminary results are discussed considering the prevailing ecological conditions in the resp. habitats, the few literature data, and some tentative electrophoresis experiments with animals from the different populations in question.

Bibliography

Avise, J. C.: Systematic value of electrophoretic data. – Syst. Zool. 23, 465–481 (1975).

Backlund, H. O.: Wrack fauna of Sweden and Finland. Ecology and chorology. – Opusc. ent. (Suppl.) **1945**, 1–237 (1945).

Giere, O.: Beitrag zur Kenntnis mariner Oligochaeten im finnischen Schärgarten. Acta zool. Fennica (in press).

Jansson, B.-O.: Salinity resistance and salinity preference of two oligochaetes *Aktedrilus monospermathecus* Knöllner and *Marionina preclitellochaeta* n. sp. from the interstitial fauna of marine sandy beaches. Oikos 13, 293–305 (1962).

Tynen, M. J.: Littoral distribution of *Lumbricillus reynoldsoni* Backlund and other Enchytraeidae (Oligochaeta) in relation to salinity and other factors. Oikos 20, 41–53 (1969).

| Mikrofauna Meeresboden | 61 | 307–308 | 1977 |

W. Sterrer & P. Ax (Eds.). The Meiofauna Species in Time and Space. Workshop Symposium, Bermuda Biological Station, 1975

Gutless Nematodes of the Deep-Sea

by

W. Duane Hope

Smithsonian Institution, Washington, D.C., USA

Abstract

In the course of studying large populations of deep-sea nematodes collected during cruises of the VEMA and Woods Hole Oceanographic Institute vessels, several species of unusual nematodes were discovered. The first to be described and named was *Rhaptothyreus typicus* Hope & Murphy, 1969 which was characterized by the absence of a vestigial stoma and esophagus, the absence of a gut, the presence of basophilic rod-like structures occupying nearly all of the body cavity, a single spiculum, and an extremely large amphid. Only males were found. However, further study of the samples yielded over a hundred females and three males, all with a vestigial stoma and esophagus, and a trophosome instead of a gut, but several species were represented and none appeared to be closely related to *R. typicus*.

The systematic relationships and mode of feeding of these nematodes was clarified in 1971 when examination of an ostracod, *Zabythrocypris* sp., from one of the same deep-sea stations revealed the presence of an endoparasitic nematode. Dissection of the nematode from its host disclosed that it was very similar to, but not identical with, some of the gutless nematodes from the sediment. It was then postulated, on the basis of morphology and parasitic existence, that these nematodes are closely related to those of the superfamily Mermithoidea, a taxon whose members are typically parasites of freshwater or terrestrial arthropods (and occasionally of other invertebrates) during the juvenile stages, and are free-living in sediment or soil during their adult life. Subsequent discovery, in another specimen of *Zabythrocypris*, of a young juvenile nematode with a stylet, further supports the proposed relationship.

Further search for these nematodes has produced several specimens from two additional species of ostracod, one specimen each in a harpacticoid copepod of the genus *Pseudotochidius* and in the remains of a tanaid, a caprellid and two isopods. Additional specimens have also been obtained from the sediment. Three specimens from sediment appear to be closely related to genera of the family Mermithidae, while all others probably cannot be justifiably assigned to that family as now defined.

Rubtzov & Platonova (1974) have proposed the family Marimermithidae to receive two new genera and four new species of marine nematodes which have structural affinities with members of Mermithoidea. One species is parasitic in a starfish while hosts in the other cases are not known. These specimens appear to differ significantly from the specimens considered here.

The discovery of nematodes apparently related to those of the Mermithoidea places in doubt the postulation by Steiner (1917) that mermithids have their closest phylogenetic affinities with the order of freshwater and soil inhabiting nematodes, Dorylaimida.

Bibliography

Hope, W. D. & D. G. Murphy: *Rhaptothyreus typicus* n.g., n. sp., an abyssal marine nematode representing a new family of uncertain taxonomic position. Proceedings of the Biological Society of Washington **82**: 81–92 (1969).

Rubtzov, I. A. & T. A. Platonova: A new family of marine parasitic nematodes. Zoologichesky
Zhurnal **53** (10), 1445–1458 (1974).

Steiner, G.: Über die Verwandtschaftsverhältnisse und die systematische Stellung der Mermithi-
den. Zool. Anz. **48**, 263–267 (1917).

Mikrofauna Meeresboden	61	309	1977

W. Sterrer & P. Ax (Eds.). The Meiofauna Species in Time and Space. Workshop Symposium, Bermuda Biological Station, 1975

Systematics and Evolution in Polycystididae (Turbellaria, Kalyptorhynchia)

by

Ernest R. Schockaert

Limburgs Universitair Centrum, Universitaire Campus, B-3610 Diepenbeek, Belgium

Abstract

Possible evolutionary trends within Polycystididae have been indicated by Karling (1956) while studying the male genital organs of several turbellarian groups. For the moment his ideas seem the most valuable basis for establishing an intrafamiliar system of Polycystididae, and have hence been applied formally (Schockaert, 1974). However, it can be demonstrated with several examples that the underlying principles prove to be unsatisfactory when one tries to apply Hennig's methodology:

— the subfamily Duplacrorhynchinae Schockaert & Karling, 1970, can be recognized as containing the most plesimorph polycystidids; a synapomorph character that would give us an indication of the monophyly of this taxon is lacking;

— the homology of some glandular elements (all derivatives of the atrial epithelium) is often doubtful;

— gutter- or tube-like cuticular stylets (derived from cirrus spines), clearly representing an apomorph condition, have undoubtedly originated more than once and do not offer a sound basis delimiting monophyletic groups . . .

We are strongly convinced that it will be absolutely necessary to abandon the genital apparatus as the nearly unique basis for taxonomy in polycystidids and other turbellarian groups. All too often we find convergence in the structure of copulatory organs in obviously remote taxa. Moreover, we can have some doubt about the selective meaning of this organ system, even about its role in reproductive isolation, i.e. in the process of speciation. Can we, under these circumstances, pay too much attention to these organs in taxonomy above the species- or genus-level? We do hope that other elements will provide us with a more solid morphological basis, if necessary with the help of the electron microscope.

Bibliography

Karling, T. G.: Morphologisch-histologische Untersuchungen an den männlichen Atrialorganen der Kalyptorhynchia (Turbellaria). Ark. Zool., Ser. 2, 9 (7), 187–279 (1956).

Schockaert, E. R.: On the male copulatory organ of some Polycystididae and its importance for the systematics of the family. In: Biology of the Turbellaria (N. W. Riser, M. P. Morse, eds.) – L. H. Hyman Memorial Volume. Mc Graw-Hill, N. Y.: 165–172 (1974).

| Mikrofauna Meeresboden | 61 | 311–312 | 1977 |

W. Sterrer & P. Ax (Eds.). The Meiofauna Species in Time and Space. Workshop Symposium, Bermuda Biological Station, 1975

Ecological Genetics and Speciation in Marine Bryozoa.

by

Thomas J. M. Schopf

Department of the Geophysical Sciences, University of Chicago, Chicago, Illinois, 60637; and Marine Biological Laboratory, Woods Hole, Massachusetts, 02543

Abstract

Morphologic features of marine species are rarely suitable as genetic markers. Accordingly the introduction of techniques to monitor variation in the enzymes and other proteins that are products of single genes, has permitted analyses that were previously impossible. Several reviews of the method and/or results have appeared (Brewer 1970; Gooch 1975; Lewontin 1974; Manwell & Baker 1970; Shaw & Prasad 1970).

For marine bryozoa, we have focused our attention on *Schizoporella errata*, a typically sessile, annual species with a short-lived larva (hours). By studying the patterns of geographic variation in these Mendelian traits, we have learned that this species routinely outbreeds, and that it seems that the size of the local population is about 200 colonies. Shallow and deeper water populations that differ by only 6 m, and shallow water populations that are 10 km apart along the coast line, can be significantly different in their gene frequencies. Populations that are 30–40 kms apart are routinely genetically differentiated, and in fact this genetic differentiation is paralleled by morphologic differentiation in a character (avicularium length) whose mean value changes the most from one place to the next, along a 100 km transect of the coast line.

All of these populations have come into existence in the past 100 years, following the rise in sea level of about 150 m in the past 15,000 years. The presently observed clines in gene frequencies, and in morphologic differentiation, parallel a temperature cline along the coast. Local reversals in the temperature owing to local circulation are paralleled by reversals in gene frequencies. These conclusions are described and summarized in Schopf (1974, 1976; Schopf & Dutton 1976).

These results show that the genetic composition of marine species can be rapidly altered as local, inshore ecologic conditions change. This phyletic evolution within local populations of *Schizoporella* suggests that the rate of evolutionary change may be quite rapid. Most morphologic characters are conservative, and eco-phenotypic explanations can be applied in cases where change occurs but where genetic data do not exist. Thus change in morphology per se may be a very poor indicator of rates of change in the underlying genome (Schopf et al. 1975). Rates of speciation for shallow water marine forms may well be on the order of only a few thousand years.

Bibliography

Brewer, G. J. & C. F. Sing: An Introduction to Isozyme Techniques. Academic Press. 186 pp. (1970).

Gooch, J. L.: Mechanisms of evolution and population genetics. In: O. Kinne, ed. Marine Ecology, vol. 2, part 1. Wiley & Sons pp. 349–409 (1975).

Lewontin, R. C.: The Genetic Basis of Evolutionary Change. Columbia University Press. 346 pp. (1974).

Manwell, C. & C. M. A. Baker: Molecular Biology and Origin of Species. Univ. Washington Press. 394 pp. (1970).

Schopf, T. J. M.: Survey of genetic differentiation in a coastal zone invertebrate: the ectoproct *Schizoporella errata*. Biological Bulletin **145**, 78–87 (1974).

— : Population genetics of Bryozoans In: R. M. Wollacott & R. L. Zimmer, eds. Biology of Bryozoa. Academic Press. (In press).

— & A. R. Dutton: Parallel clines in morphologic and genetic differentiation in a coastal zone marine invertebrate: the bryozoan *Schizoporella errata*. Paleobiology, **1**, 255–264 (1976).

—, D. M. Raup, S. J. Gould & D. S. Simberloff: Genomic versus morphologic rates of evolution: influence of morphologic complexity. Paleobiology. **1**, 63–70 (1975).

Shaw, C. R. & R. Prasad: Starch gel electrophoresis of enzymes – a compilation of recipes. Biochemical Genetics. **4**, 297–320 (1970).

| Mikrofauna Meeresboden | 61 | 313—314 | 1977 |

W. Sterrer & P. Ax (Eds.). The Meiofauna Species in Time and Space. Workshop Symposium, Bermuda Biological Station, 1975

Geographic and Reproductive Isolation in the Genus Tisbe (Copepoda, Harpacticoida)

by

Brigitte Volkmann

Istituto di Biologia del Mare, I-30122 Venezia, Italy

Abstract

The stimulus for pursuing research on reproductive isolation in *Tisbe* was given by the great taxonomic difficulties found in this genus, due to the existence of morphologically similar forms, and even sibling species groups, between which the reproductive barrier is, however, complete (Volkmann-Rocco 1971, 1972, 1973 a, b). *Tisbe* provides one of the few cases in the marine habitat where through cross-breeding experiments it is possible to estimate the degree of reproductive isolation also in species with wide geographical distribution.

1) There are species characterized by a total absence of barriers to gene flow, like *T. holothuriae*, *T. battagliai*, *T. bulbisetosa*, *T. lagunaris*, *T. cucumariae*, and *T. biminiensis*. Even crosses between Mediterranean and transatlantic populations (and a Bahama and Red Sea population in the case of *T. biminiensis*) proved to be perfectly interfertile (Battaglia & d'Avella 1964; Battaglia & Volkmann-Rocco 1973).

2) The second case, incipient reproductive isolation, is represented by *T. reticulata* and *T. dobzhanskii*. Crosses between different populations resulted in lower fecundity and viability of F_1 hybrids, and in a strong shift of the sex-ratio in favor of males (Battaglia 1957; Volkmann-Rocco & Battaglia 1972).

3) Crosses between geographic populations of *T. clodiensis* show different stages of reproductive divergence and separation (Battaglia & Volkmann-Rocco 1969; Volkmann, Battaglia & Varotto, in press). Basically 3 possibilities of relationships could be observed in crosses involving 3 Mediterranean (Venice, Ponza, Banyuls-sur-Mer), 2 Atlantic (Roscoff, Arcachon) and 1 transatlantic population (Beaufort, N.C.):

a) compatibility in both directions, resulting in normal adult offspring,

b) incompatibility in both directions, giving none or dying larvae,

c) non-reciprocal incompatibility with adult offspring in only one direction; but development of hybrids was remarkably retarded, the number of progenies per female was lower than in the controls, and a nearly complete shift of the sex-ratio toward males occurred.

Non-reciprocal or bilateral incompatibility was evident in all crosses involving the Venice or the Arcachon strain, independent of the geographic distance. The Venice population showed the strongest isolation from others when the male was utilized, the Arcachon strain, when the female was involved. The fact that this incompatibility was mostly overcome in the first backcrosses would suggest that the character "non-crossability" is controlled directly by at least one nuclear gene, but most likely by a multiple system of incompatibility genes. It was concluded that in *T. clodiensis* we are dealing with a cluster of semispecies, and that ecological factors were far more important than distance per se in the process of isolation and genetic differentiation.

Bibliography

Battaglia, B.: Ecological differentiation and incipient intraspecific isolation in marine copepods. Année biol. **33**, 259–268 (1957).

– & D. d'Avella: Differenziamento di caratteri genogeografici nel Copepode *Tisbe furcata* (Baird). La tolleranza alle variazioni di salinità. Boll. Zool. **31**, 1233–1241 (1964).

– & B. Volkmann-Rocco: Gradienti di isolamento riproduttivo in popolazioni geografiche del copepode *Tisbe clodiensis*. Atti Ist. veneto Sci. **127**, 371–381 (1969).

– & B. Volkmann-Rocco: Geographic and reproductive isolation in the marine harpacticoid copepod *Tisbe*. Mar. Biol. **19**, 156–160 (1973).

Volkmann-Rocco, B.: Some critical remarks on the taxonomy of *Tisbe* (Copepoda, Harpacticoida). Crustaceana **21**, 127–132 (1971).

–: *Tisbe battagliai* n.sp., a sibling species of *Tisbe holothuriae* Humes (Copepoda, Harpacticoida). Arch. Oceanogr. Limnol. **17**, 259–273 (1972).

–: *Tisbe biminiensis* (Copepoda, Harpacticoida) a new species of the *gracilis* group. Arch. Oceanogr. Limnol. **18**, 71–90 (1973a).

–: Étude de quatre spèces jumelles du groupe *Tisbe reticulata* Bocquet (Copepoda, Harpacticoida). Arch. Zool. exp. gen. **114**, 317–348 (1973b).

– & B. Battaglia: A new case of sibling species in the genus *Tisbe* (Copepoda, Harpacticoida). proc. 5th Eur. mar. Biol. Symp., Piccin, Padova, 67–80 (1972).

–, B. Battaglia & V. Varotto: A study of reproductive isolation within the superspecies *Tisbe clodiensis* (Copepoda, Harpacticoida). (in press).

Mikrofauna Meeresboden	61	315—316	1977

W. Sterrer & P. Ax (Eds.) The Meiofauna Species in Time and Space. Workshop Symposium, Bermuda Biological Station, 1975

An Ecophysiological Approach to the Study of Meiofauna Species in Time and Space

by

Wolfgang Wieser

Institut für Zoophysiologie der Universität Innsbruck,
Peter-Mayr-Straße Ia, A-6020 Innsbruck, Austria

Abstract

The study of adaptations has shed light on a number of problems relevant to the distribution in time and space and to the evolution of animal species. This must be particularly true for the microscopic inhabitants of an environment that is heterogeneous to such a degree that completely different faunal assemblages may live within a few mm from each other.

On a subtropical low energy beach the tolerance limits for high temperature, alkalinity, low pO_2 and low salinity of some of the major representatives of the meiofauna are closely adjusted to the microenvironment in which they live. For example, a nematode species with its center of abundance only one cm below the sediment surface displays distinctly lower resistance to high temperature and alkalinity than a typical surface-inhabiting species. This is taken to indicate that in such a habitat, which is characterized by high stability and predictability, diversification has progressed long enough to allow different niches to be filled by species which are adjusted closely to the ranges of ecological factors measured in their immediate environment. In less predictable and less stable habitats meiofauna species are expected to be less closely adjusted to the ecological conditions of the niche which they occupy, but few comparative data are available.

Thus sediment habitats with different climatic and ecological regimes may be characterized by different patterns of adaptation of their meiofauna. A comparative study of these patterns is likely to tell us something about the history and the selective pressures acting in interstitial environments, and thus will contribute to an understanding of the mechanisms that govern the distribution of species in the sea.

Bibliography

Fenchel, T.: The ecology of marine microbenthos. IV. Structure and function of the benthic ecosystem, its chemical and physical factors and the microfauna communities with special reference to the ciliated protozoa. Ophelia 6, 1—182 (1969).

— & R. J. Riedl: The sulfide system: a new biotic community underneath the oxidized layer of marine sand bottoms. Mar. Biol. 7, 255—268 (1970).

Gray, J. S.: An experimental approach to the ecology of the harpacticoid *Leptastacus constrictus* Lang. J. exp. mar. Biol. Ecol. 2, 278—292 (1968).

—: Animal-sediment relationships. Oceanogr. Mar. Biol. Ann. Rev. 12, 223—261 (1974).

Jansson, B. O.: The importance of tolerance and preference experiments for the interpretation of mesopsammon field distributions. Helgol. wiss. Meeresunters. 15, 41—58 (1967).

—: Quantitative and experimental studies of the interstitial fauna in four Swedish sandy beaches. Ophelia 5, 1—71 (1968).

Lasserre, P.: Données écophysiologiques sur la répartition des Oligochètes marins méiobenthiques. Incidence des paramètres salinité-température sur le metabolisme respiratoire de deux espèces euryhalines du genre *Marionina* Michaelsen, 1889 (Enchytraeidae, Oligochaeta). Vie Milieu **22**, 523—540 (1971).

Schmidt, P.: Die quantitative Verteilung und Populationsdynamik des Mesopsammons am Gezeiten-Sandstrand der Nordseeinsel Sylt. I. Faktorengefüge und biologische Gliederung des Lebensraumes. Int. Rev. ges. Hydrobiol. **53**: 723—779 (1968).

Slobodkin, L. B. & H. L. Sanders: On the contribution of environmental predictability to species diversity. Brookhaven Symp. Biol. **22**: 82—92 (1969).

Wieser, W.: The meiofauna as a tool in the study of habitat heterogeneity: Ecophysiological aspects. A review. Cah. Biol. Mar. **16**: 647—670 (1975).

— , J. Ott, F. Schiemer & E. Gnaiger: An ecophysiological study of some meiofauna species inhabiting a sandy beach at Bermuda. Mar. Biol. **26**: 235—248 (1974).

REIHEN DER
MATHEMATISCH-NATURWISSENSCHAFTLICHEN KLASSE

MIKROFAUNA DES MEERESBODENS

1970

1. PETER AX und RENATE AX, Das Verteilungsprinzip des subterranen Psammon am Übergang Meer-Süßwasser 51 S. mit 24 Abb., DM 13,—
2. PETER AX und RUTH HELLER, Neue Neorhabdocoela (Turbellaria) vom Sandstrand der Nordsee-Insel Sylt. 46 S. mit 20 Abb. DM 11,80

3. WILFRIED WESTHEIDE, Zur Organisation, Biologie und Ökologie des interstitiellen Polychaeten Hesionides gohari Hartmann-Schröder (Hesionidae). 37 S. mit 20 Abb., DM 10,—

1971

4. PETER AX, Zur Systematik und Phylogenie der Trigonostominae (Turbellaria, Neorhabdocoela). 84 S. mit 45 Abb., DM 26,—
5. PETER AX und KARL SCHILKE, Karkinorhynchus tetragnathus nov. spec., ein Schizorhynchier mit zweigeteilten Rüsselhaken (Turbellaria, Kalyptorhynchia). 10 S., 2 Abb., DM 6,80
6. WILFRIED WESTHEIDE, Apharyngtus punicus nov. gen. nov. spec., ein aberranter Archiannelide aus dem Mesopsammal der tunesischen Mittelmeerküste. 19 S. mit 10 Abb., DM 8,—

7. SIEGMAR HOXHOLD, Eigebilde interstitieller Kalyptorhynchier (Turbellaria) von der deutschen Nordseeküste. 43 S. mit 24 Abb., DM 13,—
8. PETER AX, Neue interstitielle Macrostomida (Turbellaria) der Gattungen Acanthomacrostomum und Haplopharynx, 14 S. mit 5 Abb., DM 6,20
9. ULRIKE MÜLLER und PETER AX, Gnathostomulida von der Nordseeinsel Sylt mit Beobachtungen zur Lebensweise und Entwicklung von Gnathostomula paradoxa Ax, 41 Seiten, 15 Abbildungen, DM 13,—

1972

10. PETER SCHMIDT, Zonierung und jahreszeitliche Fluktuationen des Mesopsammons im Sandstrand von Schilksee (Kieler Bucht). 60 S. mit 49 Abb., DM 18,—
11. ULRICH EHLERS, Systematisch-phylogenetische Untersuchungen an der Familie Solenopharyngidae (Turbellaria, Neorhabdocoela). 77 S. mit 25 Abb. DM 20,—

12. PETER SCHMIDT, Zonierung und jahreszeitliche Fluktuationen der interstitiellen Fauna in Sandstränden des Gebietes von Tromsø (Norwegen). 86 S. mit 76 Abb., DM 26,—
13. BEATE SOPOTT, Systematik und Ökologie von Proseriaten (Turbellaria) der deutschen Nordseeküste, 72 Seiten, 33 Abb., DM 18,—

1973

14. WILFRIED WESTHEIDE, Zwei neue interstitielle Microphthalmus-Arten (Polychaeta) von den Bermudas. 16 Seiten. 4 Abb., DM 6,80
15. BEATE SOPOTT, Jahreszeitliche Verteilung und Lebenszyklen der Proseriata (Turbellaria) eines Sandstrandes der Nordseeinsel Sylt. 106 Seiten, 91 Abb., DM 28,—
16. TOR G. KARLING, Anatomy und Taxonomy of a New Otoplanid (Turbellaria, Proseriata) from South Georgia. 11 Seiten, 3 Abb., DM 4,80
17. WOLFGANG MIELKE, Zwei neue Harpacticoidea (Crustacea) aus dem Eulitoral der Nordseeinsel Sylt. 14 Seiten mit 6 Abb., DM 8,20
18. EIKE HARTWIG, Die Ciliaten des Gezeiten-Sandstrandes der Nordseeinsel Sylt. I. Systematik. 69 Seiten mit 20 Abb., DM 24,—
19. ULRICH EHLERS, Zur Populationsstruktur interstitieller Typhloplanoida und Dalyellioida (Turbellaria, Neorhabdocoela). 105 Seiten mit 89 Abb. DM 32,—

20. PETER AX und PETER SCHMIDT, Interstitielle Fauna von Galapagos. I. Einführung. 37 Seiten mit 10 Abb., DM 11,20
21. EIKE HARTWIG, Die Ciliaten des Gezeiten-Sandstrandes der Nordseeinsel Sylt. II. Ökologie. 171 Seiten mit 105 Abb., DM 48,—
22. BEATE EHLERS und ULRICH EHLERS, Interstitielle Fauna von Galapagos. II. Gnathostomulida. 27 Seiten mit 13 Abb., DM 10,50
23. PETER AX und ULRICH EHLERS, Interstitielle Fauna von Galapagos. III. Promesostominae (Turbellaria, Typhloplanoida). 16 Seiten mit 5 Abb., DM 6,—
24. HORST KURT SCHMINKE, Evolution, System und Verbreitungsgeschichte der Familie Parabathynellidae (Bathynellacea, Malacostraca). 192 Seiten mit 48 Abb., DM 46,—
25. SIEVERT LORENZEN, Die Familie Epsilonematidae (Nematodes). 86 Seiten mit 23 Abb., DM 22,60

26. PETER SCHMIDT, Interstitielle Fauna von Galapagos. IV. Gastrotricha. 76 Seiten mit 29 Abb., DM 24,20

27. PETER AX und RENATE AX, Interstitielle Fauna von Galapagos. V. Otoplanidae (Turbellaria, Proseriata). 28 Seiten mit 11 Abb., DM 11,20

28. WILFRIED WESTHEIDE und PETER SCHMIDT, Interstitielle Fauna von Galapagos. VI. Aeolosoma maritimum dubiosum nov. ssp. (Annelida, Oligochaeta). 11 Seiten mit 4 Abb. DM 6,40

29. PETER AX und RENATE AX, Interstitielle Fauna von Galapagos. VII. Nematoplanidae, Polystyliphoridae, Coelogynoporidae (Turbellaria, Proseriata), 28 Seiten mit 10 Abb., DM 10,80

30. ULRICH EHLERS und PETER AX, Interstitielle Fauna von Galapagos. VIII. Trigonostominae (Turbellaria, Typhloplanoida), 33 Seiten mit 13 Abb., DM 11,80

31. WILFRIED WESTHEIDE, Interstitielle Polychaeten aus brasilianischen Sandstränden. 16 Seiten mit 6 Abb., DM 7,60

32. ANNO FAUBEL, Die Acoela (Turbellaria) eines Sandstrandes der Nordseeinsel Sylt. 58 Seiten mit 29 Abb., DM 17,20

33. DIETRICH BLOME, Zur Systematik von Nematoden aus dem Sandstrand der Nordseeinsel Sylt. 25 Seiten mit 48 Abb., DM 8,80

34. BEATE SOPOTT-EHLERS und PETER SCHMIDT, Interstitielle Fauna von Galapagos. IX. Dolichomacrostomidae (Turbellaria, Macrostomida). 20 Seiten mit 7 Abb., DM 8,—

35. HANS VOLKMAR HERBST, Drei interstitielle Cyclopinae (Crustacea, Copepoda) von der Nordseeinsel Sylt. 17 Seiten mit 32 Abb., DM 6,80

36. HELMUT KUNZ, Zwei neue afrikanische Paramesochridae (Copepoda Harpacticoidea) mit Darstellung eines Bewegungsmechanismus für die Furkaläste. 20 Seiten mit 31 Abb., DM 8,40

37. WOLFGANG MIELKE, Eulitorale Harpacticoidea (Copepoda) von Spitzbergen. 52 Seiten mit 28 Abb., DM 17,80

38. WILFRIED SCHEIBEL, Ameira divagans Nicholls, 1939 (Copepoda Harpacticoidea). Neubearbeitung aus der Kieler Bucht. 10 Seiten mit 18 Abb., DM 8,—

39. GERTRAUD TEUCHERT, Aufbau und Feinstruktur der Muskelsysteme von Turbanella cornuta Remane (Gastrotricha, Macrodasyoidea). 26 Seiten mit 20 Abb., DM 10,80

40. FRANZ RIEMANN, On hemisessile nematodes with flagelliform tails living in marine soft bottoms and on micro-tubes found in deap sea sediments. 15 Seiten mit 6 Abb., DM 6,60

41. SIEGMAR HOXHOLD, Zur Populationsstruktur und Abundanzdynamik interstitieller Kalyptorhynchia (Turbellaria, Neorhabdocoela). 134 Seiten mit 97 Abb., DM 40,60

42. REINHARD M. RIEGER and SETH TYLER, A new glandular sensory organ in interstitial Macrostomida, I. Ultrastructure. 41 Seiten mit 15 Abb. DM 15,40

43. PETER SCHMIDT, Interstitielle Fauna von Galapagos. X. Kinorhyncha. 15 Seiten mit 2 Abb., DM 6,20

44. WILFRIED WESTHEIDE, Interstitielle Fauna von Galapagos. XI. Pisionidae, Hesionidae, Pilargidae, Syllidae (Polychaeta). 146 Seiten mit 63 Abb., DM 40,80

45. ANNO FAUBEL, Macrostomida (Turbellaria) von einem Sandstrand der Nordseeinsel Sylt. 32 Seiten mit 12 Abb., DM 11,80

46. BEATE SOPOTT-EHLERS und PETER SCHMIDT, Interstitielle Fauna von Galapagos. XII. Myozona Marcus (Turbellaria, Macrostomida). 19 Seiten mit 7 Abb., DM 8,40

47. SIEVERT LORENZEN, Glochinema nov. gen. (Nematodes, Epsilonematidae) aus Südchile. 22 Seiten mit 4 Abb.. DM 8,20

48. PETER AX und ANNO FAUBEL, Anatomie von Psammomacrostomum equicaudum Ax, 1966 (Turbellaria, Macrostomida). 12 Seiten mit 2 Abb., DM 6,20

49. ULRICH EHLERS, Interstitielle Typhloplanoida (Turbellaria) aus dem Litoral der Nordseeinsel Sylt. 102 Seiten mit 45 Abb., DM 31,40

50. WERNER KATZMANN et LUCIEN LAUBIER, Le genre Fauveliopsis (Polychète sédentaire) en Méditerranée. 16 Seiten mit 4 Abb., DM 7,80

51. HERBERT MOCK und PETER SCHMIDT, Interstitielle Fauna von Galapagos. XIII. Ototyphlonemertes Diesing (Nemertini, Hoplonemertini). 40 Seiten mit 16 Abb., DM 13,40

52. WOLFGANG MIELKE, Systematik der Copepoda eines Sandstrandes der Nordseeinsel Sylt. 134 Seiten mit 87 Abb., DM 49,60

53. J. B. J. WELLS, HELMUT KUNZ, G. CHANDRASEKHARA RAO, A review of the mechanisms for movement of the caudal furca in the Family Paramesochridae (Copepoda Harpacticoida), with a description of a new species of Kliopsyllus Kunz. 16 Seiten mit 27 Abb., DM 7,80

54. BEATE SOPOTT-EHLERS, PETER SCHMIDT, Interstitielle Fauna von Galapagos. XIV. Polycladida (Turbellaria). 32 Seiten mit 13 Abb. DM 12,80

55. SIEVERT LORENZEN, Rhynchonema-Arten (Nematodes, Monhysteridae) aus Südamerika und Europa. 29 Seiten mit 11 Abb., DM 10,40

56. ANNO FAUBEL, Populationsdynamik und Lebenszyklen interstitieller Acoela und Macrostomida (Turbellaria). 107 Seiten mit 75 Abb., DM 33,–

57. PETER SCHMIDT, BEATE SOPOTT-EHLERS, Interstitielle Fauna von Galapagos XV. Macrostomum O. Schmidt, 1848 und Siccomacrostomum triviale nov. gen. nov. spec. (Turbellaria, Macrostomida) 45 Seiten mit 16 Abb., DM 15,20

58. DAVID McKIRDY, PETER SCHMIDT, MAXINE McGINTY-BAYLY, Interstitielle Fauna von Galapagos XVI. Tardigrada. 43 Seiten mit 11 Abb., DM 14,40

59. WOLFGANG MIELKE, Ökologie der Copepoda eines Sandstrandes der Nordseeinsel Sylt. 86 Seiten mit 41 Abb., DM 28,20

60. BEATE SOPOTT-EHLERS, Interstitielle Macrostomida und Proseriata (Turbellaria) von der französischen Atlantikküste und den Kanarischen Inseln. 35 Seiten mit 14 Abb. DM 12,80